NITROGEN NMR

NITROGEN NMR

Edited by

M. Witanowski

Institute of Organic Chemistry
Polish Academy of Sciences
Warsaw, Poland

and

G. A. Webb

Department of Chemical Physics
University of Surrey
Guildford, Surrey
England

Springer Science+Business Media, LLC
1973

Library of Congress Catalog Card Number: 72-95065

ISBN 978-1-4684-8177-8   ISBN 978-1-4684-8175-4 (eBook)

DOI 10.1007/978-1-4684-8175-4

Copyright © Springer Science+Business Media New York

Originally published by Plenum Publishing Company Ltd in 1973

Softcover reprint of the hardcover 1st edition 1973

Preface

To date nitrogen NMR has been discussed in research papers and review articles throughout the scientific literature. It has been our aim in preparing this book to provide a comprehensive account of the widely spread applications of nitrogen NMR. The relevant literature has been surveyed from the beginnings of NMR until early 1972. The steady annual growth in the number of references cited since 1965 is ample evidence of the ever increasing importance of the subject.

Sufficient theoretical and experimental background is given for an understanding of the applications dealt with in later chapters. The basic principles of NMR are developed with a theoretical approach to chemical shifts and spin-spin coupling constants, particular emphasis being given to nitrogen nuclei. Following this the experimental aspects of nitrogen NMR are adequately described.

Special emphasis is given to the observable effects of the nuclear quadrupole moment of the ^{14}N nucleus. It is appropriate that this topic be dealt with in depth since quadrupolar interactions frequently dominate the information available from a study of the ^{14}N nucleus and other nuclei spin-coupled to it.

The applications of nitrogen chemical shift data to organic and inorganic molecules are covered in two extensive chapters which include the effects of paramagnetism on nitrogen NMR. The continuing discussion on the choice of chemical shift scales and reference compounds is also covered. Spin-spin coupling constant values are often available from the spectra of ^{15}N nuclei and other nuclei spin-coupled to ^{15}N. This subject is amply dealt with and the correlations of coupling constants with molecular structure are discussed in some detail.

PREFACE

Finally we would like to express our appreciation to all the contributors for their dedication and forbearance during the preparation of this volume.

M. Witanowski   G. A. Webb
Warsaw, Poland.   Guildford, England.

July 1972

Contributors

T. Axenrod   Department of Chemistry, The City College of The City University of New York, New York, N.Y. 10031, U.S.A.

H. Januszewski   Institute of Organic Chemistry, Polish Academy of Sciences, Warsaw, Poland.

J. P. Kintzinger   Institute of Chemistry, Louis Pastuer University, Strasbourg, France.

J. M. Lehn   Institute of Chemistry, Louis Pasteur University, Strasbourg, France.

N. Logan   Department of Chemistry, University of Nottingham, Nottingham, England.

E. W. Randall   Department of Chemistry, Queen Mary College, London, England.

L. Stefaniak   Institute of Organic Chemistry, Polish Academy of Sciences, Warsaw, Poland.

G. A. Webb   Department of Chemical Physics, University of Surrey, Guildford, Surrey, England.

M. Witanowski   Institute of Organic Chemistry, Polish Academy of Sciences, Warsaw, Poland.

Contents

CHAPTER 1

Theoretical Background to Nitrogen NMR

G. A. Webb

Department of Chemical Physics, University of Surrey, Guildford, Surrey

and

M. Witanowski

Institute of Organic Chemistry, Polish Academy of Sciences, Warsaw.

1.1 Basic Principles of NMR

1.1.1 Introduction

There can be no doubt that nitrogen is one of the most important atoms in organic-, inorganic- and bio-chemistry. Its common molecular occurrence in a variety of valence states with various types of bonding and stereochemistry has caused the nitrogen atom to be widely studied by chemical and physical techniques.

In natural abundance nitrogen exists in two isotopic forms, the most common is ^{14}N, which is 99.635% abundant and ^{15}N which has an abundance of 0.365%. Both of these isotopes may be studied by means of nuclear magnetic resonance (NMR), additionally ^{14}N provides nuclear quadrupole resonance (NQR) data as discussed in Chapter 3.

These studies and the interpretation of the ensuing results have attracted the interests of a number of physical and theoretical chemists, for reasons which are made apparent in later chapters.

Although the ^{14}N nucleus was investigated in the early days of NMR the number of papers dealing with nitrogen NMR, published between 1950 and 1964, are very few in comparison with those on other nuclei especially ^1H and ^{19}F. This is partly due to the low sensitivity of the ^{14}N and ^{15}N nuclei, 0.00101 and 0.00104 respectively, relative to that of a proton in the same applied magnetic field. Another factor is the advent of commercially available NMR spectrometers to allow the exploitation of this sensitivity difference. The low natural abundance of ^{15}N is another reason for its neglect. This can be largely overcome by means of isotopic enrichment, which is rather expensive in practice; $^{15}NH_4^{\oplus}$ costs about U.S. $400 per gram of contained ^{15}N. However, recently molecules containing ^{15}N in natural abundance have been successfully studied with both continuous wave [1] and pulse [2] methods, together with spectrum accumulation techniques. The experimental details of these studies are discussed in Chapter 2.

Since 1964, improvements in experimental techniques and instrumentation have generated a wide and growing interest in both ^{14}N and ^{15}N NMR spectroscopy. However, the ^{14}N nucleus is inherently difficult to study not only on account of its low relative sensitivity but also because it has an electric quadrupole moment (Section 1.1.4). However, the difficulties experienced in ^{14}N NMR measurements are at least partially compensated for by the additional information, on the environment of the nucleus, which may be obtained from the quadrupolar broadening of the ^{14}N resonance lines (Section 3.4).

1.1.2   The Resonance Condition

Both common isotopes of nitrogen have a nuclear spin denoted by the quantum number I, $I = 1$ for ^{14}N and $I = \frac{1}{2}$ for ^{15}N. Since nuclei are charged bodies there is a magnetic moment μ collinear with the spin vector $\mathbf{I}$, where the length of the spin vector is $\sqrt{I(I + 1)}\hbar$.

Assuming that the nucleus can be considered as a spinning charged sphere, a simple classical calculation gives

$$\mu = \gamma I \hbar \qquad (1.1)$$

where γ is the gyromagnetic ratio of the nucleus. In the presence of an external magnetic field, B_0 defined to be in the z direction, $2I + 1$ levels are produced whose energies, relative to that in zero magnetic field E, are given by

$$E = -\gamma \hbar B_0 m_I \qquad (1.2)$$

where m_I represents the allowed component of I in the z direction and

$$m_I = I, I - 1, \ldots, -I \qquad (1.3)$$

Since the selection rule governing magnetic-dipole transitions is

$$\Delta m_I = \pm 1 \qquad (1.4)$$

and the separation between adjacent energy levels, ΔE, is given by

$$\Delta E = \gamma \hbar B_0 \qquad (1.5)$$

NMR occurs when transitions take place between these levels. This can occur by the absorption of photons from an oscillating external field, having the correct polarization and satisfying the frequency conditions:

$$h\nu = \Delta E = \gamma \hbar B_0 \qquad (1.6)$$

or more simply:

$$\nu = \frac{\gamma B_0}{2\pi} \qquad (1.7)$$

where ν is the frequency of the radiation from an external oscillating field which is absorbed by the nuclear spins.

Equation (1.7) expresses the resonance condition for NMR experiments, the practical details of which are described in Chapter 2.

Relative to that of the proton the gyromagnetic ratios of the ^{14}N and ^{15}N nuclei are $0.072236749 \pm 0.000000010$ and $-0.101330447 \pm 0.000000010$ respectively [3]. The proton has a positive gyromagnetic ratio thus in the presence of an applied magnetic field, it follows from equation (1.2) that the Zeeman splitting of the spin levels of the ^{14}N and ^{15}N nuclei

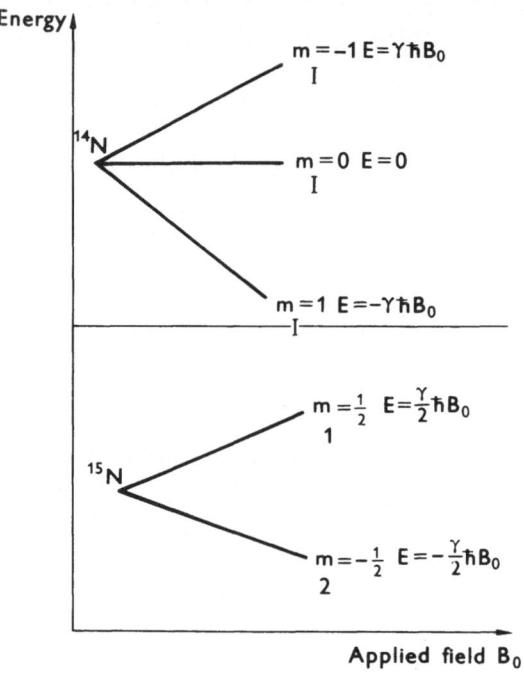

Fig. 1.1.

are as shown in Fig. 1.1. Since each of these isotopes has a different value of γ the NMR transition in a given magnetic field and in the absence of environmental effects, occurs at a different characteristic frequency for each isotope. In a field of 2.35 T (23.5 kG) the characteristic frequencies are 7.225 MHz and 10.135 MHz for ^{14}N and ^{15}N respectively, the corresponding frequency for ^{1}H is 100 MHz.

1.1.3   Nuclear Screening

In a molecule the local magnetic field, B, experienced by a nucleus differs from the applied magnetic field, B_0. This is largely due to the screening effects produced by the electrons in the molecule, expressed by

$$B = B_0(1 - \sigma) \qquad (1.8)$$

where σ is the screening constant or shielding parameter, it is usually expressed in parts per million (p.p.m.). The resonance condition for nucleus i now becomes

$$\nu_i = \frac{\gamma}{2\pi} B_0(1 - \sigma_i) \qquad (1.9)$$

If two nuclei, i and j of the same isotope, have different screening constants σ_i, σ_j, then the separation of their resonance signals in the NMR spectrum gives their relative chemical shift δ_{ij}, where

$$\delta_{ij} = \sigma_i - \sigma_j \qquad (1.10)$$

In principle a change in nitrogen chemical shifts could be expected if one nitrogen isotope in a molecule is replaced by another. This could arise from changes in the screening constant due to alterations in vibrational amplitudes and small electronic differences. Experimentally in the case of ^{14}N and ^{15}N isotopic replacement no significant change in the nitrogen screening constant has been found [4–6]. Differences between the ^{14}N and ^{15}N screening constant previously reported for some molecules [7] have been shown to be non-existent. Consequently nitrogen chemical shifts obtained from ^{14}N and ^{15}N NMR data may be used interchangeably.

Amongst the first chemical shifts to be reported was 303 p.p.m. for the separation of the ^{14}N signals arising from $NH_4^{\oplus}$ and $NO_3^{\ominus}$. This was observed by Proctor and Yu in 1950 [8].

1.1.4 Relaxation Times and Linewidth

Like the magnetic moment of a nucleus, the electric quadrupole moment is related to its spin. There are both experimental and theoretical arguments to show that only nuclei with spins greater than or equal to unity possess electric quadrupole moments [9]. Consequently the ^{14}N nucleus has an electric quadrupole moment, whereas many other nuclei, e.g. ^{15}N, ^{1}H, ^{19}F, ^{13}C, ^{31}P do not.

The presence of the electric quadrupole moment reflects a non-spherical electric charge distribution in the nucleus which can interact with any electric field gradient present at the site of the nucleus in its molecular environment. This interaction can lead to broad lines not only in the NMR spectrum of the quadrupolar nucleus but also in the spectra of neighbouring nuclei. The broadening makes the detection of the ^{14}N NMR signals more difficult and may obscure small chemical shifts and other fine details in the spectrum. The loss in spectral resolution produced by quadrupolar interaction can, in principle, be practically eliminated by the use of a sufficiently large applied magnetic field. This would simultaneously lead to a significant gain in sensitivity, especially if spectrum accumulation techniques were employed.

High resolution NMR spectra are usually obtained from non-viscous liquid samples and frequently have linewidths less than 1 Hz at a resonance frequency of several MHz. In contrast the ^{14}N lines are often *50–1000 Hz wide, and sometimes wider still.*

In order to discuss the width of an NMR line it is necessary to consider the nuclear spin–lattice and spin–spin relaxation times. In a real NMR experiment there is a distribution of nuclei between adjacent Zeeman energy levels, the equilibrium value of which depends upon a Boltzmann relationship. When the nuclear spin system is disturbed, such as occurs in an NMR experiment, a finite time is required for the Boltzmann distribution to be re-established. This is the spin–lattice relaxation time T_1.

Additionally each spin state has a finite lifetime which depends upon interactions between nuclei, the time characterizing this is called the spin–spin relaxation time T_2. For the majority of samples studied by high-resolution NMR, T_1 and T_2 are taken to be equal. As molecules move in space their nuclear spins generate microscopic fluctuating magnetic fields. These are able to interact with the nuclear spin system under investigation and produce relaxation. The relaxation processes contributing to T_1 arise only from fluctuating magnetic fields with components perpendicular to the direction of B_0. The T_2 relaxation processes are produced by components in all three mutually perpendicular directions.

Contributions to T_1 and T_2 effect the NMR linewidth, as can be seen from the Heisenberg Uncertainty Principle written as

$$\Delta E \Delta t \gtrsim \hbar \qquad (1.11)$$

where ΔE and Δt are the uncertainties in measurement of energy and time respectively.

Since the rearrangement of nuclear spins requires a finite time this produces an uncertainty in the energies of the spin states. From equation (1.6) it follows that there is a corresponding uncertainty in the frequency at which the photon is absorbed and this results in the natural linewidth which is inversely proportional to T_2.

For nuclei with $I = \frac{1}{2}$, in a chemically stable diamagnetic molecule, the most important relaxation mechanism is due to the interaction of magnetic dipoles. This is a relatively inefficient process consequently the relaxation times are long and the NMR lines are narrow. For liquids T_1 usually lies between 10^{-2} and 10^2 sec. The experimental value of T_1 usually contains three contributions arising from dipole–dipole (DD), spin-rotational (SR) and chemical shift anisotropy (SA) interactions, such that

$$\frac{1}{T_1} = \frac{1}{T_{1\,DD}} + \frac{1}{T_{1SR}} + \frac{1}{T_{1SA}} \qquad (1.12)$$

The relative magnitudes of the contributions to T_1 from these three sources have been evaluated for ^{15}N nuclei in a number of organic molecules [10] and inorganic ions [11] at a variety of temperatures. The dipole–dipole contribution is largely composed of intra- and inter-molecular interactions with protons, deuterons are much less efficient in this respect. In many cases this is found to be the dominant relaxation mechanism.

The electron distribution in a molecule can give rise to a *molecular* magnetic moment the motion of which can generate a magnetic field at the nucleus which can produce a spin-rotation interaction. For spherical molecules

$$\frac{1}{T_{1SR}} = \frac{2}{3} \frac{KT}{h^2} IC^2 \tau_{SR} \qquad (1.13)$$

Where I is the moment of inertia of the molecule, C is the spin-rotation coupling parameter, τ_{SR} is the spin-rotation correlation time, T is the absolute temperature and K the Boltzmann constant. In dealing with small symmetric molecules and ions at room temperature and above the spin-rotation interaction is found to be the more prominent one [11].

Both the chemical shift and the spin-rotation interaction depend upon the electronic structure of a molecule. If a nucleus experiences a large chemical shift in general it will also have a large spin-rotation contribution to T_1 [12].

The contribution of the chemical shift anisotropy to T_1 is usually obscured by the other interactions, however it may be estimated if both the dipole–dipole and spin-rotation contributions are small [11].

In principle a scalar spin–spin interaction between nuclei can also contribute to T_1 but the significance of this in ^{15}N NMR spectroscopy has not yet been reported.

Nuclei with electric quadrupole moments are able to relax by the more efficient quadrupolar mechanism. For most liquids the quadrupolar relaxation time T_q is given by

$$\frac{1}{T_q} = \frac{3}{8} \left(1 + \frac{\eta^2}{3}\right) \left(\frac{eq.eQ}{h}\right)^2 \tau \qquad (1.14)$$

where $(eq.eQ/h)$ is the quadrupolar coupling constant depending upon the nuclear electric quadrupole moment $eQ = 0.071$ $(e \times 10^{-28}$ m$^2)$ for ^{14}N, and the electric field gradient at the nucleus eq, η describes the deviation of the electric field gradient from axial symmetry and τ is the quadrupolar correlation time which is a characteristic of the quadrupolar mechanism. Thus the greater the quadrupolar coupling constant the broader is the ^{14}N NMR line. Typical values of T_q are about 10^{-3} sec.

When equation (1.14) is applicable the full linewidth at half-height of the ^{14}N NMR signal, Δ_N, is given by

$$\pi \Delta_N = \frac{1}{T_q} \qquad (1.15)$$

hence Δ_N can be about 10^3 Hz.

The dependence of the quadrupolar relaxation mechanism on the electric field gradient at the nucleus is shown by the dramatic increase in ^{14}N linewidth in passing from the highly symmetrical ammonium ion to the less symmetrical ammonia molecule which presents a finite field gradient at the site of the ^{14}N nucleus [13].

1.1.5   Spin–spin Coupling

The five sharp lines observed in the ^{14}N spectrum of NH$_4^{\oplus}$ and the four broader lines in the spectrum of NH$_3$ (Fig. 1.2) arise from spin–spin interaction between the ^{14}N and ^{1}H nuclei. The number of lines found to a first-order approximation is $2xI + 1$ for each set of equivalent nuclei, where x is the number of equivalent nuclei of spin I in each set, to which the nucleus in question is spin coupled. The splitting due to spin–spin coupling is a characteristic feature of high-resolution NMR spectra. For molecules containing quadrupolar nuclei the splitting may be absent both in the spectrum

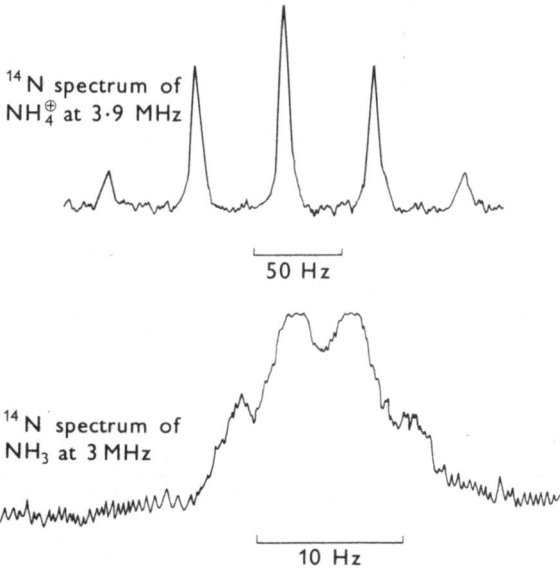

^{14}N spectrum of
NH$_4^\oplus$ at 3·9 MHz

50 Hz

^{14}N spectrum of
NH$_3$ at 3 MHz

10 Hz

Fig. 1.2.

of the quadrupolar nucleus and in the spectra of the nuclei to which it is spin coupled due to the rapid quadrupolar relaxation mechanism. The shapes of the NMR lines of ^{14}N nuclei and of nuclei with $I = \frac{1}{2}$ spin-coupled to a ^{14}N nucleus are discussed in Chapter 3.

The splitting in the nitrogen NMR spectrum due to spin–spin coupling may be modified or completely removed by intra- or inter-molecular chemical exchange of one or more of the spin coupled nuclei. In the case of ^{14}N nuclei both chemical exchange and the quadrupolar relaxation mechanism may operate as effective means of spin decoupling. The relative effects of these two contributions may be distinguished by examining the temperature dependence of the shape of the ^{14}N line (Section 3.4.3).

Nuclear spin–spin coupling is produced by an indirect interaction between the spins of neighbouring nuclei via the valence electrons in the molecule (Section 1.3). The energy of this interaction, E_J, is given by

$$E_J = hJ_{ij}\mathbf{I}_i\mathbf{I}_j \tag{1.16}$$

where J_{ij} is the spin–spin coupling constant between nuclei i and j, it is given by the separation between adjacent lines in simple NMR spectra of i and j (Fig. 1.3).

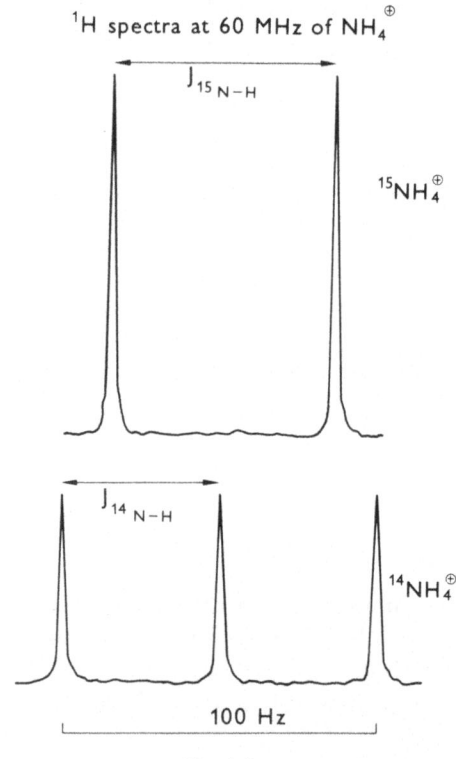

Fig. 1.3.

1.1.6   Nuclear Spin Hamiltonian and Spectral Analysis

The nuclear spin Hamiltonian operator, $\mathcal{H}$, for a system of nuclei in an external magnetic field B_0 follows from equations (1.2), (1.8) and (1.16) as:

$$\mathcal{H} = -\sum_i \gamma_i \hbar B_0 (1 - \sigma_i) I_{iz} + \sum_{i<j} hJ_{ij}\mathbf{I}_i . \mathbf{I}_j \tag{1.17}$$

where I_z is the operator corresponding to the eigenvalue m_I and the summation of the second term over $i < j$ ensures that each nuclear spin–spin interaction is only counted once.

In equation (1.17) $\mathscr{H}$ is expressed in energy units. In NMR work Hz are commonly used as units for $\mathscr{H}$, the necessary conversion involves dividing equation (1.17) by h, to give:

$$\mathscr{H} = -\frac{1}{2\pi} \sum_i \gamma_i B_0 (1 - \sigma_i) I_{iz} + \sum_{i<j} J_{ij} \mathbf{I}_i \cdot \mathbf{I}_j \qquad (1.18)$$

In order to interpret a complicated NMR spectrum it is necessary to calculate the allowed spin energy levels and the probability of a transition occurring between them in the presence of an oscillating external field [14]. The wave-functions ψ representing the spin energy levels are found by solving the eigenvalue problem

$$\mathscr{H}\psi = E\psi \qquad (1.19)$$

In order to solve equation (1.19) $\mathscr{H}$ is usually written as a sum of two terms, expressing the chemical shift and spin–spin interaction,

$$\mathscr{H} = \mathscr{H}^{(0)} + \mathscr{H}^{(1)} \qquad (1.20)$$

which represent the first and second terms of equation (1.18) respectively. Since the nuclear spin functions corresponding to m_I are eigenfunctions of the operator I_{iz} it follows that the only non-zero matrix elements of $\mathscr{H}^{(0)}$ are the diagonal ones,

$$\langle \psi_n | \mathscr{H}^{(0)} | \psi_n \rangle = -\frac{1}{2\pi} \sum_i \gamma_i B_0 (1 - \sigma_i) \langle \psi_n | I_{iz} | \psi_n \rangle \quad (1.21)$$

By matrix algebra it can be shown that the matrix elements of $\mathscr{H}^{(1)}$ are

$$\langle \psi_n | \mathscr{H}^{(1)} | \psi_n \rangle = \frac{1}{4} \sum_{i<j} J_{ij} T_{ij} \qquad (1.22)$$

and

$$\langle \psi_m | \mathscr{H}^{(1)} | \psi_n \rangle = \frac{1}{2} J_{ij} U \quad m \neq n \qquad (1.23)$$

where $T_{ij} = 1$ or -1, depending upon whether the spins i and j are parallel or antiparallel in the state corresponding to ψ_n, and $U = 1$ if ψ_m differs from ψ_n only by an interchange of spins i and j but is zero otherwise. Solution of equations (1.21), (1.22) and (1.23) yields the allowed energy levels of the nuclear spin system, the probability P_{mn} of a transition occurring between levels ψ_m and ψ_n in the presence of an oscillating external field in the y direction is given by time-dependent perturbation theory as:

$$P_{mn} \sim \left[\langle \psi_m \mid \sum_i \gamma_i I_{iy} \mid \psi_n \rangle \right]^2 \qquad (1.24)$$

The intensity of the corresponding line in the NMR spectrum is proportional to P_{mn}. Consequently the relative intensities of all permitted lines in the spectrum can be estimated from equation (1.24). Thus the solutions to equations (1.21), (1.22), (1.23) and (1.24) give the predicted positions and relative intensities of the lines in the NMR spectrum of a given nuclear spin system in terms of the empirical parameters σ_i and J_{ij}. These equations are frequently evaluated by computer for a multi-spin system and the results compared with an experimental spectrum. The matching of predicted and experimental spectra yields values for σ_i and J_{ij}. The use of computer programs of the LAOCN and NMRIT type for analysing complex spectra is now widespread and the quantum mechanics behind the analysis often tends to be forgotten. However, to gain an insight into the factors which determine the values of σ_i and J_{ij} a knowledge of quantum mechanics, valence bond theory and molecular orbital theory is necessary as shown in the next two sections.

1.2   Theory of Chemical Shifts

In general if we consider a molecule with a fixed geometry in the presence of a uniform magnetic field $\mathbf{B_0}$ in any direction, the electrons produce a secondary field $\mathbf{B'}$ at any nucleus where $\mathbf{B'}$ is not necessarily parallel to $\mathbf{B_0}$. Consequently the electronic screening of the nuclei produced by $\mathbf{B'}$ is related to $\mathbf{B_0}$ by means of a second-rank tensor or dyadic $\boldsymbol{\sigma}$ which depends upon the position of the nucleus in the molecule.

$$\mathbf{B'} = -\boldsymbol{\sigma}\mathbf{B_0} \qquad (1.25)$$

In high-resolution NMR experiments the molecules of the sample are rotating rapidly and moving randomly such that the chemical shift is determined by the mean component of

$\mathbf{B}'$ along the direction of $\mathbf{B_0}$ averaged over many rotations. As a result of the averaging the tensor $\boldsymbol{\sigma}$ can be replaced by a scalar σ which is the mean of the diagonal elements of the tensor,

$$\sigma = \frac{1}{3}(\sigma_{xx} + \sigma_{yy} + \sigma_{zz}) \qquad (1.26)$$

σ is the screening constant measured in high-resolution NMR experiments. A general expression for the tensor components σ_{xx}, σ_{yy} and σ_{zz} has been derived by Ramsey for the case of an isolated molecule [15]. He included in the electronic Hamiltonian operator of the molecule, as perturbations, terms representing the interaction of the electrons with the applied magnetic field $\mathbf{B_0}$ and with a local magnetic field due to the magnetic dipole $\boldsymbol{\mu}$ of the nucleus whose screening constant is to be evaluated.

When the electronic energy of the molecule is calculated in terms of the components of the Hamiltonian operator it is possible to collect those terms which are linear in μ and B_0.

Now the energy, E, of a magnetic dipole $\boldsymbol{\mu}$ in a field $\mathbf{B}'$ is given by

$$E = -\boldsymbol{\mu}.\,\mathbf{B}' \qquad (1.27)$$

Consequently those terms in the electronic energy expression which are linear in μ and B_0 represent the energy of interaction between the nuclear magnetic moment $\boldsymbol{\mu}$ and the secondary field $\mathbf{B}'$ due to the electron currents resulting from the application of the external field $\mathbf{B_0}$. These terms may be considered in conjunction with equations (1.25) and (1.26) to yield Ramsey's shielding formula which applies to any closed shell molecule. The formula consists of three similar expressions for σ_{xx}, σ_{yy} and $\sigma_{zz} =$

$$\frac{e^2}{2mc^2}\left\langle 0 \left| \sum_a \frac{(x_a^2 + y_a^2)}{r_a^3} \right| 0 \right\rangle - \frac{e^2\hbar^2}{4m^2c^2} \sum_{n \neq 0} (E_n - E_0)^{-1}$$

$$\left[\left\langle 0 \left| \sum_a L_{a,z} \right| n \right\rangle \left\langle n \left| \sum_a \frac{2L_{a,z}}{r_a^3} \right| 0 \right\rangle \right.$$

$$\left. + \left\langle 0 \left| \sum_a \frac{2L_{a,z}}{r_a^3} \right| n \right\rangle \left\langle n \left| \sum_a L_{a,z} \right| 0 \right\rangle \right] \qquad (1.28)$$

where $\langle 0 |$ refers to the unperturbed electronic ground state of the molecule and $\langle n |$ to the excited states with energies

E_0 and E_n respectively, r is the separation of the nucleus and electron whose coordinates are x, y, z and L_z is the z component of the electronic orbital angular momentum operator.

The operators L_z, L_z/r^3 and $(x^2 + y^2)/r^3$ in equation (1.28) are summed over all of the electrons in the molecule, not simply those of the atom to which the nucleus in question belongs. The first term in equation (1.28) is similar to Lamb's formula for atoms and becomes identical to it when averaged over all directions. This gives rise to the *diamagnetic* part of the screening constant σ^d, whereas the second term in equation (1.28) is frequently called the *paramagnetic* term σ^p, such that:

$$\sigma = \sigma^d + \sigma^p \tag{1.29}$$

To evaluate σ^p precisely it is necessary to have detailed knowledge of the eigenfunctions of all of the excited electronic states of the molecule including the continuum. This is obviously impracticable in the majority of cases, and it becomes necessary to use the quantum mechanical closure approximation in order to simplify the expression for σ^p. This involves replacing all of the electronic excitation energies $E_N - E_0$ by an average value $\Delta E_{av.}$ and using only the ground-state wavefunctions. Including this and the summations for the operators over all the electrons a, b in the molecule, σ_{zz} becomes:

$$\sigma_{zz} = \frac{e^2}{2mc^2} \left\langle 0 \left| \sum_a \frac{(x_a^2 + y_a^2)}{r_a^3} \right| 0 \right\rangle - \frac{e^2 \hbar^2}{m^2 c^2} \left(\Delta E_{av.} \right)^{-1}$$

$$\left\langle 0 \left| \sum_{a,b} \frac{(L_{z,a} L_{z,b})}{r_b^3} \right| 0 \right\rangle \tag{1.30}$$

Even in this form Ramsey's formula is only suitable for calculating σ in very small molecules. The most accurate method used is based upon perturbed Hartree-Fock expressions, but this restricts the calculations largely to diatomic molecules for which suitable wavefunctions are available [16]. For larger molecules σ^d and σ^p tend to become very large and of opposite sign, hence σ becomes the relatively small difference between large terms and its value is highly uncertain.

Saika and Slichter [17] have considered σ_j to consist of three components:

$$\sigma_j = \sigma_j^d + \sigma_j^P + \sum_{i \neq j} \sigma_{ij} \qquad (1.31)$$

where σ_j^d and σ_j^P refer to the *diamagnetic* and *paramagnetic* contributions from the electrons associated with the atom containing the nucleus and σ_{ij} represents the contribution from the other electrons in the molecule. Both σ_j^d and σ_j^P can be calculated by application of equation (1.30). In diamagnetic molecules the experimental range of nitrogen chemical shifts is about 900 p.p.m. Contributions from σ_{ij} are usually not more than a few p.p.m. and are usually considered to be unimportant when comparing the changes in electronic screening, as shown by chemical shifts, in a number of different nitrogen-containing molecules. Similarly the effects of ring currents, solvents, neighbouring electric field and shielding anisotropy are safely ignored when the sources of nuclear screening in nitrogen NMR are being discussed. Generally only changes in σ_j^d and σ_j^P are considered. σ_j^d may be expresssed as:

$$\sigma_j^d = \frac{e^2}{3mc^2} \sum_{\nu} P_{\nu\nu} \; \langle r_{\nu j}^{-1} \rangle \qquad (1.32)$$

by averaging the first term in equation (1.28) over all directions, where the summation is taken over all valence orbitals in the molecule, $P_{\nu\nu}$ is the electron charge density in orbital ν and $\langle r_{\nu j} \rangle$ is the quantum mechanical average,

$$\left\langle \phi_\nu \left| \frac{1}{r_{\nu j}} \right| \phi_\nu \right\rangle,$$

of the reciprocal distance of the electron in the νth orbital from the nucleus j. The values of $\langle r_\nu^{-1} \rangle$ for nitrogen atomic orbitals [18] may be calculated from equation (1.33).

$$\langle r_\nu^{-1} \rangle = \frac{Z_\nu}{n^2 a_0} \qquad (1.33)$$

where n is the principal quantum number of orbital ν, a_0 is its Bohr radius and Z_ν is the effective nuclear charge for electrons in this orbital. This latter term may be expressed as a function of electron charge densities $P_{\mu\mu}$:

$$Z_{1s} = 6.6 - 0.1(P_{2s2s} + P_{2px2px} + P_{2py2py} + P_{2pz2pz}) \qquad (1.34)$$

$$Z_{2s} = 5.65 - 0.35P_{2s2s} - 0.2(P_{2pz2px} + P_{2py2py} + P_{2pz2pz})$$

$$(1.35)$$

$$Z_{2p} = 5.35 - 0.5P_{2s2s} - 0.35(P_{2px2px} + P_{2py2py} + P_{2pz2pz})$$

$$(1.36)$$

In general, as the electron density around nucleus j increases σ_j^d provides an increase in screening, and from equation (1.19) it is apparent that this contribution acts towards moving the resonance signal to lower frequencies at constant applied magnetic field or to higher fields at a constant applied frequency.

It has long been realized that the range of chemical shifts observed for nitrogen nuclei cannot be accounted for by changes in σ_N^d. Experimental data on many groups of molecules show chemical shifts which are in the opposite direction to those expected from changes in the *diamagnetic* term [6].

Recent calculations on some nitrogen heterocycles have shown the variation in σ_N^d to be less than 0.5 p.p.m. whilst the chemical shifts differ within a range of about 100 p.p.m. [19]. CNDO molecular orbital results for a series of simple molecules and ions also show that contributions to the chemical shift differences arising from changes in σ_N^d are not significant [20]. Similar calculations [21] on the nitrogen chemical shift difference between pyridine and the pyridinium ion provide an estimate of -10.8 p.p.m. arising from a change in σ_N^d and of 128.8 p.p.m. from a change in σ_N^p. The total calculated difference in the screening constant between pyridine and the pyridinium ion is 118.0 p.p.m. which is in very good agreement with the observed chemical shift difference of 113–123 p.p.m. as discussed in Chapter 4. In most accounts of the theory of nitrogen chemical shifts the *diamagnetic* term has been omitted and changes in the *paramagnetic* term, σ_N^p, have been sought to rationalize experimental data.

However, recently two attempts have been made to assign a more important role to changes in the *diamagnetic* term [22, 23]. The results of one of them [22], based on CNDO molecular orbital calculations, suggest considerable contributions to the nitrogen chemical shift differences of some simple ions arising from changes in σ_N^d. A comparison of the calculated changes in σ_N^d and the experimental nitrogen chemical shifts [24] show that, with the exception of $NO_2^{\ominus}$, the calculated and observed changes are in opposite directions (Table 1.1). This must cast some doubt on the significance of the calculations. It seems very probable that the calculated data is exaggerated by an order of magnitude.

TABLE 1.1. Comparison of Calculated Changes in σ_N^d and Experimental ^{14}N Chemical Shifts for Some Simple Ions, Taking $NO_3^\ominus$ as an Arbitrary Reference

Ion	Calculated change [22] in σ_N^d (p.p.m.)	Experimental [24] ^{14}N chemical shift (p.p.m.)
$NO_3^\ominus$	0	0
$NO_2^\ominus$	−55.8	−274
$N_3^\ominus$ (central)	−66.7	+128
$N_3^\ominus$ (terminal)	−86.3	+277
$NCO^\ominus$	−91.7	+300
$NH_4^\oplus$	−154.7	+354

The second recent discussion [23] of the *diamagnetic* term is based on a simplified expression for σ_j^d which does not explicitly involve electronic wavefunctions [25].

$$\sigma_j^d = \sigma_j^d \text{ (free atom)} + \frac{e^2}{3mc^2} \sum_{i \neq j} \frac{z_i}{R_{ij}} \qquad (1.37)$$

where z_i is the charge associated with nucleus i and R_{ij} is the internuclear separation. The utility of this expression in the calculation of the *diamagnetic* component of the nitrogen screening constant is questionable [6].

As the number of electrons in the molecule increases equation (1.37) predicts a considerable increase in the screening constant. For example [23], its application predicts a shift to higher field, of the nitrogen resonance line, of about 29 p.p.m. as each hydrogen atom is replaced by a methyl group in NH_3 or $NH_4^\oplus$. Experimentally a shift of about 6 p.p.m. to lower field is observed as discussed in Chapter 4. This discrepancy between the observed chemical shifts and estimated changes in σ_N^d applies both to closely related molecules and to those containing nitrogen nuclei in very different chemical environments. Hence the attempted estimation [22, 23] of the contributions of σ_N^p to various nitrogen screening constants, by the algebraic summation of the experimental chemical shift differences and the calculated values of σ_N^d, appears to be unrealistic.

Consequently it seems that contributions to nitrogen chemical shift differences arising from changes in σ_N^d are rather small in comparison with those due to changes in σ_N^P. The former contributions may often be comparable to solvent and concentration shifts and may justifiably be dismissed in seeking an explanation of observed nitrogen chemical shift differences [18, 24, 26–34].

The second term in equation (1.28) may be evaluated by means of the LCAO molecular orbital approach to give the *paramagnetic* contribution to the shielding parameter. In order to ensure a rapid convergence gauge invariant atomic orbitals are used as a basis for the LCAO scheme. Pople [35] has shown that the resulting expression for σ_j^P is:

$$
\sigma_j^P = \frac{2e^2\hbar^2}{3m^2c^2} \langle r^{-3}\rangle_{2p} \sum_{s}^{\text{occ.}} \sum_{t}^{\text{unocc.}} (\Delta E_{s\to t})^{-1}
$$

$$
[(C_{s,xj}C_{t,yj} - C_{s,yj}C_{t,xj}) \sum_{i} (C_{s,xi}C_{t,yi} - C_{s,yi}C_{t,xi}) +
$$

$$
+ (C_{s,yj}C_{t,zj} - C_{s,zj}C_{t,yi}) \sum_{i} (C_{s,yj}C_{t,zi} - C_{s,zi}C_{t,yi}) +
$$

$$
+ (C_{s,zj}C_{t,xj} - C_{s,xj}C_{t,zj}) \sum_{i} (C_{s,zi}C_{t,xi} - C_{s,xi}C_{t,zi})] \quad (1.38)
$$

where the summation over all atoms i includes j, the C terms are the coefficients of the corresponding $2p$ orbitals in the molecular orbitals s and t and $\langle r^{-3}\rangle_{2p}$ is the mean value of the reciprocal cube of the $2p$ orbital radius, sometimes called the orbital expansion term. This latter term is usually evaluated by means of the relationship

$$
\langle r^{-3}\rangle_{2p} = \frac{1}{3} \left(\frac{Z_{2p}}{2a_0}\right)^3 \quad (1.39)
$$

which is analogous to equation (1.33).

By considering equations (1.38) and (1.39) together we can see why the experimental range of chemical shifts exhibited by a given element are approximately proportional to Z^3. Although for protons the *paramagnetic* contribution to the nuclear shielding is often negligibly small, it plays the dominant role for heavier atoms such as nitrogen.

It is apparent from equation (1.39) that an increase in the electron density on the atom containing nucleus j leads to a decrease in the value of $\langle r^{-3}\rangle_{2p}$ and thus to a decrease in the absolute magnitude of σ_j^p. Since the *paramagnetic* term makes a negative contribution to the screening parameter the resultant effect is to increase the shielding and to shift the NMR signal to lower frequencies. Thus this part of the *paramagnetic* term acts in the same sense as does the *diamagnetic* term.

The magnitude of $\langle r^{-3} \rangle_{2p}$ is often estimated from the application of Slater's rules to equation (1.39). The resulting expression for the nitrogen nucleus is:

$$\langle r^{-3} \rangle_{2p} = \frac{1}{24a_0^3} [3.90 + 0.35(x - q_N^\pi)]^3 \qquad (1.40)$$

where x is the number of electrons contributed to the molecular π electron system by the nitrogen atom and q_N^π is the π charge density at this atom. Slater atomic orbitals are known to be unreliable close to the nucleus and this is where the major contribution to $\langle r^{-3} \rangle_{2p}$ is produced. Consequently it is advisable to employ more suitable wavefunctions in estimating $\langle r^{-3} \rangle_{2p}$. By using SCF wavefunctions equation (1.40) becomes,

$$\langle r^{-3} \rangle_{2p} = 3.099 - 0.732 q_N \qquad (1.41)$$

where q_N is now the total, σ and π, charge density at the nitrogen nucleus. Equation (1.41) has been used in calculations of chemical shifts in some small molecules with reasonable success [6, 20].

It is reported that compared with the remainder of the *paramagnetic* term the variation of $\langle r^{-3} \rangle_{2p}$ is rather small in these molecules. Since equation (1.40) includes only the π electron charge density it may exaggerate the contribution of the orbital expansion term to σ_N^P.

The misinterpretation [5, 36] of nitrogen chemical shifts arising from the use of equation (1.40) has recently been criticized [6].

For nitrogen the orbital expansion term may also be evaluated from equations (1.36) and (1.39), the former of which includes all valence electron contributions.

By substituting equation (1.36) into (1.39), partially differentiating with respect to charge density and then dividing by Z_{2p}^3, we get:

$$\frac{1}{Z_{2p}^3} \cdot \frac{\partial [(Z_{2p})^3]}{\partial P_{2s2s}} = \frac{1.5}{Z_{2p}} \qquad (1.42)$$

and

$$\frac{1}{Z_{2p}^3} \cdot \frac{\partial |(Z_{2p})^3|}{\partial P_{2p2p(x,y,z)}} = \frac{1.05}{Z_{2p}} \qquad (1.43)$$

Equations (1.42) and (1.43) may be used in considering relative differences in σ_N^P arising from changes in the oribital expansion term since the latter is proportional to $(Z_{2p})^3$. For structures containing hybridized orbitals the changes in $(Z_{2p})^3$ may be estimated as follows:

Type of hybridization at the nitrogen atom	Bond system	Relative change in $(Z_{2p})^3$ for nitrogen per unit electron charge
sp	sigma	$-1.275/Z_{2p}$
	pi	$-1.05/Z_{2p}$
sp^2	sigma	$-1.2/Z_{2p}$
	pi	$-1.05/Z_{2p}$
sp^3	sigma	$-1.1625/Z_{2p}$

If in a structure such as that of the nitro group (I)

$$R-\overset{\oplus}{N} \Big\langle \begin{array}{c} O^{-1/2} \\ O^{-1/2} \end{array}$$

(I)

an electronegative substituent R removes the electron charge from the nitrogen atom through the σ-bond and this 'inductive' loss is compensated for by a gain of an equal charge from the π-bond system then the net change in $(Z_{2p})^3$ will be $+0.15/Z_{2p}$ which constitutes a 4% increase per unit charge exchanged. At the same time, the corresponding relative change in the remaining factor in the *paramagnetic* term [24] is not linear with the charge transferred and may lead to a 5–24% decrease *per unit charge* depending on the magnitude of the charge migration, and on the reference structure used. It is obvious that such a simple model indicates that changes in $\langle r^{-3} \rangle_{2p}$ may be much smaller than those resulting from other factors if the entire electron system is considered. This is probably the reason why the nitrogen chemical shifts within the $R-NO_2$, $R-NCO$, $R-NCS$, $R-NC$, $R-N_3$, and related groups of molecules [24, 28, 32] may be explained without any reference to changes in the orbital expansion term. For structures where the electron charge variations at the nitrogen atom result from effects which may operate in the same direction along both the σ-bond and π-bond systems, the significance of changes in the $\langle r^{-3} \rangle_{2p}$ term may be increased. Consideration of the π-orbital system alone may lead to reasonable correlations of nitrogen chemical shifts, as has been found for six-membered [33] and five-membered [34] hetero-aromatic rings.

Equation (1.38) shows that the other important contribution leading to changes in σ_j^P is the summation expression. Equation (1.38) has been used as such, or in similar form, for the computation of theoretical nitrogen chemical shifts in some simple nitrogen-containing molecules [20], pyridine and the pyridinium ion [21], and in simple nitrogen–oxygen–halogen compounds [37]. In the first case [20] reasonable agreement with the experimental shifts was obtained only within groups of molecules with the same hybridization of the valence orbitals of the nitrogen atom. The other two cases [21, 37] give good agreement with the observed values.

The summation in equation (1.38) contains contributions from excitation energies $\Delta E_{s \to t}$ and from orbital coefficients C representing changes in the orbital angular momentum. Generally, a reduction in $\Delta E_{s \to t}$ increases the contribution of the corresponding component to σ_j^P, whilst a general decrease in the magnitude of the coefficients acts in the opposite sense. Usually the terms involving the coefficients are positive so that they act towards increasing the *paramagnetic* term, which results in a deshielding of the nucleus. The relative significance of the excitation energy values and the angular momentum in the differences of nitrogen chemical shifts may vary depending on the structures considered and, as was shown [37] for nitrogen–oxygen–fluorine compounds, it can be dangerous to ascribe changes in nitrogen shifts to one or other of these factors exclusively. A good example of this is the consideration [37] of the relative nitrogen chemical shifts in the following pairs of molecules:

NF_3 123 p.p.m. to lower field than ONF_3

FNO   181 p.p.m. to lower field than FNO_2

$NO_2^{\ominus}$ 250 p.p.m. to lower field than $NO_3^{\ominus}$

where the structural differences consist of the replacement of the so-called lone electron pair by an O—N bond. The calculations show that the low-field shift of NF_3 relative to ONF_3 is caused mainly by a considerable relative increase in the orbital angular momentum for the former, since the presence of the additional oxygen atom in the latter reduces the values of the p-orbital coefficients at the nitrogen atom. The effect of changes in the number and magnitude of contributing excitations is not important here, since there is no appreciable change in the magnitude of the excitation energies and the increase in the number of the excitations for ONF_3 is not enough to counterbalance the angular momentum change. The chemical shift difference between FNO and FNO_2 results from a combination of effects, a substantial increase of the angular momentum for FNO and a substantial decrease in energy of at least one excitation in FNO compared with FNO_2, weakly counteracted by the increased number of excitations in FNO_2. The downfield shift of the nitrogen resonance signal of $NO_2^\ominus$ relative to that of $NO_3^\ominus$ arises mainly from a very low energy excitation which involves the lone electron pair in $NO_2^\ominus$. The contribution of this excitation dominates the low field shift. It is an extreme case of a lone pair effect on nitrogen chemical shifts. One should note that $\pi \to \pi^*$ excitations do not contribute to the summation term in equation (1.38) which is apparent from the ordering of the coefficients of the $2p$ orbitals.

The summation term in equation (1.38) is often expressed in a simplified form by means of the closure approximation as in the case of equation (1.30), which leads to the so-called average excitation energy (AEE) approximation.

In this approximation

$$\sigma_j^P = -\frac{e^2\hbar^2}{2m^2c^2\,\Delta E_{av.}}\,\langle r^{-3}\rangle_{2p}\sum_i Q_{ij} \qquad (1.44)$$

where the summation over i includes atom j, and

$$Q_{ij} = \tfrac{4}{3}\delta_{ij}(Px_ix_j + Py_iy_j + Pz_iz_j) - \tfrac{2}{3}(Px_ix_jPy_iy_j + Px_ix_jPz_iz_j +$$

$$+ Py_iy_jPz_iz_j) + \tfrac{2}{3}(Px_iy_j + Px_jy_i + Px_iz_jPx_jz_i + Py_iz_jPy_jz_i) \quad (1.45)$$

where δ_{ij} is the Kroenecker delta ($\delta_{ij} = 1$ if $i = j$, otherwise $\delta_{ij} = 0$), and the P's are the elements of the bond order matrix for the atomic orbitals comprising the molecular orbitals.

$$P_{\mu\nu} = 2 \sum_{s}^{\text{occ.}} C_{s\mu} C_{s\nu} \tag{1.46}$$

where $C_{s\mu}$ are the coefficients of the atomic orbitals χ_μ in the LCAO description of the molecular orbital ψ_s, i.e.

$$\psi_s = \sum_{\mu} C_{s\mu} \chi_\mu \tag{1.47}$$

In equation (1.44) $\Delta E_{\text{av.}}$ is the average excitation energy. The usefulness of this approximation depends on whether $\Delta E_{\text{av.}}$ may be assumed constant for a group of molecules [6]. This is possible in many cases of structurally related molecules, and calculations based on equation (1.44) give reasonable correlations with the observed nitrogen chemical shifts for nitroalkanes, nitramines and alkyl nitrates [24], nitriles and isonitriles [28], alkyl isocyanates, isothiocyanates and azides [24], nitrocarbanion systems [32], linear triatomic molecules and ions [27], six-membered hetero-aromatic rings [33], and azoles [34].

There have been attempts [29, 30, 36, 38–40] to explain nitrogen chemical shifts in terms of changes in $\Delta E_{\text{av.}}$ which are assumed to follow changes in the lowest-frequency transitions, these are usually those involving an excitation of the lone electron pair. This is a very crude assumption, and the detailed discussion of the theoretical shifts in nitrogen–oxygen–fluorine compounds [37] indicates that only in special cases of extremely low-lying excited states may this be valid. In general a lack of correlation is found for amines [41] and heterocyclic aromatic rings [33]; also there are obvious discrepancies between the shifts and the trends in the lowest excitation energies. Even in the case of azo-compounds [40] where an explanation of the shifts has been claimed on the basis of the lowest-energy transitions, there is no clear relationship between these experimental quantities, as can be seen from Table 1.2.

The only exception is the nitroso group [38] and similar structures where the longest-wavelength transitions may reasonably be correlated with the nitrogen chemical shifts. However this is an extreme case with an unusually low energy excitation which dominates the summation term in equation

TABLE 1.2. Comparison of Nitrogen Chemical Shifts, Relative to Pyridazine as an Arbitrary Reference, and the Longest Wavelength Absorption Band for a Series of Azo Compounds [40].

Compound	Nitrogen chemical shift (p.p.m.)[a]	Longest wavelength absorption band (Å)
Pyridazine	0	3400
trans-FN=NF	38	<2000
$^{\ominus}ON=NO^{\ominus}$ in NaOH	52	3000
HON=NOH in $HClO_4$	55	2430
trans-Azobenzene	52–102	4200–4400
$CF_3N=NCF_3$	109	3550
$CH_3CH_2N=NCH_2CH_3$	124	3550

[a] The shifts are all to low field of the pyridazine resonance.

(1.38) and probably makes the other contributions appear rather insignificant.

Changes in $\sum_i Q_{ij}$ have been widely considered [24, 27–30, 32–34] as an important contributor to nitrogen chemical shift differences. The electronic charge densities and bond orders used in equation (1.45) are readily available from molecular orbital calculations. Computer programs for performing approximate molecular orbital calculations to various degrees of sophistication, are now widely used.

The summation term, $\sum_i Q_{iN}$, for nitrogen in equation (1.44) may be generally expressed for various types of hybridization of nitrogen valence orbitals and those of directly bonded atoms in the approximation that divides the electronic structure of a molecule into a π-orbital delocalized system and polarized σ-bonds. The latter may be represented by approximate wavefunctions $\psi_{N,i}^{\sigma}$ as

$$\psi_{N,i}^{\sigma} = \frac{1}{\sqrt{2}}(\sqrt{1+a_i}\phi_N + \sqrt{1-a_i}\phi_i) \qquad (1.48)$$

where a_i is the corresponding polarization parameter of the bond between the nitrogen atom and atom i and the ϕ's are the corresponding hybridized orbitals. The summation term becomes a function of the polarization parameters a_i, the π-bond orders $p_{N,i}^{\pi}$, and the π-charge density at the nitrogen atom q_N^{π}. If there is a lone electron pair, which is not part of the π-orbital system, then it may be represented as a 'σ-bond' with the corresponding $a_i = 1$.

For sp^3 hybridization of the nitrogen valence orbitals [41] in structure (II),

$$R_1 \diagdown \; \diagup R_2$$
$$N$$
$$R_3 \diagup \; \diagdown R_4$$

(II)

which may represent for example ammonium ions and amines, the expression is:

$$\sum_i Q_{iN} = 2 - \tfrac{1}{3}(a_1 a_2 + a_1 a_3 + a_1 a_4 + a_2 a_3 + a_2 a_4 + a_3 a_4)$$

(1.49)

For the sp^2 hybridization of the nitrogen valence orbitals [33] as in structure (III)

$$R_1 \diagdown \; \diagup R_2$$
$$N$$
$$|$$
$$R_3$$

(III)

the expression becomes:

$$\sum_i Q_{iN} = 2 - \tfrac{4}{9}(a_1 + a_2 + a_3)(q_N^\pi - 1) - \tfrac{2}{9}(a_1 a_2 + a_1 a_3 + a_2 a_3)$$

$$+ \tfrac{4}{9}(\sqrt{1 - a_1^2}\, p_{N,1}^\pi + \sqrt{1 - a_2^2}\, p_{N,2}^\pi + \sqrt{1 - a_3^2}\, p_{N,3}^\pi)$$

(1.50)

Equation (1.50) is valid if either sp^2 or sp^3 hybridized orbitals are used for the R's. This includes structures with lone electron pairs on the nitrogen atom, hence it may be applied to pyridine-type nitrogen atoms and the corresponding pyridinium ions, to pyrrole-type nitrogen atoms, to nitro and nitroso groups, azo-compounds and analogous structures. It should be noted that in the approximation used, the general form of the expressions for ΣQ_{iN} depends on the type of hybridization at the nitrogen atom, but the hybridization at the neighbouring atoms only affects the general expression when there is a non-zero π-bond order between the nitrogen and the atom in question. Thus, equations derived for sp^2 or sp hybrids at R may be used if we substitute a sp^3 hybrid at R, but not *vice versa*, also sp^2 and sp hybrids are not mutually interchangeable in this way. Therefore, equation

(1.50) is not applicable to structures with sp-hybridized orbitals at any group R, such as in alkyl or aryl isocyanates, isothiocyanates and azides where the orbital system is approximately of the type (IV)

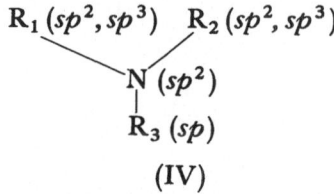

$$R_3 \, (sp)$$

(IV)

In this case the equation is different from equation (1.50) in that the $p_{N,3}^\pi$ value in the latter should be multiplied by a factor of $2\sqrt{3}$. The presence of a lone electron pair on the nitrogen atom in these molecules requires that a_1 (or a_2) = 1.

For sp hybridization at the nitrogen atom in a linear structure (V)

$$R_1 - N - R_2$$

(V)

the general expression is

$$\sum_i Q_{iN} = \tfrac{8}{3} + \tfrac{1}{3}(2 - a_1 - a_2)(q_N^{\pi'} + q_N^{\pi''} - 2) - \tfrac{2}{3}q_N^{\pi'} \, q_N^{\pi''}$$

$$+ k_1\sqrt{1 - a_1{}^2}\,(p_{N,1}^{\pi'} + p_{N,1}^{\pi''}) + m_1 p_{N,1}^{\pi'} \, p_{N,1}^{\pi''}$$

$$+ k_2\sqrt{1 - a_2{}^2}\,(p_{N,2}^{\pi'} + p_{N,2}^{\pi''}) + m_2 p_{N,2}^{\pi'} \, p_{N,2}^{\pi''} \qquad (1.51)$$

where the superscripts π' and π'' refer to the two π-orbital systems with mutually perpendicular nodal planes, whilst the k_i and m_i factors depend on the type of hybridization of R_i as follows:

Type of R_i orbital used for sigma-bonds	k_i	m_i
sp^3	0	0
sp^2	$\dfrac{2}{3\sqrt{3}}$	0
sp	$\tfrac{1}{3}$	$-\tfrac{2}{3}$
p	$\dfrac{\sqrt{2}}{3}$	$-\tfrac{2}{3}$

Equation (1.51) may be applied to nitriles, isonitriles, nitrilium ions, azides (central and terminal atoms), and similar molecules.

Finally it is possible to consider unhybridized $2s$ and $2p$ orbitals on the nitrogen atom and assume that the lone electron pair is in the $2s$ orbital whilst one of the $2p$ orbitals participates in the molecular σ-bond orbital, as in structure (VI).

$$(s, \text{lone pair}) \text{ N } (p)-R_1 (sp)$$

$$(VI)$$

This represents an alternative to the preceding structures for nitriles, the cyanide ion, and terminal atoms in azides and the azide ion, in which case equation (1.51) becomes:

$$\sum_i Q_{iN} = \tfrac{8}{3} + \tfrac{2}{3}(1 - a_1)(q_N^{\pi'} + q_N^{\pi''} - 2) - \tfrac{2}{3}q_N^{\pi'}q_N^{\pi''}$$

$$+ \frac{\sqrt{2}}{3}\sqrt{1 - a_1{}^2}(p_{N,1}^{\pi'} + p_{N,1}^{\pi''}) - \tfrac{2}{3}p_{N,1}^{\pi'}p_{N,1}^{\pi''} \qquad (1.52)$$

It should be noted that if the nitrogen atom attains either the $1s^2$ structure, by losing five electrons from the valence shell (this corresponds to $q_N^{\pi} = 0$ and all $a_i = -1$), or the $1s^2 2s^2 2p^6$ structure by accepting three electrons (this corresponds to all $q_N^{\pi} = 2$ and $a_i = +1$) then equations (1.49), (1.50), (1.51) and (1.52) yield a zero value for the summation term so that the *paramagnetic* term vanishes, as is expected for the noble gas configuration. These conditions are not met by an equation derived in the literature [29] and subsequently employed [30, 31] in calculations of nitrogen chemical shifts. This is due to the erroneous neglect of non-zero cross terms of the type $P_{x_i y_j}$, etc. in the final bracket of equation (1.45), as well as the fact that this bracket is given the wrong sign. In the case of calculations of the chemical shifts of ^{13}C nuclei it is possible, by a judicious choice of axes, to set the cross terms of the charge-density and bond-order matrix equal to zero but this is not generally applicable. For an sp^2-hybridized atom, by chosing the z-axis to be parallel to the π system arising from the unhybridized p orbitals, then terms of the type $P_{x_N z_j}$ and $P_{y_N z_j}$ are zero for a nitrogen nucleus bonded to nucleus j. However, the $P_{x_N y_j}$ and $P_{x_j y_N}$ terms only vanish if either the x- or y-axis coincides with the direction of the $N-j$ bond. If these terms are neglected for any other orientation of the bond then the rotational invariance of the expression for Q_{Nj} is violated. Consequently the equation reported by Wu [29] is seriously in error, and the information obtained by the use of his equation should be treated accordingly.

There have been some attempts at a compromise between the summation over all excited states, as in equation (1.38) or similar expressions, and the average excitation energy approximation in the calculation of the *paramagnetic* term. A 'different excitation energy' (DEE) approximation has been proposed [19], this incorporates the lowest known excited states according to equation (1.38) but uses an average value for the remaining excited states. This method has been employed with some success in discussing the pyridine–pyridinium ion nitrogen chemical shift [18, 26] and on a larger scale to the ^{13}C chemical shifts of pyridine and some azines [19]. However, its application to the nitrogen chemical shifts of azines [19] has given less satisfactory results. It has been reported [33], for a larger number of azines which include structures with fused aromatic rings, that the DEE approximation, but not the AEE approximation, appears to fail because of exaggerated contributions of the lowest excited states to the *paramagnetic* term.

Apart from the choice of the method in the calculation of the *diamagnetic* term and the *paramagnetic* term, which have been discussed here, it should be noted that all of them involve the so-called semi-empirical or plainly empirical approach from the point of view of molecular orbital theory. Since the results of such methods may depend on subjective parametrization, they offer much less satisfaction from a purely theoretical point of view than do the so-called *ab initio* calculations where electronic wavefunctions are used explicitly. An attempt at such *ab initio* calculations of C, N, O, and F chemical shifts for some polyatomic molecules has been made [42], but the results for nitrogen are not very encouraging. The results may be of interest with respect to the discussion of the relative share of the *paramagnetic* and *diamagnetic* terms in nitrogen chemical shift differences. Combinations of Gaussian functions were used for the $1s$, $2s$, $2p_x$, $2p_y$ and $2p_z$ orbitals of each first-row atom and $1s$ orbitals for hydrogen. In the STO-5G set, Slater-type orbitals were constructed with either a standard parametrization (St. scale) or the minimization of the total calculated molecular

energy (Opt. scale). The LEMAO-5G sets (least-energy minimal atomic orbitals) were obtained from Gaussian functions
by the minimization of the calculated energies of isolated
atoms, and here again a standard or optimized set of scaling
factors for the valence shells were used. The fifth variation
was an extended basis set, 4-31G, where the valence shells (1s
for H, 2s and 2p for heavy atoms) were divided into inner
and outer parts, described by sums of three and one Gaussian
functions respectively. The results of the calculations including changes in the *diamagnetic* term, the *paramagnetic* term
and the total screening constant for some simple molecules
are compared here with experimental nitrogen chemical shifts
quoted from more recent sources (Chapter 4) than those used in
the original work [42] (Table 1.3).

The calculated shifts are generally inconsistent with the
observed ones, except that the large downfield shift for
CH_3CN is qualitatively reproduced. There is even a systematic
discrepancy between the directions of the calculated and
experimental values for the series NH_3, CH_3NH_2, $(CH_3)_2NH$
with the exception of the LEMAO-5G (Opt.) set. However, a
detailed inspection of the calculated contributions to the
shifts reveals that the source of the lack of even qualitative
agreement is the calculated *diamagnetic* term contribution
which is large and opposed to the observed trend in the shifts,
except for CH_3CN. The *paramagnetic* term contributions,
calculated for each of the basis sets, do reproduce qualitatively
the experimental sequence of the nitrogen chemical shifts and
this leads in the case of the LEMAO-5G (Opt.) set to the
same order of the calculated values but not to any quantitative reproduction of the measured differences. It seems that
here again the experimental data indicate a relatively small
contribution from changes in the *diamagnetic* term to the
total difference in the screening constants for nitrogen nuclei,
and that the present methods of calculation of the latter term
are quite unsatisfactory. So far, the best results in the theory
of nitrogen chemical shifts have been obtained by considering
the *paramagnetic* term alone. This is the case for both the full
treatment of magnetic screening in the different structures of
nitrogen–oxygen–fluorine compounds as well as the far-
reaching approximations in calculations of relative nitrogen
chemical shifts within groups of structurally similar molecules.

The variety of molecular electronic environments of nitrogen nuclei provides through nitrogen chemical shifts a rigorous

G. A. WEBB and M. WITANOWSKI

TABLE 1.3. Comparison of Some Experimental Nitrogen Chemical Shifts with Values Calculated by *ab initio* Procedures [42]

Molecule	Experimental N chemical shift (p.p.m.)[a]	Calculated N chemical shifts (*diamagnetic* + *paramagnetic* components) = total shift (p.p.m.)				
		STO-5G (St.)	STO-5G (Opt.)	LEMAO-5G (St.)	LEMAO-5G (Opt.)	4-31G
NH_3	0	0	0	0	0	0
CH_3NH_2	-5	(26 - 25) = +1	(26 - 32) = -6	(26 - 17) = +9	(26 - 27) = -1	(25 - 22) = +3
$(CH_3)_2NH$	-12	(70 - 42) = +28	(69 - 51) = +18	(71 - 55) = +16	(70 - 79) = -9	(68 - 47) = +21
$HCONH_2$	-115	(28 - 58) = -30	(29 - 62) = -33	(29 - 99) = -70	(30 - 91) = -61	(29 - 80) = -51
CH_3CN	-246	(-7 - 261) = -268	(-7 - 266) = -273	(-8 - 416) = -424	(-7 - 427) = -434	(-11 - 320) = -331

[a] NH_3 is taken as an arbitrary standard.

test of the theories of chemical shifts, and of the methods of calculation of electron charge distribution. The importance of this is amplified by the fact that ^{14}N nuclei provide an independent means of estimating the values of electron charge density from ^{14}N NQR measurements and from ^{14}N NMR linewidths.

1.3   Theory of Spin–Spin Coupling

The rotationally averaged nuclear spin–spin coupling arises from an interaction between the nuclei and their surrounding electrons.

The first nucleus perturbs the electrons and these produce a magnetic field at the second nucleus. The nuclear magnetic moment may interact either with the spin or orbital motion of the electrons. For the former there are two types of interaction, thus the total interaction energy may be written as the sum of three terms.

The first term represents the interaction between the nuclear magnetic moment and the field produced by the orbital motion of the electrons. It is included in the electronic Hamiltonian operator of the molecule as $\mathscr{H}^{\mathrm{I}}$, where:

$$\mathscr{H}^{\mathrm{I}} = 2\beta\hbar \sum_a \sum_i \gamma_i \frac{\mathbf{L}_{ai} \cdot \mathbf{I}_i}{r_{ai}{}^3} \qquad (1.53)$$

where $\beta = (e\hbar/2mc)$ is the Bohr magneton, $\mathbf{L}_{ai}$ is the orbital angular momentum operator about nucleus i, r_{ai} is the separation between nucleus i and electron a, and the summation is taken over all electrons and nuclei.

The second contribution to the Hamiltonian, $\mathscr{H}^{\mathrm{II}}$ accounts for the dipole–dipole interaction between the nuclear and electron spins.

$$\mathscr{H}^{\mathrm{II}} = 2\beta\hbar \sum_a \sum_i \gamma_i \left[\frac{3(\mathbf{S}_a \cdot \mathbf{r}_{ai})(\mathbf{I}_i \cdot \mathbf{r}_{ai})}{r_{ai}{}^5} - \frac{\mathbf{S}_a \cdot \mathbf{I}_i}{r_{ai}{}^3} \right] \qquad (1.54)$$

The electron spin-nuclear spin interaction described by equation (1.54) is only applicable to extra-nuclear electrons. If the electron concerned is in an orbital which has a finite density at the nucleus it is necessary to consider the Fermi contact term $\mathscr{H}^{\mathrm{III}}$, where

$$\mathscr{H}^{\mathrm{III}} = \frac{16\pi\beta\hbar}{3} \sum_a \sum_i \gamma_i \delta(r_{ai}) \mathbf{S}_a \cdot \mathbf{I}_i \qquad (1.55)$$

The Dirac delta function $\delta(r_{ai})$ represents the probability of the electron, a, being present at the site of the nucleus, i.

It is noteworthy that equations (1.53), (1.54) and (1.55) are all linear in I_i, whereas the spin–spin coupling term in the nuclear spin Hamiltonian, equation (1.16), depends upon the product $I_i . I_j$. Hence, if J_{ij} is to be calculated by means of perturbation theory it is necessary to consider second-order terms in the energy expression. The perturbation approach has been most widely employed to date in calculations of J_{ij}.

The effects of nuclear spin–spin coupling are usually found by taking the total electronic Hamiltonian operator of the molecule and concentrating on the perturbation term $\mathscr{H}'$ contained in it, where $\mathscr{H}'$ is given by

$$\mathscr{H}' = \mathscr{H}^{I} + \mathscr{H}^{II} + \mathscr{H}^{III} \tag{1.56}$$

It follows from equations (1.53), (1.54), (1.55) and (1.56) that $\mathscr{H}'$ is an operator which depends upon the nuclear spin and electron spin and space coordinates. However, the nuclear spin Hamiltonian, equation (1.18) depends only upon the nuclear spin coordinates. Hence by calculating the effect of $\mathscr{H}'$ on the electronic states of a molecule it is possible to obtain an energy expression dependent upon the nuclear spin operators. This can be equated with $\mathscr{H}^{(1)}$ in equation (1.20), whence the dependence of J_{ij} upon the electronic structure of the molecule is found.

Ramsey [43] has shown that when the second-order correction to the energy of the electronic ground state due to $\mathscr{H}'$ is averaged over all molecular orientations, there is no contribution from cross terms between $\mathscr{H}^{I}$, $\mathscr{H}^{II}$ and $\mathscr{H}^{III}$. Hence the three contributions to the perturbation $\mathscr{H}'$ can be considered separately. Usually the Fermi contact expression, $\mathscr{H}^{III}$ is the most important contributor, $\mathscr{H}^{I}$ and $\mathscr{H}^{II}$ are almost always neglected [44]. Assuming that $\mathscr{H}^{III}$ is the only perturbation to be considered, the resulting second-order correction to the electronic energy, E^{III}, of the ground state is given by:

$$E^{III} = -\left(\frac{16\pi\beta\hbar}{3}\right)^2 \sum_n \sum_i \sum_j \gamma_i\gamma_j (E_n - E_0)^{-1} \langle o | \sum_a \delta(r_{ai}) S_a . I_i | n \rangle \times$$
$$\langle n | \sum_b \delta(r_{bj}) S_b . I_j | 0 \rangle \tag{1.57}$$

where $\langle n |$ refers to electronic excited states of the molecule and $\langle 0 |$ to the ground state with energies E_n and E_0 respectively.

Since the matrix elements in equation (1.57) involve integration only over electronic coordinates the nuclear spin operators can be taken outside the summation over n. In order to relate equations (1.16) and (1.57) it is necessary to average the latter over all possible molecular orientations because J_{ij} is obtained from solution spectra. The resulting value of E^{III} is,

$$E^{III} = -\frac{1}{3}\left(\frac{16\pi\beta\hbar}{3}\right)^2 \sum_i \sum_j \gamma_i \gamma_j \, \mathbf{I}_i . \mathbf{I}_j \times$$

$$\sum_n (E_n - E_0)^{-1} \langle 0| \sum_a \delta(\mathbf{r}_{ai})\mathbf{S}_a |n\rangle \langle n| \sum_b \delta(\mathbf{r}_{bj})\mathbf{S}_b |0\rangle \quad (1.58)$$

The coefficient of $\mathbf{I}_i . \mathbf{I}_j$ in equation (1.58) can now be equated to J_{ij}. After dividing by h, the Fermi contact contribution to J_{ij} is expressed in Hz as:

$$J_{ij}^{III} = -\frac{2}{3h}\left(\frac{16\pi\beta\hbar}{3}\right)^2 \gamma_i \gamma_j \sum_n (E_n - E_0)^{-1}\langle 0| \sum_a \delta(\mathbf{r}_{ai})\mathbf{S}_a |n\rangle .$$

$$\langle n| \sum_b \delta(\mathbf{r}_{bj})\mathbf{S}_b |0\rangle \quad\quad\quad (1.59)$$

The factor 2 appears in equation (1.59) because the summations over i and j are independent.

J involves an average over the molecular vibrational levels of the electronic states. Consequently isotopic substitution of ^{14}N by ^{15}N could give an additional contribution to J, in practice this is found to be insignificant. However, $\gamma_{^{15}N}$ and $\gamma_{^{14}N}$ have opposite signs, thus the replacement of ^{14}N by ^{15}N in a molecule will change the sign of J for the spin–spin coupling between the nitrogen nucleus and neighbouring nuclei. The relationship between the ^{14}N and ^{15}N coupling constants to another nucleus X is,

$$J(^{14}N - X) = -0.7129 J(^{15}N - X) \quad\quad (1.60)$$

which is useful for calculating ^{14}N coupling constants from ^{15}N NMR data and *vice versa*. In order to understand the relationship between nuclear spin–spin coupling and electronic structure it is convenient to use a reduced coupling constant K. The sign of K depends only upon the electronic environment of the nucleus. The usual convention is that a positive value of K denotes a stabilization of antiparallel nuclear spins.

K_{ij} is defined by:

$$K_{ij} = \frac{4\pi^2 J_{ij}}{h\gamma_i\gamma_j} \qquad (1.61)$$

In electromagnetic units K is given as cm^{-3}, in SI units it has dimensions of $NA^2 \ m^{-3}$, where $1 \ NA^2 \ m^{-3}$ is equivalent to $10 \ cm^{-3}$.

By substituting equation (1.61) into equation (1.59), the Fermi contact contribution to K_{ij} is given by:

$$K_{ij}^{III} = -\frac{512\pi^2\beta^2}{27} \sum_n (E_n - E_0)^{-1} \langle 0| \sum_a \delta(r_{ai})S_a |n\rangle$$

$$\langle n| \sum_b \delta(r_{bj})S_b |0\rangle \quad (1.62)$$

The problems encountered in evaluating K_{ij}^{III} from equation (1.62) are similar to those experienced in calculating σ^P and they lead to similar approximations.

By means of the closure approximation equation (1.62) may be written in terms of ground-state electronic wavefunctions as:

$$K_{ij}^{III} = -\frac{512\pi^2\beta^2}{27} (\Delta E_{av.})^{-1} \langle 0| \sum_a \sum_b \delta(r_{ai}) \delta(r_{bj})S_a \cdot S_b |0\rangle \quad (1.63)$$

If the ground state is a spin singlet then the only excited states giving non-zero matrix elements in equation (1.62) are spin triplets. Thus $\Delta E_{av.}$ is usually taken to be an average of the excitation energies of the spin triplet states.

Equation (1.63) has been used with molecular orbital ground state functions and with valence bond functions. Reviews dealing with the relative merits of calculations of K_{ij} involving these functions for various molecular environments have appeared in recent years [44, 45, 46].

Equation (1.63) may be written in terms of the ground-state probability density $\rho_0(i, j)$ which describes the probability of one electron being at nucleus i and a second electron at nucleus j.

$$K_{ij}^{III} = -\frac{256\pi^2\beta^2}{9} (\Delta E_{av.})^{-1} \rho_0(i, j) \qquad (1.64)$$

For both molecular orbital and valence bond wave functions the major contribution to $\rho_0(i, j)$ comes from the s electron density at the nuclei.

In the general molecular orbital theory equation (1.64) is written as:

$$K_{ij}^{\mathrm{III}} = -\frac{256\pi^2\beta^2}{9} (\Delta E_{\mathrm{av.}})^{-1} \sum_{s}^{\mathrm{occ.}} \sum_{t}^{\mathrm{occ.}} \psi_s(i)\psi_t(i)\psi_s(j)\psi_t(j) \quad (1.65)$$

where the summations are taken over all of the molecular orbitals occupied in the ground state. By using the LCAO expression, equation (1.47), for the molecular orbitals, assuming that only s orbitals centred on the nuclei have a significant density at the nucleus and that there is only one s orbital on each centre in the basis set of atomic orbitals, we get:

$$K_{ij}^{\mathrm{III}} = -\frac{256\pi^2\beta^2}{9} (\Delta E_{\mathrm{av.}})^{-1} \sum_{s}^{\mathrm{occ.}} \sum_{t}^{\mathrm{occ.}} C_{si}C_{ti}C_{sj}C_{tj}\chi_\mu^2(0)\chi_\nu^2(0)$$
$$(1.66)$$

where $\chi(0)$ is the value of the atomic orbital at the nucleus, χ_μ is centred on nucleus i and χ_ν on nucleus j. Equation (1.66) may be rewritten in terms of the bond orders given by equation (1.46):

$$K_{ij}^{\mathrm{III}} = -\frac{64\pi^2\beta^2}{9} (\Delta E_{\mathrm{av.}})^{-1} p_{\mu\nu}^2 \chi_\mu^2(0)\chi_\nu^2(0) \quad (1.67)$$

Application of the McConnell [47] equation (1.67) can only give rise to positive values of K_{ij}. This is a limitation of the closure approximation which does not acknowledge the possibility of terms in equation (1.62) taking either sign.

The most recent work has been mainly based on

$$K_{ij}^{\mathrm{III}} = -\frac{256\pi^2\beta^2}{9} \sum_{s}^{\mathrm{occ.}} \sum_{t}^{\mathrm{unocc.}} (^3\Delta E_{s\to t})^{-1} \psi_s(i)\psi_t(i)\psi_s(j)\psi_t(j)$$
$$(1.68)$$

where the summations are taken over all the s occupied molecular orbitals and the t unoccupied ones contributing to the excited spin triplet states whose excitation energies are $^3\Delta E_{s\to t}$.

Pople and Santry [48] have successfully calculated a number of coupling constants using only an incomplete set of excited states in conjunction with an expression based on equation (1.68). By analogy with equation (1.66), equation (1.68) may be written as:

$$K_{ij}^{\mathrm{III}} = -\frac{256\pi^2\beta^2}{9} S_i^2(0)S_j^2(0) \sum_{s}^{\mathrm{occ.}} \sum_{t}^{\mathrm{unocc.}} (^3\Delta_{s\to t})^{-1} C_{si}C_{ti}C_{sj}C_{tj}$$
$$(1.69)$$

The Pople-Santry approach embodies the zero-overlap approximation in which only one-centre terms are used for calculating the electron density at the nucleus. Hence in equation (1.69) the electron density at the nucleus is assumed to be due only to one-centre terms involving valence shell s orbitals, e.g. 2s for nitrogen and 1s for hydrogen. $S_i(o)$ is the value at the nucleus of the valence shell s orbital associated with nucleus i.

By considering a series of closely related molecules, empirical relationships based upon equation (1.69) have been developed between $N-H$ coupling constants and the amount of S electron character in the $N-H$ hybrid orbitals. This has led to experimental differences in $K(N-H)$ being attributed to changes in the type of hybridization experienced by the nitrogen atom [49]. Although the results are intuitively reasonable it has to be remembered that they are based upon the assumption that the summations in equation (1.69) are roughly constant throughout the series of molecules considered.

In the case of the tetrahedral $NH_4^\oplus$ ion the only molecular orbitals having a non-zero density at the nucleus are the totally symmetric ones. These are constructed from the nitrogen 2s and 2p atomic orbitals and the hydrogen 1s orbitals.

There are two totally symmetric molecular orbitals, one bonding ψ_1 and one antibonding ψ_2 which have the form,

$$\psi_1 = \alpha 2S_N + \frac{\beta}{2} (1S_H + 1S_H + 1S_H + 1S_H) \qquad (1.70)$$

$$\psi_2 = \beta\, 2S_N - \frac{\alpha}{2} (1S_H + 1S_H + 1S_H + 1S_H) \qquad (1.71)$$

where $$\alpha^2 + \beta^2 = 1 \qquad (1.72)$$

Only one excited state can now contribute to equation (1.69), which becomes:

$$K_{N-H}^{III} = \frac{64\pi^2\beta^2}{q} (^3\Delta E_{1\to 2})^{-1} 2S_N^2(0) 1S_H^2(0)\alpha^2\beta^2 \qquad (1.73)$$

from which $K_{N-H}^{III} = +7.5 \times 10^{20}$ NA^2 m^{-3}, the experimental value is 5.9×10^{20} NA^2 m^{-3} but the sign has not been determined.

The Pople-Santry theory predicts positive values for $K(N - H)$, $K(N - B)$ and $K(N - N)$ and a negative one for $K(N - F)$. This awaits experimental confirmation. In general equation (1.69) used in conjunction with reliable wavefunctions has been found to give satisfactory estimates of K for small molecules [44].

The variables in equation (1.69) have been obtained for larger molecules using approximate molecular orbital procedures of the CNDO/2 and INDO types, for which computer programs are now widely available [50]. These procedures are more flexible than a non-empirical SCF theory in that most of the difficult multi-centre integrals are not directly evaluated.

A further technique discussed by Pople and co-workers is finite perturbation theory [51-53]. This involves adding the Fermi contact perturbation term to the SCF Hamiltonian operator and evaluating the resulting SCF orbitals. The coupling constants calculated in this way are considered to be very similar to those obtained using SCF orbitals and a complete set of excited spin triplet states. It is anticipated that this approach will stimulate further interest in nitrogen spin-spin coupling constants.

References

1. R. L. LICHTER and J. D. ROBERTS, *J. Amer. Chem. Soc.*, **93**, 3200 (1971).
2. J. M. BRIGGS, L. F. FARNELL, and E. W. RANDALL, *Chem. Comm.*, **68** (1971).
3. J. D. BALDESCHWIELER, *J. Chem. Phys.*, **36**, 152 (1962).
4. E. D. BECKER, R. D. BRADLEY, and T. A. AXENROD, *J. Mag. Res.*, **4**, 136 (1971).
5. E. W. RANDALL and D. G. GILLIES in *Progress in NMR Spectroscopy*, Vol. 6 (J. W. EMSLEY, J. FEENEY, and L. H. SUTCLIFFE, Eds.), Pergamon Press, London, 1970, p. 119.
6. M. WITANOWSKI and G. A. WEBB, in *Annual Reports on NMR Spectroscopy*, Vol. 5a (E. F. MOONEY, Ed.), Academic Press, London, 1972, p. 395.
7. J. D. RAY, *J. Chem. Phys.*, **40**, 3440 (1964).
8. W. G. PROCTOR and F. C. YU, *Phys. Rev.*, **77**, 717 (1950).
9. N. F. RAMSEY, *Nuclear Moments*, J. Wiley and Sons Inc., New York, 1953.
10. T. SALUVERE and E. T. LIPPMAA *Eesti NSV Tead. Akad. Toim. Füüs.-Matem.*, **20**, 91 (1971).
11. T. SALUVERE and E. T. LIPPMAA *Chem. Phys. Letts.*, **7**, 545, (1970).

12. A. A. MARYOTT, T. C. FARRAR, and M. S. MALMBERG, *J. Chem. Phys.*, **54**, 64 (1971).
13. R. A. OGG and J. D. RAY, *J. Chem. Phys.*, **26**, 1339 (1957).
14. J. A. POPLE, W. G. SCHNEIDER, and H. J. BERNSTEIN, *High Resolution NMR*, McGraw-Hill, New York, 1959.
15. N. F. RAMSEY, *Phys. Rev.*, **78**, 699 (1950).
16. D. E. O'REILLY in *Progress in NMR Spectroscopy*, Vol. 2 (J. W. EMSLEY, J. FEENEY, and L. H. SUTCLIFFE, Eds.), Pergamon Press, London, 1967, p. 1.
17. A. SAIKA and C. P. SLICHTER, *J. Chem. Phys.*, **22**, 26 (1954).
18. J. W. EMSLEY, *J. Chem. Soc.*, A, 1387 (1968).
19. T. TOKUHIRO and G. FRAENKEL, *J. Amer. Chem. Soc.*, **91**, 5005 (1969).
20. A. VELENIK and R. M. LYNDEN-BELL, *Mol. Phys.* **19**, 371 (1970).
21. H. KATO, H. KATO and T. YONEZAWA, *Bull. Chem. Soc. Japan*, **43**, 1921 (1970).
22. A. J. SADLEJ, *Org. Mag. Res.*, **2**, 63 (1970).
23. R. GRINTER and J. MASON, *J. Chem. Soc.*, A, 2196 (1970).
24. M. WITANOWSKI, *J. Amer. Chem. Soc.*, **90**, 5683 (1968).
25. W. H. FLYGARE and J. GOODISMAN, *J. Chem. Phys.*, **49**, 3122 (1968).
26. V. M. S. GIL and J. N. MURRELL, *Trans. Faraday Soc.*, **60**, 248 (1964).
27. J. E. KENT and E. L. WAGNER, *J. Chem. Phys.*, **44**, 3530 (1966).
28. M. WITANOWSKI, *Tetrahedron*, **23**, 4299 (1967).
29. TING KAI WU, *J. Chem. Phys.*, **49**, 1139 (1968).
30. TING KAI WU, *J. Chem. Phys.*, **51**, 3622 (1969).
31. D. N. HENDRICKSON and P. M. KUZNESOF, *Theoret. Chim. Acta.*, **15**, 57 (1969).
32. M. WITANOWSKI and S. A. SHEVELEV, *J. Mol. Spectrosc.*, **33**, 19 (1970).
33. M. WITANOWSKI, L. STEFANIAK, H. JANUSZEWSKI, and G. A. WEBB, *Tetrahedron*, **27**, 3129 (1971).
34. M. WITANOWSKI, L. STEFANIAK, H. JANUSZEWSKI, Z. GRABOWSKI, and G. A. WEBB, *Tetrahedron*, **28**, 637 (1972).
35. J. A. POPLE, *Disc. Faraday Soc.*, **34**, 7 (1963).
36. J. MASON and W. VAN BRONSWIJK, *J. Chem. Soc.*, A, 1763 (1970).
37. F. AUBKE, F. G. HERRING, and A. M. QURESHI, *Can. J. Chem.*, **48**, 3504 (1970).
38. L. O. ANDERSSON, J. MASON, and W. VAN BRONSWIJK, *J. Chem. Soc.*, A, 296 (1970).
39. J. B. LAMBERT and J. D. ROBERTS, *J. Amer. Chem. Soc.*, **87**, 4087 (1965).
40. J. MASON and W. VAN BRONSWIJK, *J. Chem. Soc.*, A, 791 (1971).
41. M. WITANOWSKI and H. JANUSZEWSKI, *Can. J. Chem.*, **47**, 1321 (1969).
42. R. DITCHFIELD, D. P. MILLER, and J. A. POPLE, *J. Chem. Phys.*, **54**, 4186 (1971).

43. N. F. RAMSEY, *Phys. Rev.*, **91**, 303, (1953).
44. J. N. MURRELL, in *Progress in NMR Spectroscopy*, Vol. 6 (J. W. EMSLEY, J. FEENEY, and L. H. SUTCLIFFE, Eds.), Pergamon Press, London, 1970, p. 1.
45. M. BARFIELD and D. M. GRANT, in *Advances in Magnetic Resonance*, Vol. 1 (J. S. WAUGH, Ed.), Academic Press, New York, 1969, p. 149.
46. M BARFIELD and B. CHAKRABORTI, *Chem. Rev.* **69**, 757 (1969).
47. H. M. McCONNELL, *J. Chem. Phys.*, **24**, 460 (1956).
48. J. A. POPLE and D. P. SANTRY, *Mol. Phys.*, **8**, 1 (1964).
49. T. AXENROD, M. J. WIEDER, G. BERTI, and P. L. BARILI, *J. Amer. Chem. Soc.*, **92**, 6066 (1970).
50. J. A. POPLE and D. L. BEVERIDGE, *Approximate Molecular Orbital Theory*, McGraw-Hill, New York, 1970.
51. J. A. POPLE, J. W. McIVER, and N. S. OSTLUND, *Chem. Phys. Letts*, **1**, 465 (1967).
52. J. A. POPLE, J. W. McIVER, and N. S. OSTLUND, *J. Chem. Phys.*, **49**, 2960, 2965 (1968).
53. G. E. MACIËL, J. W. McIVER, N. S. OSTLUND, and J. A. POPLE, *J. Amer. Chem. Soc.*, **92**, 1, 11, 4151, 4497, 4506 (1970).

CHAPTER 2

Experimental Aspects of Nitrogen NMR

E. W. Randall

Department of Chemistry, Queen Mary College, London

2.1 Scope

The main aim of this chapter is to enumerate and discuss the experimental methods and operational procedures for nitrogen magnetic resonance studies devoid of theoretical details on the one hand and details of electronic circuitry on the other. Since the required techniques are obviously not restricted to nitrogen nuclei solely, they have been individually, but not collectively, covered in detail elsewhere, as will be pointed out.

An outline approach has been adopted and nitrogenous examples, where these are available, are employed in a hopefully illustrative way. The latest reviews [1, 2, 15] of nitrogen NMR contain discussions of experimental techniques, but such is the rapidity of change that they do not contain examples of the latest useful accoutrements; the pulse box and the Fourier transform device.

2.2   Introduction

The experimental requirements for the detection and accurate measurement of nitrogen resonance signals are set by the characteristics of each of the two isotopes of magnetic interest, namely, ^{14}N and ^{15}N. These are principally:—the magnetic moments, the abundances, the relaxation characteristics and the shift ranges. Of these, the shift ranges are identical, at least to the current limit of detection, i.e. isotope effects on the screening constant are negligible, [3, 4] and the magnetic moments are almost equal in magnitude although they differ in sign [5, 6]:

$$\mu(^{14}N) = +0.40357; \quad \mu(^{15}N) = -0.28304 \text{ nuclear magnetons}$$

These two factors produce no difference in single resonance technique between the isotopes, and the basic resonance frequencies at the same field strength are of comparable magnitude (Table 2.1).

TABLE 2.1

Nucleus	Relative resonance frequencies (MHz) (at $B_0 = \sim 2.3$ T)	Natural abundance (%)
^{14}N	7.224	99.64
^{15}N	10.133	0.36
^{1}H	100.000	99.98

The main experimental differences between ^{14}N and ^{15}N NMR studies arise from the widely different abundances and the crucially different relaxation times of the two nuclei.

The 300-fold superiority in abundance of ^{14}N over ^{15}N, shown in Table 2.1, obviously gives ^{14}N studies a considerable advantage from the point of view of sensitivity. This has been the case to such an extent that ^{15}N NMR studies have been very few, and until very recently were confined to ^{15}N-enriched samples [1, 2].

On the other hand, the commonly short relaxation times for ^{14}N nuclei (Chapters 1 and 3) give rise to very broad lines, and linewidths of a 100 Hz or more are the rule, whereas ^{15}N nuclei characteristically give lines of 1 Hz width or less. The linewidth problem for ^{14}N nuclei therefore restricts high resolution work on nitrogen, except in a few special cases, to the ^{15}N isotope. Work at natural abundance then presents problems of very high sensitivity, and successful studies have been accomplished only since 1971 by bringing the techniques of proton noise-decoupling, spectral accumulation and pulsed Fourier NMR into conjunction [7, 8].

The order of presentation of these techniques in this chapter can be most readily understood from a consideration of the general characteristics of a high sensitivity spectrometer. Most modern spectrometers employ three main types of radiofrequency and a constant magnetic field, B_0. One frequency, f_0, is used to stimulate a 'lock' signal in a stabilization channel to keep B_0 constant to a very high order of accuracy. Another frequency, f_1, is detected to monitor the spectrum. The third frequency, f_2, can be used in a double resonance channel to perturb the signals detected at f_1 in a favourable way. Thus f_0 and f_1 each have a transmitter and a receiver, whereas f_2 has only a transmitter. The three frequencies may be derived from the same synthesizer or be effectively locked together.

Experiments fall into two categories: continuous wave experiments in which the frequencies are always switched on, and pulsed experiments in which normally one frequency is gated.

Two general approaches are possible, the DIRECT MODE of observation in which f_1 is set at the resonance frequency of interest, and the INDIRECT MODE in which the nucleus of interest has a resonance frequency near f_2 but is monitored by observation of f_1 via a scalar coupling of the two types of spins.

In the DIRECT mode the f_1 frequency is set in the ^{14}N or ^{15}N spectral regions and is either swept, frequency sweep continuous wave case, or is pulsed. Of the various pulsed experiments we shall be concerned here only with the simplest—giving normally a single pulse with a pulse angle of about $90°$ (Fourier transform method) which is sufficient to give the normal spectrum. Other pulse experiments, which we shall not consider here, can give valuable results for relaxation times. In either arrangement, the continuous wave or Fourier mode case, weak signals from ^{15}N nuclei in natural abundance further weakened by coupling (between say ^{15}N and protons), may be collapsed to a single more intense line by irradiation of the sample with f_2 under suitable conditions. The coupling information is sacrificed in order to gain the precious signal. This is the complete decoupling version of the double resonance experiment. In a well-established notation [9] these double resonance experiments are conveniently denoted by ^{15}N (^1H) and ^{14}N (^1H) for fixed field experiments, and by ^{15}N $\{^1H\}$ and ^{14}N $\{^1H\}$ for field swept experiments.

The alternative use of the double resonance technique is to provide the INDIRECT method. In this case f_2 is set near the frequency of the nucleus being studied and the effect of varying f_2 on the signal observed at f_1 is monitored either by sweeping f_1 through the region of interest for discreet steps of f_2, or by confining f_1 accurately to a transition frequency in the f_1 region and sweeping f_2 (INDOR technique). This may be repeated for different f_1 transitions. These experiments are denoted ^1H (^{14}N) or ^1H (^{15}N). The corresponding field swept experiments are denoted by ^1H $\{^{14}N\}$ and ^1H $\{^{15}N\}$. The indirect method involves building a montage of spectra if the full spectrum of line positions is to be recorded. The technique does not detect any f_2 transitions which do not have energy levels in common with f_1 transitions.

The INDIRECT method is thus time consuming and not of general utility. Nevertheless it can be accurate and offers some sensitivity gain, as can be seen in Section 2.7.

The background details for all these particular techniques as applied to nitrogen NMR are discussed below, after a consideration of the locking, referencing and spectral accumulation procedures, as well as some general theoretical aspects which are common to all the techniques.

A useful account of spectrometer design has recently been given by Gillies [10].

2.3 Lock and Referencing Procedures

The favoured way of referencing in NMR has been to use an internal standard containing an isotope which is homonuclear with the isotope being studied. Tetramethylsilane (TMS) is the familar proton standard, and it has been adopted by a large number of people for ^{13}C work also.

The reasons for using internal standards are: (i) susceptibility effects on shifts are absent, or at least are the same for the sample and the reference; (ii) the stability requirement on the spectrometer is maintained and not broken as it is when sample substitution is employed.

External referencing is useful where solvent shifts are being studied, in this case the reference substance cannot interfere with the solution characteristics. It carries the concomitant disadvantage that susceptibility corrections need to be estimated and applied. Normally these should be only about 0.2 p.p.m or less.

Homonuclear referencing was the initial fashion because the first spectrometers were single frequency f_1 instruments. In double channel developments, where the second channel handled a lock signal f_0, it became possible to use this lock signal as the reference. The second frequency was obtained in the early 'locked' instruments by audio side-banding techniques. The small audio frequencies thus restricted the lock to the homonuclear case.

Many of the early double resonance experiments were heteronuclear by contrast, and involved two different radiofrequencies [9]. This allowed the use of heteronuclear referencing in the INDIRECT METHOD. For example a proton signal split by a nitrogen-15 coupling could be monitored while the nitrogen f_2 frequency was altered until the proton signal was perturbed in a known way. Simultaneous measurement of the two radiofrequencies gives the essential frequency calibration. The observed proton resonance can be related to a preferred proton standard, e.g. tetramethylsilane by a secondary, homonuclear, shift comparison [11].

Each of these corrections of course introduces an error, so that there is considerable advantage in avoiding the last correction to the homonuclear standard by using the heteronuclear result for the highest precision. Thus Price [4], and Becker *et al.* [3], in studies of the isotope effect on the screening constants of ^{14}N and ^{15}N nuclei used this method. It requires a radiofrequency counter to measure to sufficient accuracy the heterofrequencies arising from the heteronuclei.

It is obvious, however, that a homonuclear standard is more convenient, and, provided the convenience is worth the loss of accuracy, homonuclear standards are used.

For the case of low-abundance isotopes like ^{13}C and ^{15}N homonuclear locking is not feasible unless enriched material is used, either internally, which is expensive, or externally, which gives the susceptibility error. Strong heteronuclear resonances are therefore to be preferred. The proton was the normal nucleus first used in this way, but it became inconvenient with the advent of the noise decoupling of protons, since the lock channel is then irradiated with high-intensity noise. The modern preference for lock nuclei is either ^{19}F or ^{2}D. The former has the advantage of high sensitivity, 0.83 relative to the proton. However, it has the disadvantage that its resonance frequency is relatively close to the proton resonance frequency and the associated noise-modulated frequency f_2 so that noise pick-up is a problem unless careful filtering is employed. The nucleus ^{2}D does not present this difficulty, its resonance frequency is 15.35 MHz at 2.3 T but it does have the disadvantage of low sensitivity (9.6×10^{-3} relative to ^{1}H). Fortunately for most solutions a deuterated solvent may be used; this normally ensures that sufficient lock signal is present. A possible nuisance in high-resolution work is that for some solvents the linewidth of the lock signal may be rather large.

For the latest spectrometers, which allow heteronuclear locking and which have very high stability, the use of external locking is growing. Just as for external referencing, one advantage is that the lock compound does not interfere with the solution being investigated, except through the bulk susceptibility effect, and conversely the lock signal is not affected either.

One interesting development for improving lock stability, which has been slow to become commercially available, is the so called 'time-shared' technique [12-14]. As seen in Section 2.5.1. 'time-sharing' may be used to advantage on f_1 and f_2 channels as well. For f_0 and f_1 the basic detection problem concerns the interaction between the transmitter and the receiver of each frequency. The receiver must be set to receive no signal from the transmitter except via the sample at resonance. This is accomplished in single coil instruments by bridge balancing. In double coil instruments the coils are set so as to be effectively orthogonal by the use of controls which minimize the leakage between the coils. Bridge balancing and leakage are each rather unstable with time, probably due to the sensitivity of their components to temperature and vibration effects. A simple but elegant solution is to gate the transmitter and receiver so that one is off while the other is on. Interaction is thus minimized by time displacement. The gating must be fast enough on the lock channel to deal with the field fluctuations for the system in question, and the time must be shared in such proportion to give the highest sensitivity. Although the signal detected is expected to drop because of the sharing of time *per se*, the noise is cut to such an extent that overall a sensitivity gain is realized. Factors of about 10 may be reached. This technique should prove most useful for ^{2}D locks for which the inherent sensitivity is low.

The problems of reference compounds for nitrogen NMR and sign conventions for chemical shifts are discussed in Chapter 4.

2.4   General Methods of Increasing Sensitivity

The general principles of sensitivity enhancement are:

  (i)   to maximize the total available magnetization,
  (ii) to monitor it effectively,
  (iii) to repeat the monitoring procedure and accumulate the results if necessary.

The question of how to monitor the magnetization is dealt with in some detail in Sections 2.5 and 2.7. The maximization and repetition of the monitoring are considered here.

2.4.1   Maximization of The Available Magnetization

The sensitivity of the experiment is obviously increased if the total magnetization available can be increased. This may be accomplished by developing different facets of the magnetic field: its field strength, B_0, and its homogeneity over large sample volume. This is evident on consideration of the expression for the signal strength S:

$$S \propto \frac{nB_0{}^2\gamma^3}{T} \qquad (2.1)$$

where n is the total number of nuclei and T is the absolute temperature. The effects of n, B_0 and T are considered below.

Increase of the Number of Nuclei
  n may be increased by enlarging the operational volume of the instrument. This volume is defined by the dimensions of

the f_1 coil. One of the main limitations to expansion of the volume is the homogeneity requirement, since for high resolution work there is a toleration limit towards degradation of field homogeneity and the associated linewidth.

Modern magnets can give sufficient homogeneity, about 1 Hz linewidth, across cylindrical tubes of 10–13 mm diameter. The approximate gain in the signal-to-noise (S/N) ratio compared with tubes of 5 mm diameter is thus between $(10/5)^2$ and $(13/5)^2$, i.e. by a factor of about 5. The time saving is thus nearly 25-fold (Section 2.4.2).

For samples not critically restricted in amount the increased sensitivity can be used to shorten the time of the experiment. Alternatively, high dilution effects may be studied either deliberately, e.g. in hydrogen bonding studies on amides enriched in ^{15}N at very low concentration, or with reluctance, e.g., in the case of inconveniently low solubility. Natural abundance studies of low-abundance nuclei like ^{15}N are also rendered easier. However, for samples which may be obtained only in restricted amounts, such as specifically ^{15}N labelled compounds prepared only with difficulty, the potential gain cannot of course be realized.

Field Strength

The signal may be improved by raising the field strength, B_0. The limit for electromagnets is about 2.3 T (23,000 G). The use of superconducting solenoids allows this to be more than doubled, so that the theoretical gain should be a factor of 4 [equation (2.1)]. This may be realized only if the resonance frequency has not been raised to a difficult electronic region. Although difficulty is experienced for protons, promoted in the frequency scale from 100 MHz to 200 MHz or more, this is hardly the case either for ^{14}N or ^{15}N nuclei.

One problem with superconducting solenoids for ^{15}N studies until recently was the low homogeneity of the field. This restricted the tube diameter to 5 mm so that, for samples which are sufficiently available to fill the 13-mm tubes of electromagnets, the use of the superconducting solenoids gave no gain. Currently (1972), however, larger bore superconducting solenoids with good homogeneity are being developed and should prove very useful indeed for ^{15}N and ^{13}C studies in particular. Higher fields of course also allow better separation of chemically shifted peaks. This is of some considerable utility for ^{1}H but of less, perhaps only marginal, utility for nuclei like ^{15}N and ^{13}C endowed as they are with large shift ranges. However, higher magnetic fields can also be very useful for ^{14}N studies, since the spectral resolution can be controlled by the natural linewidth in this case.

The Effect of Temperature

The inverse temperature effect arises from the Boltzman term which describes the population difference between the energy states and is the origin of the Curie Law dependence of paramagnetic susceptibilities in both the nuclear and electronic cases. With normal solvents having a restricted temperature range the gain from this effect is small and is most likely to be offset in a large number of cases by decreased solubility at lower temperatures. However, in noise decoupling experiments, where the f_2 power used may be as high as 10 W, sample heating to 100°C may occur and cooling the sample may give significant improvements. There is also a temperature effect on the relaxation times (Section 2.6).

2.4.2   Spectral Accumulation

A familiar way of improving the S/N ratio in any spectroscopic experiment is to record many spectra under the same conditions and to add them. The random noise accumulates less rapidly than the coherent signals. The spectrometer must be stable in tuning, field strength, resolution and sweep-width for the total time of the experiment, and the spectra must be correctly registered. These conditions are easily, if expensively, realized in modern spectrometers with locked fields and related f_0, f_1 and f_2 frequencies.

The S/N gain over a single experiment is proportional to $\sqrt{n}$, where n is the number of scans, provided the sample is allowed to recover between scans. Note that the time required is proportional to the square of the S/N gain desired.

The technique may be used in conjunction with either the continuous wave or Fourier direct methods, and even with the indirect methods, most commonly the INDOR version [10].

2.5.   Direct Methods of Studying Resonance

The direct methods may be defined as those in which the observing frequency f_1 is set in the nitrogen region. The version which is more familiar to the average chemist from the standard texts [25, 26] is the continuous wave method. Because of its familiarity it will command little more than cursory attention here. The more unfamiliar Fourier pulse method will be outlined in slightly more detail; this method has been reviewed recently [27, 28].

In using either the continuous wave or pulse methods decoupling and Overhauser effects are of considerable help, and these are discussed in Section 2.5.3.

2.5.1 Continuous Wave Method

The frequency f_1 may be swept through the region of interest in two ways: either by varying f_1 itself (frequency sweep), or by varying the field B_0 (field sweep). Most modern spectrometers are operated at a fixed field determined by the frequency of the lock, f_0, and the resonance condition for the lock compound used, as well as by the frequency of the particular modulation side-band used, in the normal event that f_0 is audiofrequency modulated. The important variables in continuous wave measurements are the f_1 power and the sweep rate, and their relation to the relaxation characteristics of the sample.

One seemingly obvious way of increasing the sensitivity of the NMR experiment is to improve the monitoring of the total magnetization by increasing the radiofrequency power. The object is to increase the rate at which transitions are induced between the spin levels, i.e. the rate of power absorption. The net power absorbed is, however, the differential effect between upward and downward transitions, and it depends upon the difference in population between the spin levels, since the transition probabilities are the same. This population difference is maintained at the equilibrium Boltzman value by relaxation processes in the z or longitudinal direction, and are governed by T_1.

The population difference, and the resulting signal, is decreased if the relaxation processes cannot cope with the net upward pumping effect of a high radiofrequency field. We shall see later that double resonance effects on the spin populations may help to offset the pumping action. In other words the sensitivity increase consequent on increasing B_1 is limited by the relaxation time T_1. In the limit when the populations are equated the system is said to be saturated, the magnetization is reduced to zero and no signal is obtained.

The centre of the absorption saturates first, so that increasing the power may boost the wings of a resonance while saturating the centre—in other words the line may broaden.

At the centre of the line the absorption is decreased by the so-called *saturation factor* given by $(1 + \gamma^2 B_1^2 T_1 T_2)^{-1}$, in which B_1 is the field strength of the applied radiation and T_2 is the transverse relaxation time which describes the decay of magnetization in directions perpendicular to B_0. T_2 may be thought to enter the expression because it gives the linewidth. In liquids, $T_1 \simeq T_2$.

One way around the saturation problem is to monitor transient effects, i.e. to conduct the experiment in times shorter than the relaxation times. For example one may sweep rapidly through the resonance condition so that there is not enough time for saturation to occur. This has the limitation that the line position is shifted and the line shape distorted.

This effect is shown in Fig. 2.1 for the field sweep case. The three curves are absorptions for different values of $\sqrt{aT_2}$, where a is the rate of change of the Larmor condition

$$\left(a = |\gamma| \frac{dB_0}{dt} \right)$$

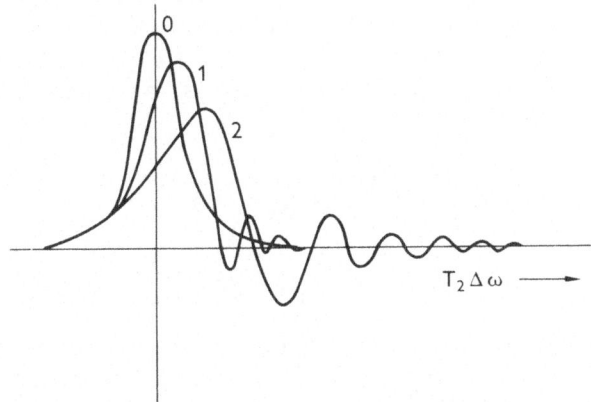

Fig. 2.1   Line shapes for various sweep speeds showing the presence of wiggle beats for the higher sweep speeds given by $\sqrt{a\,T_2}$ = 0, 1 and 2, where

$$a = |\gamma| \frac{dB_0}{dt}$$

(After Jacobsohn and Wangsness [29]).

and $\Delta\omega$ is the difference between the Larmor precession frequency and the fixed frequency, f_1.

The rate at which the resonance is swept through is thus a most important quantity, and the behaviour of the total magnetization depends markedly upon it. The behaviour of the spin system described above under continuous wave conditions and in the pulsed Fourier mode may be regarded as the two extremes.

In the case of ^{14}N nuclei the relaxation times are usually governed by quadrupolar interactions which may vary by two or three orders of magnitude for different nuclei. Consequently the degree of saturation at a given power level may change considerably for different signals in a spectrum. This difficulty may be circumvented by making use of the modulation sidebands resulting from an audiofrequency of a few kHz [16]. The frequency modulation technique is often used for baseline stabilization and for providing analytical and control channels in the field—or frequency—sweep modes of operation. The saturation term for the sidebands is much smaller than for the central band. Consequently, a single sweep experiment extending over the central band and both sidebands at a high power level results in a good S/N ratio for broad signals in the centreband and narrow undistorted resonances in the sidebands. It is most convenient experimentally to adjust the phase reference detector such that the centreband appears in the upright absorption form whilst the sidebands appear in the form of inverted absorption. For unsaturated signals in this form of display the total signal shape $F(f_1)$ may be represented as a combination of Lorentzian lines [17]:

$$F(f_1) = A + Bf_1 + \sum_n C_n b_n^2 \left(\frac{1}{b_n^2 + (f_n - f_1)^2} + \right.$$

$$\left. + \frac{-K}{b_n^2 + (f_n + M - f_1)^2} + \frac{-K}{b_n^2 + (f_n - M - f_1)^2} \right) \quad (2.2)$$

where f_1 is the observing frequency, M is the modulation frequency, $A + Bf_1$ represents the starting background, C_n is the maximum height of a separated signal from nucleus n relative to the true background, $2b_n$ is its half-height width, f_n is the chemical shift of nucleus n, and K is a proportionality factor for the sideband-to-centreband intensity.

The pulse mode constitutes a second way of using high

power. It is applied for a very short time in a pulse of radiofrequency at the resonance condition. The magnetization is now not destroyed but is turned through an angle in a plane at right angles to z. The total magnetization is being effectively used, and this is one view of the sensitivity enhancement obtained in pulse methods.

A useful comparison of these cases may be gained by considering the detailed motion of the spins. For this purpose it is useful to consider the motion in the rotating frame, i.e. in a frame rotating at the applied frequency, ω, whether fixed and pulsed, or swept.

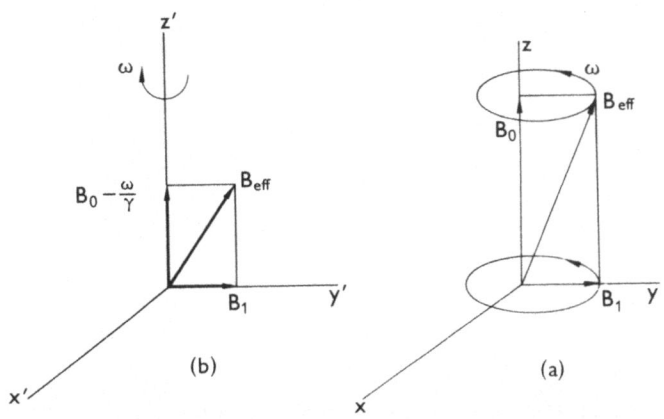

Fig. 2.2 The two applied fields, B_0 (static) and B_1(oscillatory), produce an effective field, B_{eff}. (a) in the laboratory frame, and (b) in a frame rotating at a frequency ω.

The rotation of the frame around the z direction at an angular frequency, ω, produces a magnetic field along this direction opposed to B_0 of magnitude ω/γ shown in Fig. 2.2(b). If the frame is rotated at exactly the Larmor frequency the effect of B_0 is entirely cancelled by ω/γ.

We then have the total magnetization M along the z axis and the individual magnetic moments distributed around the z axis in two cones corresponding to the two spin states (assuming $I = \frac{1}{2}$). Whereas these spins in the laboratory frame are rotating at the Larmor frequency, in the rotating frame they are stationary. We shall use primes (z', y', x') to signify that the rotating frame is being used [Fig. 2.3(a)].

If a field B_1 is suddenly applied along the x' axis in the rotating frame, the magnetization M will immediately precess around the new field and will tip from z' towards y' and eventually to $-z'$ [Fig. 2.3(b)]. The precession around the x' axis is given by the Larmor condition appropriate to the new field B_1, viz., $\omega = \gamma B_1$. The angle, θ, through which the magnetization precesses is given by $\theta = \gamma B_1 t_p$, where t_p (the pulse length) is the time measured from when B_1 is switched on. B_1 and t_p may be chosen so that $\theta = 90°$ ($90°$ pulse).

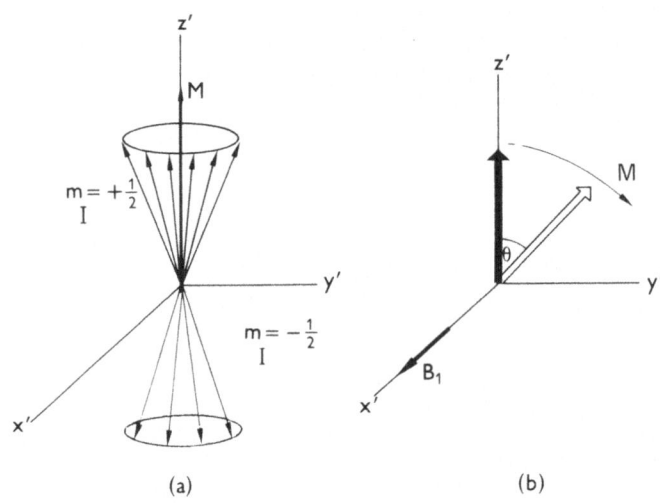

(a)   (b)

Fig. 2.3 (a) The individual spin magnetic moments quantized into two spin states for the $I = \frac{1}{2}$ case, e.g. ^{15}N. (The state of lower energy for ^{15}N has $m_I = +\frac{1}{2}$, whereas for ^{14}N the lowest level of the three appropriate to the $I = 1$ case has $m_I = -1$. The sign change is due to the different signs of the magnetogyric ratios of ^{14}N and ^{15}N.) (b) The effect on M of a field B_1 suddenly applied along the X' axis.

Consider the situation not at resonance in the continuous wave approach shown in Fig. 2.2(a). The field arising from rotation of the frame does not now cancel B_0 and there is developed an effective field, $B_{\text{eff.}}$, whose vector components are

$$\left(B_0 - \frac{\omega}{\gamma}\right) \quad \text{and} \quad B_1$$

In the field sweep experiment it is B_0 which is varied, whereas in the frequency sweep experiment the variable is ω.

Away from resonance the individual spins in the laboratory frame precess around B_0 and the magnetization is along z. Approaching resonance slowly, the spins precess around B_{eff}.

and given time the resulting total magnetization aligns itself along $B_{eff.}$. At resonance $B_{eff.} = B_1$ and the magnetization M is aligned along the x' direction.

This differs from the pulse situation where the magnetization M again is initially along z' and is *suddenly* subjected to the effective field B_1. As before, the individual spins instantaneously precess around $B_{eff.}$, i.e. B_1, but now their resultant magnetization, M, is also precessing around B_1 and is not aligned along it as for the slow passage experiment in the continuous wave approach.

Alignment of M with $B_{eff.}$ is maintained if the sweep rate, either B_0 or ω, is slow according to the expression:

$$\frac{d}{dt}\left(B_0 - \frac{\omega}{\gamma}\right) \ll \gamma B_1^2 \qquad (2.3)$$

This is the so-called adiabatic theorem.

If the sweep is rapid M will not follow $B_{eff.}$. In the laboratory frame M and $B_{eff.}$ will be precessing around z at different frequencies and these will beat together. This is the origin of the so-called wiggle beats in continuous wave operation (Fig. 2.1).

The components of M perpendicular to z decay with a time constant T_2. If the slow passage spectrum contains more than one peak the wiggles from a given peak will beat with the wiggles from another peak. Peak separations may be measured from the resulting beat pattern, the period being $1/\delta$, where δ is the frequency separation of the peaks. As we shall see in Section 2.5.2, similar behaviour occurs in the pulse experiment.

Time Sharing [10]

The device of time sharing may be used on the f_1 channel as usefully as has already been described for f_0 [12–14]. The experiment is simply to switch the f_1 transmitter and receiver on and off alternately. Time sharing is a form of modulation technique in which the wave form is square. There are two variables, the frequency of the switching and the proportion of the time devoted to the transmitter and the receiver—the duty cycle.

The frequency of switching should be high enough so that the associated sidebands do not interfer with the spectrum. Normal modulation frequencies, 2–20 kHz, therefore are appropriate. The switching time is thus short compared with the relaxation times, except possibly for ^{14}N, so that relaxation effects are not important.

The effect of the duty cycle on the S/N ratio is obviously vital. Disregarding questions of leakage, balance and associated noise, signal is lost if the receiver is on for only a proportion, P, of the time. The S/N loss is proportional to $\sqrt{P}$. P can, however, be made large by ensuring that the transmitter is used at reasonably high power just near the saturation limit.

2.5.2   Fourier Pulse Method [27, 28]

As already discussed a short radiofrequency pulse at the resonance frequency will turn the magnetization M through an angle, called the pulse angle, towards the x', y' plane. A 90°

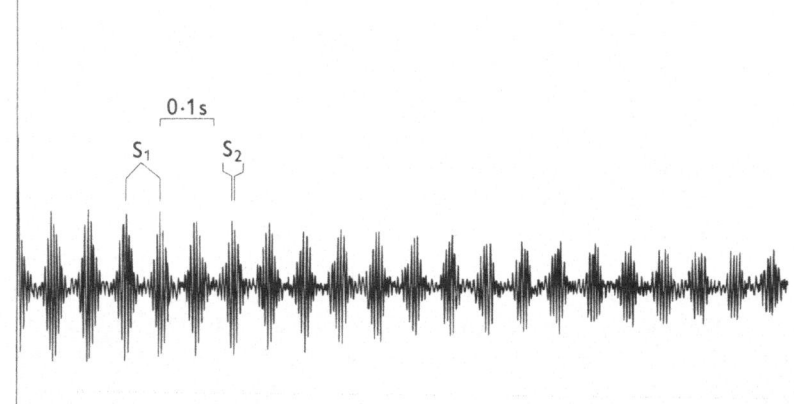

Fig. 2.4   Accumulated ^{15}N free induction decay from a 5 molar sample of 95% enriched ^{15}NH$_4$Cl in 2N-HCl at 9.120 MHz (1024 pulses).

pulse places M along the y' axis. This magnetization may be detected. When the pulse is switched off the system takes some time to recover, i.e. the transverse magnetization decays at a finite rate, governed by the relaxation time T_2. This free induction decay (FID) contains all the normal spectral information. For example, if the frequency of the pulse is not exactly at the resonance condition a beat is observed between the carrier frequency and the Larmor frequency. This is superimposed on the decay If more than one peak is present there will be two superimposed decays.

Figure 2.4 shows part of the free induction decay of the ^{15}N resonance at 9.120 MHz for the ^{15}NH$_4^{\oplus}$ ion. The two main spacings in the decay S_1 and S_2 are determined by the two frequency differences which characterize the spectrum, the chemical shift relative to the carrier frequency and the coupling constant. For the conditions shown the shift is larger than the coupling and corresponds to the smaller time, S_2.

The reciprocal relation between the frequencies in the frequency-dependent spectrum and the time intervals holds for other important characteristics of the spectrum. The largest frequency separation is the spectral width, W. This corresponds in the time domain to the smallest time interval. If the decay is stored in a multichannel analyser or computer this time is the interval between adjacent channels, i.e. it is the dwell time, t_d. The smallest frequency difference of

TABLE 2.2.   Frequency and Time Equivalents

Frequency domain	Time domain
Resolution	Acquisition time (T_a)
Spectral width monitored	Dwell time (t_d)
Spectral width excited	Pulse length (t_p)

interest in the spectrum is the resolution, this corresponds to the largest time interval—the acquisition time T_a. The verbal transformation of phrases in the time domain into the frequency domain is given in Table 2.2.

An important point concerning the FID is that each point contains information from every part of the spectrum. It is apparent from Fig. 2.4 that S_1 and S_2 are repeated, and that observation of all of the FID is not necessary for their measurement. Discarding the last part of the decay does not *prevent* the measurement of S_1 and S_2 but it does reduce the *quality* of the measurement. It is tantamount to decreasing the acquisition time. This worsens the resolution and, if the signal discarded is not recouped, the S/N ratio.

The Fourier transform of the FID gives the normal frequency spectrum. This is true even if the pulse angle is not 90°: other angles merely reduce the S/N ratio by reducing the $M_{x'y'}$ magnetization.

Fast Fourier transformations may now be accomplished with small digital computers. A block diagram of a Fourier spectrometer is shown in Fig. 2.5.

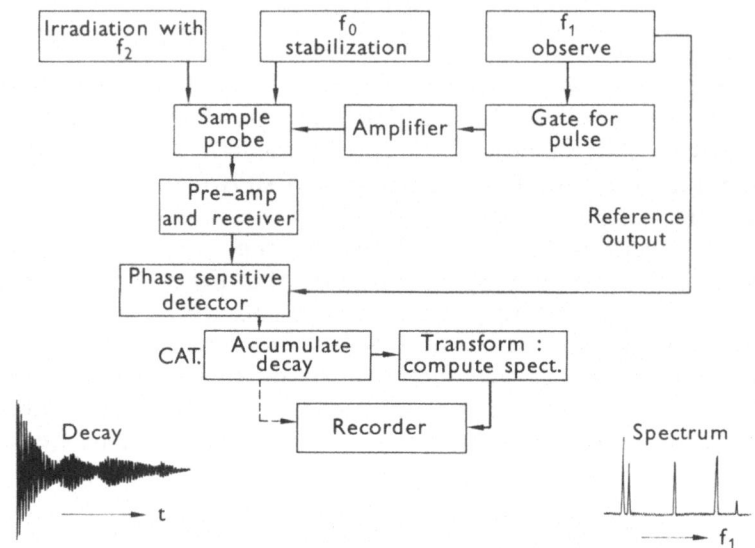

Fig. 2.5   Block diagram of a three-channel NMR spectrometer including Fourier attachments.

Sensitivity

One simple view of sensitivity enhancement using the Fourier mode is to regard the pulsing as giving a large number of sidebands of equal spacing around the carrier frequency. The spacing is governed by $1/T_a$ where T_a is the acquisition time. This spacing determines the resolution. The sidebands extend over a frequency range determined by $1/t_p$, where t_p is the pulse length. The envelope of the sidebands is given by the function $\sin x/x$ where x is $n\pi t_p/T_a$. t_p should be short enough to ensure that the sidebands which cover the spectrum are in the narrow region in the centre of the envelope and have the same intensity.

These frequencies are applied simultaneously to the spin system and not successively as in say a digitized sweep experiment, and the responses of all the spins are also gathered simultaneously. The rate at which information is garnered is thus greater, so that in a given time more signal is obtained.

Ernst and Anderson [30] have compared the S/N ratios for slow passage continuous wave and Fourier experiments taken in the same total time with accumulation, presuming that equilibration occurs between pulses in the Fourier mode [31]. The Fourier spectrum is expected to be superior by a factor of $[W/\Delta]^{1/2}$, where W is the spectral width and Δ is a typical linewidth.

If $W = 1$ kHz and $\Delta = 1$ Hz then the S/N gain is $(1000)^{1/2}$ or alternatively, the time gain for an equivalent S/N is 1000. In practice the gains obtained from utilizing Fourier methods are less than this, for two reasons. Firstly, the continuous wave mode as normally employed corresponds more closely to intermediate passage rather than the slow passage presumed above. Secondly, as we shall see below, some saturation effects may occur in the Fourier mode. Practical S/N gains exceeding a factor of 10 can, however, be obtained.

It may be noted that in the limit where a single line of known position is being monitored at narrow sweep-width there is no advantage in the Fourier method. It follows that for a spectrum with many lines of known position a frequency sweep experiment programmed to omit the regions containing no signal would be equivalent to the Fourier method. Such an experiment might be of some utility for analytical purposes on known samples. Most experiments, however, involve the elements of search even if the spectrum contains only one line, so that the Fourier mode is still superior.

The feasibility of natural abundance ^{15}N Fourier work at high resolution has been demonstrated only recently [7, 8]. For liquid organic samples of low molecular weight contained in 10-mm tubes, S/N ratios of about 20 : 1 from nitrogen signals enhanced by the proton Overhauser effect were obtained in about 1 hr utilizing 4 K data points.

Solubility and exchange problems can produce severe handicaps as exemplified by work on amino acids even when enriched by either continuous wave [32], or Fourier techniques [33]. The use of the methyl esters in acid solution, however, increases the solubility to about 5–9 M and overcomes the exchange problem to such an extent that good spectra are obtained in about 5 hr for samples with a natural abundance of ^{15}N [8]. Two typical examples are given in Fig. 2.6.

Line Intensities

Different spins in the system decay at different rates according to the individual values of T_2. For a single pulse, which has a pulse length much shorter than the shortest relaxation time, differential relaxation effects between the various spins are not manifest in the spectrum except that the linewidths may be different. The spins do not relax appreciably during the pulse and so they are all turned through the same angle.

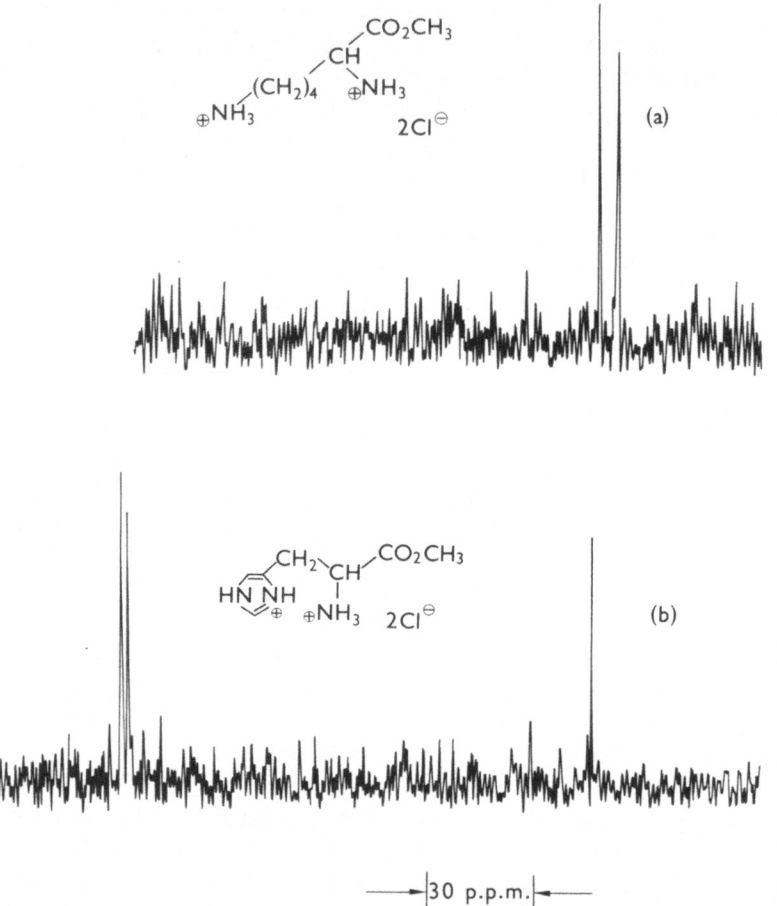

Fig. 2.6   Natural abundance [15]N Fourier spectra at 9.120 MHz of (a) lysine methylester dihydrochloride, and (b) histidine methylester dihydrochloride in aqueous solution at pH between 1 and 2. The spectra are the result of 20,000 pulses each, (4 hr total acquisition time) the samples were contained in 10-mm tubes with 5-mm concentric tubes containing C_6F_6 as lock compound.

In a multipulse experiment, the above holds true provided the magnetization returns to its equilibrium value between the pulses. This will be so if the pulse repetition time is long compared with any relaxation time in the system. This can be accomplished by having a long acquisition time, T_a, or by putting in a delay between the end of an acquisition and the start of the next pulse.

If, however, the repetition is too rapid for some of the spins then the dynamic steady state giving the average pulse angle will be less than for the other spins, and the intensities of the lines from the spins with long T_1 values will suffer relative to the spins with shorter T_1's. This effect is comparable to differential saturation effects in the continuous wave mode. One difference, however, is that in the Fourier experiment the linewidth does not increase from this cause.

In cases where there is a very wide difference in relaxation times it is conceivable that some resonances may not be detected. There is thus a parallel dichotomy in the continuous wave and Fourier modes: lines may be missed because they are broad due to short relaxation times, or because they are narrow due to long relaxation times and saturated. ^{14}N resonances characterized by very rapid, electric quadrupole induced relaxation might be an example of the first case, and ^{15}N resonances from quaternary nitrogens, which might be expected to have very long relaxation times, might constitute examples of the second type. The use of T_1 reagents in the latter case is discussed in Section 2.6.

Resolution

For a given number of channels or data points, increasing the dwell time, i.e. the time interval between channels, increases proportionately the acquisition time, T_a. In other words, decreasing the spectral width, W, increases the resolution. The proportionality constant is $N/2$, where N is the number of data points. Thus:

$$2W \cdot T_a = N \qquad (2.4)$$

If T_a is made long in order to gain resolution then W, the resulting spectral width, will be small. Of course, increasing N by using a larger computer enables higher resolution to be obtained for a given spectral width.

Since the signal in the time domain decays, the later parts of the decay do not add greatly to the sensitivity except for any spins with very long relaxation times. The requirement for high resolution therefore is opposed to the sensitivity requirement. Again, this is similar to the situation in continuous wave experiments in which fairly rapid passage through the spectrum may be used to gain signal but results in line distortion and spectral shifts.

2.5.3   Decoupling and Overhauser Effects

The direct methods of measurement of chemical shifts can be aided by decoupling the nitrogen spins from nuclei to which they are coupled. The most common couplings involve hydrogen, so that proton noise decoupling finds useful application in nitrogen as in ^{13}C NMR spectroscopy. In noise decoupling, f_2 is modulated randomly or pseudo-randomly so that all transitions in the f_2 region are effectively excited [34–36]. In this way all the nitrogen nuclei in a given sample may be decoupled simultaneously from each type of hydrogen in the molecule [7, 8].

In the case of ^{14}N for which the sensitivity problem is not so great such decoupling may prove to be unnecessary, since ^{14}N is usually effectively decoupled from other spins anyway because of the normally fast relaxation rate (Chapters 1 and 3). For ^{15}N, especially in natural abundance, decoupling can be crucially important. From the effect of multiplet collapse alone significant S/N gains may be obtained. For doublets the expected gain is two-fold, the same as it is for triplets relative to the central peak of the 1 : 2 : 1 pattern. For quartets and quintets this gain is 8/3 compared to the largest peaks in the multiplets.

Additionally one may expect significant enhancement due to the redistribution of spin population among the spin levels under the action of the strong decoupling conditions. This is the well-known nuclear Overhauser effect—NOE [9]. The importance of the NOE enhancement of ^{13}C resonances in natural abundance is well known [34–36]. Since ^{15}N work in natural abundance is even more difficult, the NOE effect is expected to be increased in importance for ^{15}N studies. Unfortunately, complications arise for nitrogen, which are not normally present for ^{13}C.

If dipole–dipole interactions dominate the relaxation mechanism in an AX spin case, the S/N gain, G, from the Overhauser effect is a maximum:

$$G_{\text{max.}} = 1 + \frac{\gamma_X}{2\gamma_A} \tag{2.5}$$

where γ_X is the magnetogyric ratio for the second nucleus.

For ^{13}C and protons, γ_X/γ_A is 3.97 and the value of $G_{max.}$ is 2.99. For ^{15}N the magnetogyric ratio is negative and $\gamma_{^1H}/\gamma_{^{15}N} = -9.867$, and the value of $G_{max.}$ is -3.93. The negative sign means that the decoupled signal should be inverted (Fig. 2.7) [7, 32]. For ^{14}N, $\gamma_{^1H}/\gamma_{^{14}N} = +13.88$ consequently $G_{max.}$ is 7.94. This positive value implies that the decoupled signal is not inverted.

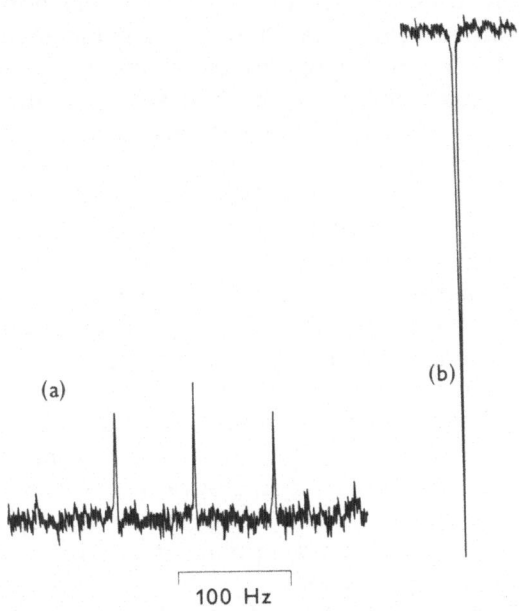

, Fig. 2.7 ^{15}N spectra of 5M-NH$_4$Cl in 2N-NCl 95% enriched in ^{15}N, at 9.120 MHz. The spectra were obtained under single passage continuous wave conditions. Spectrum (a) is undecoupled, and (b) is wide-band noise decoupled at the proton resonance frequency (90 MHz).

If dipole–dipole interactions are not dominant then the term $\gamma_X/2\gamma_A$ in equation (2.5) is multiplied by a factor representing the contribution of these interactions to the total relaxation time. For ^{14}N nuclei this leads to values of $G_{max.}$ in the range from 1, representing no gain in intensity, to 7.94. The corresponding ranges for ^{13}C and ^{15}N nuclei coupled to protons are from 1 to 2.99 and from 1 to -3.93 respectively. Since the $G_{max.}$ range for ^{15}N covers both positive and negative values, the NOE may lead to a decrease in signal intensity or even to a complete disappearance of the signal.

It is interesting to reflect that the low values of the nitrogen magnetogyric ratios relative to the proton value, which reduce the sensitivity of direct NMR experiments, nevertheless can in certain circumstances ensure large compensatory Overhauser effects. This facet of the double resonance experiment is also useful in the indirect double resonance experiments described in Section 2.7.

For cases more complex than the simple AX, differential enhancements can occur because of long range dipole–dipole interactions (Section 2.6). If however, the one bond dipole–dipole mechanism dominates then the enhancements should be the same for all nitrogen atoms which are bonded to at least one hydrogen. Thus in phenylhydrazine the intensities of the ^{15}NH and $^{15}NH_2$ nitrogens are found to be the same [7].

If relaxation effects other than the dipole–dipole interaction contribute to the relaxation mechanism then $G_{max.}$ is not obtained. This provides a useful tool for the study of the proportion which the dipole–dipole mechanism makes to the total relaxation. Other mechanisms are electric quadrupole, spin-rotation and chemical anisotropic mechanisms (Section 2.6). Thus in the $^{15}NH_4^{\oplus}$ ion the observed value for G, taken in acid solution so as to cut down the exchange rate, is -2.2 (pH ~ 0; $50°C$). This reduction from $G_{max.}$ is probably due to significant contributions to the relaxation from the spin-rotation effect [32]. Other deviations from $G_{max.}$ have been noted by Lippmaa et al., for a variety of ^{15}N-containing molecules [24].

For $^{14}NH_4^{\oplus}$ in acid solution, the collapsed peak is indeed positive but little Overhauser gain is found [39]. The differential effect between the $^{15}NH_4^{\oplus}$ and $^{14}NH_4^{\oplus}$ cases shows that the electric quadrupole effect of ^{14}N is dominant even in the highly symmetrical case of tetrahedral symmetry.

In the case of labile N–H systems an additional possible effect on $G_{max.}$ is due to proton exchange [33]. Exchange modulation of the scalar ^{15}N–H coupling does contribute to the total relaxation.

Lichter and Roberts have reported pH effects upon G for the $^{15}NH_4^{\oplus}$ ion [32]. Since pH effects upon the proton exchange are well established it is likely that the scalar modulation mechanism is significant. As Lichter and Roberts are careful to point out, however, this is not definitively established, since the resultant value of G has not been observed to exceed $+1$.

The Overhauser effect may be negated purposely by the

addition of paramagnetic material—T_1 reagents. This technique may be of considerable use where exchange constitutes a problem.

The general conclusion concerning the NOE for ^{15}N systems is that because of the effect of the negative magnetogyric ratio, its utility for signal enhancement may be seriously impaired in situations where the dipole–dipole mechanism is not dominant. The compensation may be of increased utility in the study of relaxation processes.

For ^{14}N proton noise decoupling experiments are likely to be of little use as far as enhancement is concerned. Normally the ^{14}N lines are very broad and the spins are already effectively decoupled. No Overhauser effect is expected in general, since none was found for the most favourable case of ^{14}NH$_4^{\oplus}$ [37].

An interesting modification of the normal Overhauser enhancement experiment has been described for ^{13}C by Feeney et al. [38]. The idea is to conduct an observation with f_1 immediately after switching off f_2. Since the decoupling effect is established (and broken) instantaneously, the coupling is restored as soon as f_2 is switched off. On the other hand, the Overhauser effect on the spin populations is governed by the relaxation effects which restore the populations to their equilibrium values. It follows that a measurement made reasonably quickly relative to the relaxation times will give an enhanced spectrum without decoupling. This has been demonstrated by Feeney et al. using continuous wave monitoring. The gain is marginal. Obviously, the faster Fourier sampling is advantageous, and an automated sequence to allow accumulation has been described for ^{13}C [39]. The S/N gain is improved to 2.4–3.0. This is yet another version of 'time sharing' in which the gating involves f_1 and f_2 alternately. An extension to ^{15}N studies is feasible [37].

Off-Centre Double Resonance [40, 41]

It is possible to choose the power and frequency conditions, B_2 and f_2, so that the multiplet peaks are not completely collapsed but exhibit essentially the correct multiplicity arising from the largest, normally one bond, couplings with a splitting smaller than the proper J value. The so-called 'effective coupling' $J_{\text{eff.}}$ is governed by the expression

$$J_{\text{eff.}} \propto \Delta f_2 J \qquad (2.6)$$

where Δf_2 is the difference between f_2 and the resonance condition, i.e. it is the offset parameter.

This technique is generally useful in a number of ways. Firstly, it allows the signal multiplicity to be determined, so that N, NH, NH_2 and $^\oplus NH_3$ situations can be distinguished formally. In the case of ^{13}C spectra, which are normally very rich in resonances, the technique is very useful. For nitrogen one may anticipate less utility if only because the extent of the initial ambiguity is likely to be less, there being fewer resonances to worry about. It will be of little value for ^{14}N for which spin-spin effects are not normally observed.

Secondly, it allows some retention of the Overhauser effect, as in the time-shared experiment mentioned above. Complications may arise here for ^{15}N, however, because of the Overhauser inversion: i.e. reduction of the effect as in the case of exchange modulation may lead to small signals, and to differential effects between different lines in the spin multiplet.

Thirdly, if the relative values of Δf_2 and $J_{eff.}$ for different lines in the spectrum are found, they enable one to calculate the relative J values.

2.6 Relaxation Times

Relaxation phenomena are crucial to the NMR experiment either in the continuous wave or Fourier mode as we have seen. The ^{14}N nucleus possesses an electric quadrupole moment and relaxation effects for this isotope are dominated by the quadrupoler mechanism. The details are covered in Chapter 3 and are not considered further here.

For ^{15}N, Lippmaa and co-workers have investigated the various contributions to the spin lattice relaxation time T_1 by using isotopic substitution of ^2D for ^1H, temperature and magnetic field effects on T_1, and temperature variation of the NOE [24].

The possible relaxation mechanisms, excluding electric quadrupole effects which are not relevant for ^{15}N, and exchange effects which have been considered in Section 2.5.3, are: the dipole–dipole effect, $T_1(DD)$; the spin rotation effect, $T_1(SR)$ and the effect of chemical shift anisotropy.

The dipole–dipole effect for two spins A and X is in general given by

$$T_1(DD) = R^6/h^2 \gamma_A^2 \gamma_X^2 \tau \qquad (2.7)$$

where R is the distance between the A and X spins and τ is a correlation time [42].

$T_1(DD)$ may be conveniently divided into three categories which in decreasing importance are (i) the one-bond intramolecular effect, (ii) intermolecular effects, (iii) larger-range intramolecular effects. The appropriate correlation times are different for (i) and (iii) on the one hand and (ii) on the other. The relaxation rate for the one-bond contribution in an AX_n case is normally assumed to be n times larger than for the AX case [43]. Note that substitution of ^2D for ^1H lengthens the relaxation time of an attached spin, if the quadrupole effects of ^2D are small.

In general, the largest effect is from the one bond, $T_1(DD)$, mechanism. For ^{15}N with protons attached, relaxation times are of the order of 10 secs. at normal probe temperatures. The contribution to the relaxation time from this cause increases with temperature. Cooling the sample as noted in Section 2.4.1 may therefore increase the sensitivity.

For nitrogen with no attached protons values of 10^2-10^3 sec, depending on the temperature, are obtained and $T_1(DD)$ is lengthened to about the spin-rotation value, $T_1(SR)$. As shown by equation (1.13) $T_1(SR)$ unlike the dipolar contribution, decreases with increasing temperature. It is larger for smaller molecules like $^{15}NH_4^\oplus$ which have short correlation times. For nitrobenzene at $10°C$ the values of the various contributions are: $T_1(SR) = 800$ sec, $T_1(DD, H)$ intermolecular $= 1000$ sec, and $T_1(DD, H)$ intramolecular $= 3300$ sec [24].

The proportion (K) of the relaxation arising from the dipole–dipole mechanism can be found by measurement of the Overhauser effect (Section 2.5.3) which is dominated by the dipole–dipole term. A smaller value than the theoretical, $K = 1$, shows the importance of other terms. Thus, whereas $K = 1$ for aniline over the temperature range 60 to $200°C$ for proton decoupling experiments, compounds which do not contain N—H bonds e.g. nitrobenzene, azobenzene, benzonitrile, quinoline, have values of K between 0.5 and 0.2 at $60°C$ decreasing ot 0.1 or less at higher temperatures [24].

It is of obvious advantage to shorten these long relaxation times from the sensitivity point of view, at least to values compatible with the required resolution. If this is of the order of 1 Hz then the relaxation times could be shortened usefully to about 1 sec. The addition of paramagnetic substances, T_1

reagents, can thus dramatically improve the sensitivity. The Overhauser effect is of course lost. This is obviously of considerable advantage for nitrogen atoms with no attached protons, and even for N–H cases the T_1 reagent may circumvent the difficulties of the incomplete inversion of the resonance [44].

One restriction on the use of these reagents is that the effect on the shifts should be either negligible or measurable. If there is a shift then a number of measurements may be required at different concentrations of the reagent so that the shifts may be extrapolated to zero concentration of the reagent. This in itself will take time which unless productive in the sense of giving geometrical information, as in the case of lanthanide reagents with ^{13}C [45], may offset the time gained.

2.7   Indirect Double Resonance, Methods of Studying Resonance

The double resonance technique has been extensively reviewed [9, 46, 47]. It is a technique which is essential for a modern NMR spectrometer in more than one way [10]. We have already seen that, for two hetero or two chemically shifted homonuclei which are not coupled, the double resonance experiment may be used for a field-locking purpose. Additionally for coupled nuclei, say A and X, the use of high power in f_2 in the X region can aid direct detection of the f_1 signal in the A region by causing decoupling and nuclear Overhauser effects as discussed in Section 2.5.3. In this section the adaptation of the technique for the INDIRECT study of nuclei, i.e., the X nuclei resonating at the frequency f_2, is briefly outlined, where X is either ^{14}N or ^{15}N.

The field associated with f_2 is B_2 and this through the Larmor expression relevant to this field, namely $\gamma_X B_2 / 2\pi$, defines an effective frequency spread in the X region over which the irradiation effects are felt. The resultant changes in the A region depend upon two factors: the exact relation of f_2 to the frequencies of the transitions in the X region and the power or amplitude in f_2 ($\gamma_X B_2 / 2\pi$ factor). The various cases may be conveniently classified according to the value of $\gamma_X B_2 / 2\pi$, let this be F, say. If F is large, many or all of the nitrogen transitions are perturbed and the A spins will be decoupled from the N spins—Complete Decoupling Case. This can be useful in removing undesireable broadening of ^{1}H transitions by the effects of the nuclear electric quadrupole of ^{14}N (Chapter 3).

At lower powers of F (but still say larger than some J's) smaller numbers of transitions are perturbed—Intermediate Case. The exact nature of the effects then depends upon the particular grouping of frequencies near the value of f_2 used. If the group of perturbed frequencies arises from one spin state of a third nucleus, say M, then the decoupling effects in the ^{1}H region are selective to this spin state—Selective Decoupling. Observation of which group of lines collapses and measurement of the frequencies f_1 and f_2 gives the relative signs of the couplings J(H–M) and J(M–N).

The limit of selectivity and the highest accuracy is reached when the perturbing effect is of the order of the linewidth $(1/T_2^*)$ in the nitrogen region since then individual lines may be irradiated—Spin Tickling.

2.7.1   High and Intermediate Powers

Shift Measurement

In a ^{15}N–H system both the ^{1}H and ^{15}N regions should be doublets in which the separations are $J(^{15}$N–H).

If the f_2 is set at the centre of the doublet in the ^{15}N region with a power such that $\gamma B_2 \gg 2\pi J$ the proton doublet will be collapsed to a single line.

This is easily accounted for in the rotating frame having the Larmor precession of the ^{15}N spin: the magnetization of the ^{15}N spins at resonance lies along the effective field direction which is in the xy plane, at right angles to B_0 whereas the ^{1}H spins are still quantized along the B_0 direction. The spin–spin coupling energy, given by equation (1.16), is thus reduced to zero. Figure 2.8 shows a field sweep H $\{^{15}$N$\}$ result [48].

If the power is reduced the decoupled line looses intensity to two equally spaced satellites which have a separation which increases as B_2 decreases until in the single resonance limit as $B_2 = 0$, the separation is J, and the central line has zero intensity. If f_2 is moved 'off centre', as we have already seen, similar effects occur. The optimum frequency for decoupling thus gives the shift of the nitrogen resonance.

The technique has been used extensively for both ^{14}N and ^{15}N cases [1].

Figure 2.9 shows the α-protons of pyridine-^{15}N without

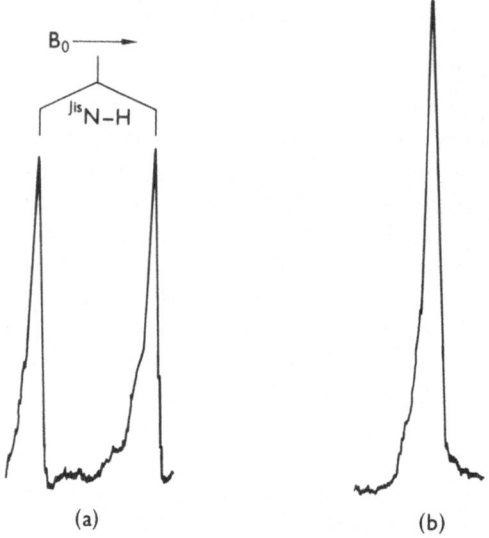

Fig. 2.8   Spectra of the ^{15}NH proton in $(CH_3)_3 Si\,^{15}NH\,C_6H_5$ at 40 MHz, (a) without and (b) with high-power irradiation at the ^{15}N resonance frequency.

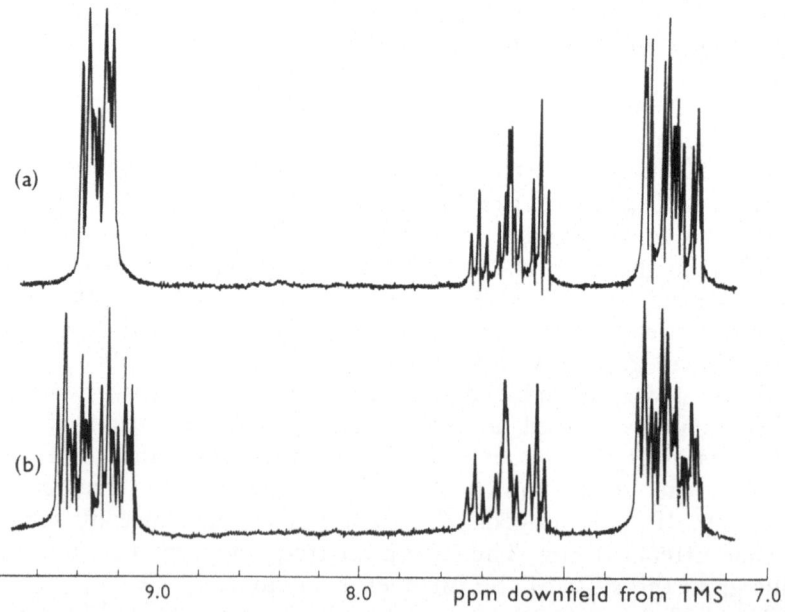

Fig. 2.9   Spectra at 100 MHz of the α protons of pure pyridine-^{15}N: (a) irradiated with high power at the ^{15}N resonance frequency of 10.13 MHz, and (b) normal single resonance spectrum.

and with decoupling. The residual splittings in Fig. 2.9(a) are due to H—H couplings. The method gives accurate results ($\sim$ 1 p.p.m.) in the case of ^{15}N, but it requires greatly enriched compounds—otherwise one is attempting to monitor the ^{1}H lines decoupled from ^{15}N superimposed on the (probably broadened) lines of the ^{14}N isotopomer. In natural abundance the method is therefore virtually unapplicable.

For ^{14}N the accuracy is generally much less than it is for ^{15}N, since normally the proton resonances are almost decoupled from the ^{14}N spins anyway. Thus the accuracy for the pyridine-^{15}N case given above ($\pm$ 1.5 p.p.m.) may be contrasted with the original accuracy for the ^{1}H $\{^{14}$N$\}$ experiment of $\pm$ 10 p.p.m. [9].

Intermediate Power

If the power is reduced and selected groups of nitrogen transitions are irradiated in turn not only can the shift be determined more accurately but additional detail may be obtained, namely the frequency separations of the groups corresponding to coupling constants, and the relative signs of the couplings. This may be illustrated with unpublished work on N-methylformamide-^{15}N (75% w/w in H_2O) [see (I)], for

$$O{\Large\diagdown}_{C-N}{\Large\diagup}^{Me\ (2)}$$
$$(1)\ H{\Large\diagup}\qquad{\Large\diagdown}H\ (3)$$
$$(I)$$

which the N region should be four groups of 1 : 3 : 3 : 1 quartets [48]. The quartet separations arise from the methyl proton coupling J^{15}(N—Me) = 1 Hz. Two doublet splittings $J(^{15}$N—H$_3$) = 92.6 Hz and $J(^{15}$N—H$_1$) = 15.6 Hz give the four groups. If

$$F = \frac{\gamma B_2}{2\pi} \sim 1\ \text{Hz}$$

then the four groups may be irradiated selectively and in each case should produce a collapse of two lines in the methyl region which should consist of eight lines (2 x 2 x 2) arising from three doublet splittings. The four H $\{^{15}$N$\}$ experiments conducted at proton frequencies of 40 MHz are shown in Fig. 2.10(b)–(e). The frequencies shown are the optimum nitrogen frequencies corrected to the field at which TMS resonates at 40 MHz (+52.1 Hz). The differences between them give the couplings J_{N1} and J_{N3} to within 1 Hz, which agree with the values taken from the proton region. The *mean* value gives the nitrogen shift. It is $4,053,628.9 \pm 0.3$ Hz. This agrees well with the result $4,053,629.5$ Hz obtained by Gillies, using a

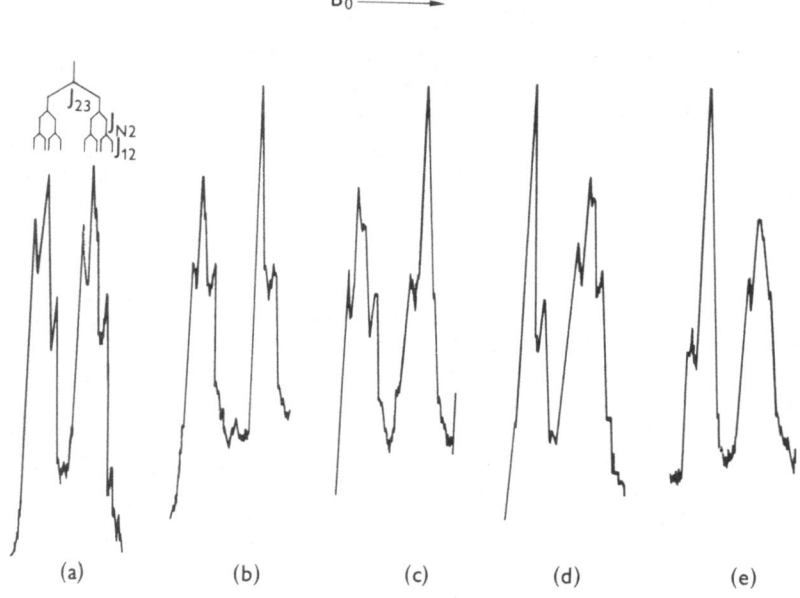

Fig. 2.10 Proton spectra at 40 MHz of the methyl region of N-methyl formamide-^{15}N. (a) normal spectrum showing couplings J_{12}, J_{23}, and J_{N2}; (b)–(e) spectra resulting from irradiation in the nitrogen region at intermediate powers with frequencies of (b) $4,053,465.5$ Hz, (c) $4,053,480.7$ Hz, (d) $4,053,558.0$ Hz, and (e) $4,053,753.1$ Hz.

slightly different f_1 value. A second experiment relates this shift to $^{15}NH_4^{\oplus}$ in ammonium nitrate. The final shift value is 92.3 ± 0.2 p.p.m.

The details of the particular lines which collapse give the information that the reduced couplings K_{N3} and K_{12} are of opposite sign.

2.7.2   Spin Tickling

If the irradiating power is reduced such that $(2\pi)^{-1}\gamma_2 B_2$ is small then individual lines may be irradiated. The effect in the f_1 region is that any transition having an energy level in common with the irradiated transition is split. For ^{15}N $(I = \frac{1}{2})$

the splitting is into a doublet, whereas for ^{14}N $(I = 1)$ the splitting is triplet in character.

The method is the most refined of the double resonance experiments and may be used to detect all the nitrogen lines which are coupled to, say, protons. The signs of couplings and accurate shifts may be obtained as before. The method has been used on formamide-^{15}N [11]. The relative splittings for different lines at constant f_2 power give the relative peak heights. Some tickling experiments have been performed on ^{14}N in those cases where $^{14}N{-}H$ couplings are well

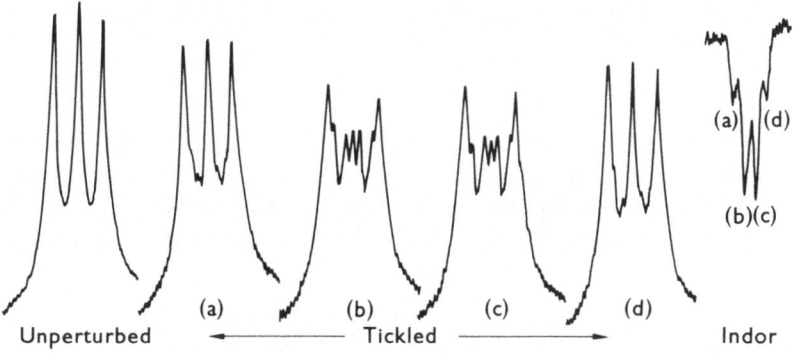

Fig. 2.11. The proton f_1 region at 100 MHz of $CH_3{}^{14}NC$. The unperturbed triplet is a single resonance spectrum in which the splitting arises from $^{14}N{-}H$ coupling. Spectra (a)–(d) are spin tickled double resonance spectra corresponding to irradiation of the four lines of the ^{14}N region with variation of f_1. An INDOR version (f_2 swept) is also shown with the four lines labelled (a) to (d) as in the f_1 frequency swept case.

resolved [18–21]. Figure 2.11 shows the case of a ^1H (^{14}N) experiment with methyl isocyanide.

The method is disadvantageous for nitrogen coupled to a number of equivalent spins since degeneracies then occur. Thus only part of an irradiated line may arise from transitions having energy levels in common with the line being monitored so that only part of a line may be split [11]. If the splitting is not resolved the effect may be detected by a broadening of the line and a reduction in intensity. This is evident from the cases (a) and (d) in Figure 2.11 [4].

The precision of the spin tickling experiment has been used by Price to find ^{14}N and ^{15}N chemical shifts very accurately related to proton frequencies [4]. This allows an estimate to be made of the isotope effect on the screening constant. In the case of the $NH_4^{\oplus}$ ion Price has carried out a series of very careful measurements on a number of solutions first using ^{1}H {N} tickling experiments at 40 MHz and then using ^{1}H (N) at 100 MHz. Accurate counting of the nitrogen and proton frequencies in each case gives, for the ratio of the ^{15}N and ^{14}N resonance, frequencies 1.402,757,13 ± 0.000,000,2 and 1.402,757,10 ± 0.000,000,08, whereas Becker *et al.* [3] have obtained a value of 1.402,756,95 ± 0.000,000,08. These results disagree with one of the two values reported earlier [5].

Further work by Becker *et al.* has employed frequency sweep decoupling on other compounds and allows the conclusion that the isotope effect is less than 0.2 p.p.m.

2.7.3　INDOR Measurements

The technique here is to set the field at a value corresponding to a transition in the f_1 region and then to sweep f_2. The results depend critically on the power used. As before, the most discriminating results are obtained in the

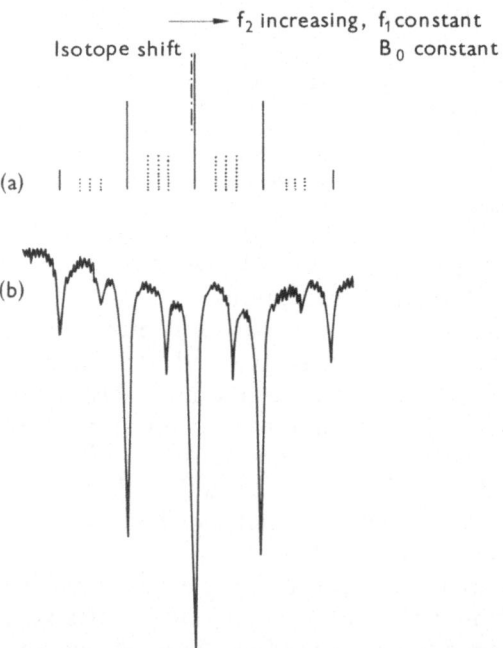

Fig. 2.12.　^{15}N theoretical (a) and experimental (b) INDOR spectra of a mixture consisting mainly of ^{15}NH$_4^\oplus$ and ^{15}NH$_3$D$^\oplus$.

spin tickling limit [18, 20, 22, 23]. Some of Price's precision work has involved the use of the INDOR technique, e.g. on $CH_3-^{14}NC$ (Fig. 2.11). Some beautiful INDOR spectra on a mixture of deuterated ammonium ions are shown in Fig. 2.12 [4]. In this work the proton frequency f_1 has been chosen to correspond to a degenerate line in the proton region consisting of accidentally overlapping lines from the two ^{15}N-containing species, both of which are therefore detected. It should be noted that not all of the lines predicted by the theoretical spectrum, Fig. 2.12(a), are observed in Fig. 2.12(b). The remaining lines can be monitored by changing f_1 to match other proton transitions, and then repeating the experiment.

Further mention of the INDOR technique is made in Chapter 6.

References

1. E. W. RANDALL and D. G. GILLIES, in *Progress in Nuclear Magnetic Resonance Spectroscopy*, Vol. 6 (J. W. EMSLEY, J. FEENEY, and L. H. SUTCLIFFE, Eds.), Pergamon Press, Oxford, 1971, p. 119.
2. R. L. LICHTER, in *Determination of Organic Structures by Physical Methods*, Vol. 4 (F. C. NACHOD and J. J. ZUCKERMAN, Eds.), Academic Press, London, New York, 1972, p. 195.
3. E. D. BECKER, R. B. BRADLEY, and T. AXENROD, *J. Mag. Res.* 4, 136 (1970).
4. R. PRICE and E. W. RANDALL, unpublished work; R. PRICE, Ph.D. Thesis, London University, 1969.
5. M. R. BAKER, C. H. ANDERSON, and N. F. RAMSEY, *Phys. Rev.*, 133, A1533 (1964).
6. W. G. PROCTOR and F. C. YU, *Phys. Rev.*, 77, (1950); *Phys. Rev.*, 81, 20 (1951).
7. J. M. BRIGGS, L. F. FARNELL, and E. W. RANDALL, *Chem. Comm.* 399 (1971).
8. P. S. PREGOSIN, E. W. RANDALL, and A. I. WHITE, *Chem. Comm.*, 1602 (1971).
9. J. D. BALDESCHWIELER and E. W. RANDALL, *Chem. Rev.*, 63, 81 (1963).
10 D. G. GILLIES, in *Chemical Society Specialist Report on NMR* (R. K. HARRIS, Ed.) (in the press).
11. R. J. CHUCK, D. G. GILLIES, and E. W. RANDALL, *Mol. Phys.*, 16, 121 (1962).
12. W. A. ANDERSON, *Rev. Sci. Instr.*, 33, 1160 (1962).
13. E. B. BAKER, L. W. BURD, and G. N. ROOT, *Rev. Sci. Instr.*, 36, 1495 (1965).

14. E. T. LIPPMAA, J. PAST, A. OLIVSON, and T. SALUVERE, *Eesti N.
    S. V. Tead, Akad. Toime Füüs. Mat.*, **58**, 15 (1966).
15. M. WITANOWSKI and G. A. WEBB, in *Annual Reports on NMR
    Spectroscopy*, Vol. 5 (E. F. MOONEY, Ed.), Academic Press,
    London, New York, 1972, p. 395.
16. M. WITANOWSKI, *J. Amer. Chem. Soc.*, **90**, 5683 (1968).
17. M. WITANOWSKI, L. STEFANIAK, H. JANUSZEWSKI, Z.
    GRABOWSKI, and G. A. WEBB, *Tetrahedron*, **28**, 637 (1972).
18. W. McFARLANE and D. H. WHIFFEN, *Mol. Phys.*, **17**, 603 (1969).
19. W. McFARLANE and R. R. DEAN, *J. Chem. Soc., A*, 1535 (1968).
20. W. McFARLANE and R. R. DEAN, *J. Chem. Soc., A*, 1187 (1968).
21. W. McFARLANE, *Mol. Phys.*, **10**, 603 (1966).
22. G. A. OLAH and T. E. KIOVSKY, *J. Amer. Chem. Soc.*, **90**, 4666
    (1968).
23. W. McFARLANE in *Annual Reports on NMR Spectroscopy*, Vol. 1
    (E. F. MOONEY, Ed.) Academic Press, London, New York, 1968,
    p. 135.
24. E. T. LIPPMAA, T. SALUVERE, and S. LAISAAR, *Chem. Phys.
    Letts.*, **11**, 120 (1971).
25. J. A. POPLE, W. G. SCHNEIDER, and H. J. BERNSTEIN, *High
    Resolution Nuclear Magnetic Resonance*, McGraw-Hill, New York,
    1959.
26. J. W. EMSLEY, J. FEENEY, and L. H. SUTCLIFFE, *High
    Resolution Nuclear Magnetic Resonance Spectroscopy* Pergamon,
    Oxford, 1966.
27. T. C. FARRAR and E. D. BECKER *Pulse and Fourier Transform
    NMR*, Academic Press, London, New York, 1971.
28. D. G. GILLIES and D. SHAW, in *Annual Reports on NMR
    Spectroscopy*, Vol. 5 (E. F. MOONEY, Ed.), Academic Press, London,
    New York, 1972 (in the press).
29. B. A. JACOBSOHN and R. K. WANGSNESS, *Phys. Rev.*, **73**, 942
    (1948).
30. R. R. ERNST and W. A. ANDERSON, *Rev. Sci. Instr.*, **37**, 93
    (1966).
31. R. R. ERNST, In *Advances in Magnetic Resonance*, Vol. 2 (J. S.
    WAUGH, Ed.), Academic Press, London, New York, 1966, p. 1.
32. R. L. LICHTER and J. D. ROBERTS, *J. Amer. Chem. Soc.*, **93**,
    3200 (1971).
33. W. A. GIBBONS, P. S. PREGOSIN, J. A. SOGN, E. W. RANDALL,
    and A. I. WHITE, unpublished results.
34. R. R. ERNST, *J. Chem. Phys.*, **45**, 3845 (1966).
35. F. J. WEIGERT, M. JANTELAT, and J. D. ROBERTS, *Proc. Nat.
    Acad. Sci., U.S.*, **66**, 1152 (1968).
36. L. F. JOHNSON and M. E. TATE, *Can. J. Chem.*, **47**, 63 (1969).
37. L. F. FARNELL and E. W. RANDALL, unpublished results.
38. J. FEENEY, D. SHAW, and P. J. S. PAUWELS, *Chem. Comm.*, 544
    (1970).
39. O. A. GANSOW and W. SCHITTENHELM, *J. Amer. Chem. Soc.*,
    93, 4294 (1971).
40. E. WEINKERT, A. O. CLOUSE, D. W. COCHRAN, and D.
    DODDRELL, *J. Amer. Chem. Soc.*, **91**, 6879 (1969).

41. H. J. REICH, M. JANTELAT, M. T. MESSE, F. J. WEIGERT, and J. D. ROBERTS, *J. Amer. Chem. Soc.*, **91**, 7445 (1969).
42. I. SOLOMON, *Phys. Rev.*, **99**, 559 (1965).
43. K. F. KUHLMANN, D. M. GRANT, and R. K. HARRIS, *J. Chem. Phys.*, **52**, 3439 (1970).
44. J. C. CAROL and E. W. RANDALL, unpublished results.
45. J. BRIGGS, F. A. HART, G. P. MOSS, and E. W. RANDALL, *Chem. Comm.*, 364 (1971).
46. R. A. HOFFMAN and S. FORSEN, in *Progress in Nuclear Magnetic Resonance Spectroscopy*, Vol. 1 (J. W. EMSLEY, J. FEENEY and L. H. SUTCLIFFE, Eds.), Pergamon Press, Oxford, 1966, p. 15.
47. W. McFARLANE, in *Determination of Organic Structures by Physical Methods*, Vol. 4 (F. C. NACHOD and J. J. ZUCKERMAN, Eds.), Academic Press, London, New York, 1971, p. 139.
48. R. J. CHUCK, Ph.D. Thesis, London University, 1966.

CHAPTER 3

Nitrogen-14 Nuclear Quadrupole Effects

J. M. Lehn and J. P. Kintzinger

Institute of Chemistry, Louis Pasteur University, Strasbourg, France

3.1   Introduction

In the present days of high resolution NMR spectroscopy, broad spectral lines are often considered as a mere manifestation of life's everyday embarassments. One may, however, also take up the converse opinion and consider linewidths as a new piece of information about a property of the system studied, the nuclear relaxation times, which together with chemical shifts and coupling constants† make up the main information content of a high resolution NMR spectrum. Nuclear relaxation times are characteristic time constants representing the rate of change of the longitudinal (longitudinal or spin-lattice relaxation time T_1) or of the transverse (transverse or spin-spin relaxation time T_2) components of the macrosopic nuclear magnetization. They are determined by the interactions existing within the spin-system and between the spins and the lattice.

Nuclear relaxation may occur through several mechanisms depending on the type of interactions **A** existing in a given system. The modulation of these interactions by the molecular motions, described by a characteristic time function **B**, causes fluctuations and leads to a relaxation time T_r given by:

$$T_r^{-1} = \mathbf{AB} \tag{3.1}$$

In the case of dipolar nuclei (spin $\frac{1}{2}$) the main interaction is generally the magnetic dipole–dipole interaction between the nuclei.

Nuclei with spin $>\frac{1}{2}$ such as the nitrogen-14 nucleus (spin 1), also possess a *nuclear electric quadrupole moment eQ* which interacts with the molecular *electric field gradient eq*, the interaction being measured by the *nuclear quadrupolar coupling constant* (NQCC) χ, where

$$\chi = \frac{e^2 qQ}{h} \text{ (in frequency units)} \tag{3.2}$$

The quadrupolar interaction being generally much larger than the dipolar one, *quadrupolar relaxation* is the dominant term for such nuclei in the sum over the contributions from the various mechanisms i:

$$T_r^{-1} = \sum_i T_r(i)^{-1} \tag{3.3}$$

† Relative areas of signals represent another piece of information, which is, however, not of fundamental nature with respect to the NMR phenomenon itself.

Equation (3.1) (with **A** being the quadrupolar term) shows that if the quadrupolar relaxation time T_q can be measured either **A** or **B** may be studied. In the present chapter we shall consider in turn the nature of and structural or medium effects on ^{14}N quadrupolar interactions, and the description of molecular motions.

We shall then analyse the resulting ^{14}N relaxation times, describe their effects on the ^{14}N NMR spectra and on the spectra of dipolar nuclei spin–spin coupled to the nitrogen nucleus and discuss their use in the study of structural and dynamic properties of molecules.

3.2   Nitrogen-14 Nuclear Quadrupole Interactions

3.2.1   Elementary Theory [1–3]

The electric interaction between the nuclear and electronic charge distributions in an atom may be represented by the Hamiltonian $\mathscr{H}_E$

$$\mathscr{H}_E = \sum_m \sum_l \mathscr{N}_l^m \mathscr{E}_l^{-m} \tag{3.4}$$

$\mathscr{N}_l^m$ and $\mathscr{E}_l^{-m}$ are respectively the nuclear and electronic tensor operators of order l:

$$\mathscr{N}_l^m = \left(\frac{4\pi}{2l+1}\right)^{1/2} \sum_j e_j R_j^l Y_l^m (\Theta_j, \Phi_j) \tag{3.5}$$

$$\mathscr{E}_l^{-m} = \left(e\,\frac{4\pi}{2l+1}\right)^{1/2} \sum_k r_k^{-(l+1)} Y_l^{-m} (\theta_k, \varphi_k) \tag{3.6}$$

The functions Y are the spherical harmonics of order l; R_j, Θ_j, Φ_j, and r_k, θ_k, φ_k are respectively the polar coordinates of the A nucleons and of the Z electrons. The charge $e_j = e$ for a proton and 0 for a neutron. The interaction energy is

$$W_E = \sum_l \sum_m N_l^{-m} E_l^{-m} \tag{3.7}$$

where N_l^{-m} and E_l^{-m} are respectively the expectation values of $\mathscr{N}_l^{-m}$ and $\mathscr{E}_l^{-m}$ over the nuclear and electronic wave functions, $\Psi_N(R_1, R_2, \ldots, R_A)$ and $\Psi_E(r_1, r_2 \ldots, r_Z)$.

N_l^{-m} is the nuclear multipole moment of order l. Because R_j is an even and $Y_l^m (\Theta, \Phi)$ an even or odd (when l is even or odd) function of the nuclear coordinates, the integrals of the expectation values vanish unless l takes up even values. In addition one may show that $l \leqslant 2I$. Thus only nuclei with spin $\geqslant 1$ can have an electric *quadrupole moment*, which corresponds to $l = 2$, in addition to the monopole $l = 0$. Treating the magnetic interactions, it can also be shown that all nuclei with spin $\geqslant \frac{1}{2}$ have a magnetic dipole moment.

Quadrupolar nuclei possess an ellipsoidal distribution of nuclear charges. The components of the nuclear quadrupole tensor $\mathcal{N}_2^m$ may be expressed in terms of a parameter eQ, the *nuclear electric quadrupole moment*, which is colinear with the nuclear magnetic dipole moment. eQ is positive or negative depending on whether the principal axis of the nuclear charge ellipsoid is parallel or perpendicular to the nuclear spin axis; it is zero when the charge distribution has spherical symmetry. The components of $\mathcal{N}_2^m$ are then given by:

$$\mathcal{N}_2^0 = Q^0 = Q'2[3I_z^2 - I(I + 1)] \qquad (3.8)$$

$$\mathcal{N}_2^{\pm 1} = Q^{\pm 1} = Q'\sqrt{6}[I_z I_\pm + I_\pm I_z] \qquad (3.9)$$

$$\mathcal{N}_2^{\pm 2} = Q^{\pm 2} = Q'\sqrt{6}I_\pm^2 \qquad (3.10)$$

where

$$Q' = \frac{eQ}{4I(2I - 1)}$$

The components of the expectation values E_2^{-m} of the electronic tensor $\mathscr{E}_2^{-m}$ over the electronic wavefunction may be written in terms of the *electric field gradient components* V_{ij}:

$$V_{ij} = \frac{\delta^2 V}{\delta_i \delta_j} = eq_{ij} \qquad (3.11)$$

where $V(x,y,z,)$ is the electrostatic potential produced by the electrons at point (x,y,z). One has:

$$E_2^0 = V^0 = \tfrac{1}{2}V_{zz} \qquad (3.12)$$

$$E_2^{\pm} = V^{\pm 1} = \frac{1}{\sqrt{6}}(V_{zx} \pm iV_{yz}) \qquad (3.13)$$

$$E_2^{\pm 2} = V^{\pm 2} = \frac{1}{2\sqrt{6}}(V_{xx} - V_{yy} \pm 2iV_{xz}) \qquad (3.14)$$

Thus, unlike a dipole moment, the nuclear quadrupole moment does not interact with a uniform electric field but with an electric field gradient, the interaction being given by the Hamiltonian:

$$\mathscr{H}_Q = \sum_m Q^m V^{-m} \tag{3.15}$$

Diagonalization of the tensor V^{-m} leads to the three principal components V_{XX}, V_{YY}, and V_{ZZ} in the electric field gradient principal axis system X, Y, Z. According to Laplace's equation:

$$V_{XX} + V_{YY} + V_{ZZ} = 0 \tag{3.16}$$

One may then express the field gradient tensor in terms of two quantities

$$eq = V_{ZZ} = eq_{ZZ} \tag{3.17}$$

and the *asymmetry parameter* η

$$\eta = \frac{q_{XX} - q_{YY}}{q_{ZZ}} \tag{3.18}$$

with the conventional choice of axes such that $0 \leqslant \eta \leqslant 1$ and,

$$|q_{XX}| \leqslant |q_{YY}| \leqslant |q_{ZZ}|. \tag{3.19}$$

The components of the electric field gradient [equations (3.12–3.14)] now become:

$$V^{\circ}_{XYZ} = \tfrac{1}{2}eq \tag{3.20}$$

$$V^{\pm 1}_{XYZ} = 0 \tag{3.21}$$

$$V^{\pm 2}_{XYZ} = \frac{1}{2\sqrt{6}}\,\eta_{eq} \tag{3.22}$$

For $\eta = 0$ the environment of the nucleus has axial symmetry. All components are zero for spherical symmetry. The quadrupolar interaction Hamiltonian in the principal axis system may then be written as:

$$\mathscr{H}_Q = \frac{e^2 qQ}{8I(2I-1)}\left[Q^{\circ} + \frac{\eta}{\sqrt{6}}(Q^{\pm 2} + Q^{-2})\right]$$

$$\mathscr{H}_Q = \frac{e^2 qQ}{4I(2I-1)}\,[3I_z^2 - I(I+1) + \tfrac{1}{2}\eta(I_+^2 + I_-^2)] \tag{3.23}$$

In the case of a spin 1 nucleus the eigenvalues of $\mathscr{H}_Q$ are the three energy levels W_0 and $W_{\pm 1}$ of the quadrupolar interaction:

$$W_0 = \frac{-e^2 qQ}{2} \qquad (3.24)$$

$$W_{\pm 1} = \frac{e^2 qQ}{4} \ (1 \pm \eta) \qquad (3.25)$$

In a molecule, the field gradient V_{ii} contains, in addition to the electronic term, a nuclear term arising from the effect of the charges of the other nuclei in the molecule at the site of the nucleus considered as shown by equation (3.34).

The nitrogen nuclear quadrupolar coupling constant $\chi_N = (e^2 qQ/h)$, equation (3.2) and the asymmetry parameter η_N equation (3.18) both effect the quadrupolar relaxation time of the ^{14}N nucleus; they depend on the electronic distribution and on the nuclear framework of the molecule.

3.2.2   Experimental Determination of Nuclear Quadrupole Coupling Constants (NQCC)

We shall briefly mention here the experimental methods which allow measurement of χ_N and η_N values. A detailed discussion may be found in reference 2.

The ^{14}N Nuclear Quadrupole Moment

First of all, it should be pointed out that there is at present no accurate experimental value available for the nuclear quadrupole moment of ^{14}N itself. Atomic beam studies do not yield any information, the ^4S ground state of atomic nitrogen being spherically symmetrical. The only experimental value (based on microwave data of the paramagnetic NO molecule) is quite inaccurate $1.6 \pm 0.7e \times 10^{-26}$ cm^2 [4].

Accurate calculations of the field gradient eq from high quality electronic wavefunctions combined with experimental values of $(e^2 qQ/h)$ may also yield information about eQ. A very extensive study of the N_2 molecule has yielded a value of $1.5e \times 10^{-26}$ cm^2 [5], other calculations on other systems have led to the following values: 1.47 [6], 1.56 [7] and 1.66 [8] ($e \times 10^{-26}$ cm^2). It thus appears that a value of $(eQ)_N = 1.55 \ e \times 10^{-26}$ cm^2 should probably be within 5–10% of the exact value.

Microwave Spectroscopy

χ_N may be measured in the gas phase from the spectral splittings due to the coupling of the nuclear spin angular momentum with the molecular rotation angular momentum via the interaction of the nuclear quadrupole moment with the molecular field gradient. These measurements generally yield the magnitude and the sign of the components $\chi_{\alpha\alpha}$, $\chi_{\beta\beta}$ and $\chi_{\gamma\gamma}$ of the quadrupolar coupling constant in the principal axis system of inertia; however except in some favourable cases, the off diagonal elements which would allow the transformation to the field-gradient principal axis system, are too small to be measurable. This problem vanishes in symmetrical molecules where the inertial and field gradient principal axis systems coincide. In addition, because of the complexity of the spectral patterns only relatively small or simple molecules may be studied, like for instance XCN (X = H, F, Cl, Br, I) NH_3, NF_3, $N(CH_3)_3$, NSF_3, NOF_3, nitroethylene, pyridine, pyrrole, aziridine, etc. Finally, the resolution of the spectrometers does not allow measurement of χ values below about 0.5 MHz. Table 3.1 gives a list of nitrogen NQCC's obtained from microwave data.

Nitrogen Nuclear Quadrupole Resonance (NQR) [2, 3, 9, 10]

We have seen above that the ^{14}N nuclear quadrupole interaction gives rise to three energy levels. In the solid state, where the field gradient axes are fixed in space, transitions between these levels may be induced by a radio frequency field. Three transitions are possible:

$$\Delta m = \pm 1$$

$$\nu_\pm = \frac{3}{4}\frac{e^2qQ}{h}\left(1 \pm \frac{\eta}{3}\right) \qquad (3.26)$$

$$\Delta m = 0$$

$$\nu_0 = \frac{e^2qQ}{h}\eta \qquad (3.27)$$

Since χ_N is usually small (below 6 MHz) the resonance lines of the NQR spectrum are of low intensity and their detection may become very tedious. Generally only the $\nu_\pm$ resonances are observed, but this is enough to yield the *principal values* of

TABLE 3.1. Nitrogen-14 Nuclear Quadrupolar Coupling Constants and Asymmetry Parameters obtained from Microwave Data[a]

Compound	e^2qQ/h (in MHz)	η_N %	Ref.
NH_3	-4.084	0	157
NF_3	-7.07	0	158
$N(CH_3)_3$	-5.47	0	159
$HN(CH_3)_2$	-5.05	20	160
NHF_2	-8.9	44	161
NF_3O	-1.52	0	162
CN	-5 ± 5	—	163
HCN	-4.58	0	164
FCN	-2.67	0	165
ClCN	-3.63	0	166a
BrCN	-3.83	0	166a
ICN	-3.80	0	166b
CH_3CN	-4.214	0	167
C_2H_5CN	-4.14	0	168
$HC{\equiv}CCN$	-4.28	0	165
	-4.20	0	169
$H_2C{=}CCH_3CN$[c]	-4.18	—	245
C_6H_5CN	-4.10	0	246
CF_3CN	-4.70	0	170
GeH_3CN	-5.0	0	171
SiH_3CN	-4.7	0	172
$S(CN)_2$[c]	-3.48	30	173
CH_3NC	$+0.483$	0	167
	0.488	0	250
CH_3CNO	$+0.495$	0	174
HCNO	$\lesssim\lvert 0.30 \rvert$	—	175
CH_3SCN	-4.18	44.5	176
$CH_2{=}CHNO_2$[c]	-1.25	39.2	177
$HN{\scriptstyle N}N$[b]	$+4.85$	—	178
$HN{\scriptstyle N}N$[b]	$\leqslant\lvert 0.7 \rvert$	—	178
$HNN{\scriptstyle N}$[b]	-1.35	—	178
F_3SN	$+1.19$	0	179
ethyleneimine	-3.689	62.9	180
transpropyleneimine	-3.71	62.8	181
pyridine	-4.88	41.4	182
γ-picoline	-4.82	57	183
pyrrole	-2.66	9	184
N-methylpyrrole	$+2.05$	64.4	185
	-2.16	82.4	185
cyanamide			
$\quad NH_2$[b]	-4.90	24.5	186
$\quad CN$[b]	-3.30	73	186
pyridazine	-5.65	15.7	247, 248
thiazole	-4.41	17.1	248
1,3,4-thiadiazole	-4.64	9	248, 249
	-4.81	12.3	

TABLE 3.1—*Continued*

Compound	e^2qQ/h (in MHz)	η_N %	Ref.
1,3,4-oxadiazole	−4.83	26↑3	249
pyrazole $\begin{cases} \text{NH} \\ \text{N} \end{cases}$	−3.02 −4.48	53.6 64.7	256

[a] Only values of the NQCC in the field gradient axis system are listed. χ_N and η_N values in the inertial axis systems are omitted.

[b] Values in the inertial axis system which should be near to the field gradient system for HN*NN* and H$_2$N*CN* but not for H*N*NN and H$_2$*N*CN.

[c] Values in the bond axis system.

the components of χ_{ii}. Their sign is not obtained. The directions of the field gradient principal axis system with respect to the molecular framework could be obtained from studies on single crystals of known crystallographic structure. Although the NQR method is in principle applicable to any compound which gives well formed single crystals or crystalline powders (at normal or at low temperature) there are severe experimental limitations due mainly to the difficulty of detecting the weak resonance lines (the practical lower limit in frequency being *ca.* 1 MHz) and to the problems arising from the solid state properties of the compound (crystallization; piezoelectricity; existence of several sites in the crystal). It is, however, the NQR method which has provided by far the largest number of the χ_N and η_N values presently available [2] (Table 3.2) (see also Section 3.2.4).

Solid State NMR Measurements.

χ_N and η_N can also be determined by high-field NMR measurements in a rigid lattice when the coupling constant is small with respect to the NMR frequency. Thus this method complements zero field NQR measurements in the region of small χ_N values. Quadrupole effects may be observed in the NMR spectra of solids as line splittings or specific powder patterns [2, 9, 11]. For a spin 1 nucleus, a splitting into $2I = 2$ transitions is predicted, the doublet separation being:

$$\Delta_{\text{solid}} = \frac{3}{2}\left(\frac{e^2qQ}{h}\right)(1 - \eta) \qquad (3.28)$$

TABLE 3.2.   Nitrogen-14 Nuclear Quadrupolar Coupling Constants and
Asymmetry Parameters obtained from NQR Spectroscopy[a]

Compound	e^2qQ/h (in MHz)	η_N %	Ref.
NR₃ Group			
ammonia	3.161	0	43
ammonia-d₃	3.231	0	43
NF₃ (0K)	7.068	0	188
trimethylamine	5.194	0	187
(4.2K)	5.238	0	189
triethylamine	5.02	0	41
4-dimethylaminopyridine	4.799	2.5	33
N-methyl-pyrrolidine	5.014	0.9	190
N-methyl-piperidine	4.986	3.6	190
N-methyl-morpholine	5.054	2.1	190
N-methyl-piperazine	5.064*	1.8*	190
triethylenediamine	4.925	0	191
hexamethylenetetramine	4.543	0	192
OP(N(CH₃)₂)₃	4.951	23.3	193
	4.791	19.3	
(CH₃)₂NNH₂	5.943	36.2	254
	5.931	36.9	254
1-chloroaziridine	5.1	29.5	194
1-chloropiperidine	6.251	83.2	251
1-chloro dimethylamine	6.333	78.8	251
NH₂ Group			
methylamine (4.2K)	4 018	37	189
formamide	2.274	37.8	44
urea	3.507	32.3	195, 196, 197
thiourea	3.110*	39.4*	198
aniline	3.933*	26.9*	199
aniline *o*-NH₂	3.733*	33.1*	33, 34, 199
aniline *p*-NH₂	3.910	26.4	195, 199
aniline *p*-Cl	4.117	24.3	195, 199
aniline *p*-Br	4.135	23.1	195, 199
aniline *p*-I	3.766	33.1	199
aniline *p*-OCH₃ (293K)	4.006	28.2	193
aniline *p*-CH₃ (293K)	4.103	14.2	193
	4.050	10.9	
2,6-diaminotoluene (294K)	3.742	31.7	193
	3.707	31.8	
2-aminopyridine	3.550	34.6	33, 34
3-aminopyridine	3.710	38.35	33, 34
4-aminopyridine	3.506	38.6	33, 34
2-amino-3-methylpyridine	3.629	31.4	33
2-amino-5-methylpyridine	3.642	34.3	33
hydrazine	4.821	78.4	200
	4.818	82.8	

TABLE 3.2—*Continued*

Compound	e^2qQ/h (in MHz)	η_N %	Ref.
CH_3NHNH_2	4.766	76.8	254
$(CH_3)_2NNH_2$	4.973	76.1	254
	4.941	77.6	
$C_6H_5NHNH_2$	5.134	69.8	254
$(NH_2NH_3)Br$	5.223	93.4	254
$(NH_2NH_3)I$	5.169	94.2	254
ethylenediamine	3.996	31.2	201
	3.997	31.3	
hexanediamine	4.03	35	202
2-aminopyrimidine	3.270	40.6	253
4,6-dimethyl-2-amino pyrimidine	3.622	25.7	253
NH Group			
dimethylamine (4.2K)	4.681	33.4	189
	4.645	32.3	
diethylamine	4.493	26.7	41
CH_3NHNH_2	5.481	58.8	254
$C_6H_5NHNH_2$	5.431	61.6	254
aziridine	3.581*	53*	203, 190, 194
2,2-dimethylaziridine	3.53	58	194
azetidine (183K)	4.250*	26.4*	190
	4.53*	30*	194
pyrrolidine	4.360	33.8	190
	4.34	34.5	194
piperidine	4.431	31	190, 194, 41
morpholine	4.482	26.8	190
	4.47	26.4	194
4-methylpiperidine	4.413	30.7	190
imidazole	3.271	12.8	252
piperazine (300K)	4.413	26.6	204
N-methylpiperazine	4.452*	30*	190
Nitrogen Heterocycles			
Substituted pyridines			
Pyridine	4.584*	39.6*	202, 205
2-methyl	4.459	33.1	31, 34
3-methyl	4.620	39.4	34
4-methyl	4.414	34.2	202, 205
3,5-dimethyl	4.640	40.6	31
2,4-dimethyl	4.321	28.4	31
2,4,6-trimethyl	4.250*	24.1*	31
2-chloro	4.452	25.9	31, 32, 34
3-chloro	4.628*	37.9*	31, 32, 34
4-chloro	4.562*	36.3*	31, 32
3,5-dichloro	4.630*	36.7*	31

TABLE 3.2—*Continued*

Compound	e^2qQ/h (in MHz)	η_N %	Ref.
Pyridine			
2,6-dichloro	4.256	10.6	31, 34
2,6-dibromo	4.330	11.9	31
2-cyano	4.718	34.7	31, 32, 34
3-cyano	4.634	36.7	31, 32, 34
4-cyano	4.772	43.7	31, 32, 34
2-nitro	4.585	32.4	35
3-nitro	4.629	32.9	35
4-nitro	4.786	48.6	35
2,6-dimethyl-3-nitro	4.395	24.8	35
2-chloro-3-nitro	4.465	21.6	35
2-acetyl	4.689	40.9	32
3-acetyl	4.625	37.3	32
4-acetyl	4.762	43.5	32
2,6-diacetyl	4.815	43.4	32
2-carbomethoxy	4.783*	38.5*	32
3-carbomethoxy	4.653*	38*	32
4-carbomethoxy	4.778	43.05	32
2-amino	3.745	3.5	33, 34
3-amino	4.497	39	33, 34
4-amino	3.781	8.4	33, 34
2-amino-3-methyl	3.942	5.9	33
	3.731	5.8	
2-amino-4-methyl	3.706	6.2	33
2-amino-5-methyl	3.773	3.9	33
2-amino-6-methyl	3.719	5.3	33
4-dimethylamino	4.022	14.95	33
2,6-dimethoxy	3.856	11.4	31
2-trimethylsilyl	4.687	38.2	35
2-fluoro	4.202	23.6	32
2-vinyl	4.257	30.9	32
imidazole	1.425	98	252
pyridazine	5.189*	8.5*	31, 32
3,6-dichloropyridazine	4.966	30	31
Substituted Pyrimidine			
pyrimidine	4.436*	38.6*	31, 32
4-methyl: N_1	4.398	36.05	253
N_3	4.277	32.5	253
5-methyl	4.443*	39.8*	253
4,6-dimethyl	4.225*	30.4*	253
2-amino	3.756	2.99	253
	3.710	6.7	
4,6-dimethyl-2-amino	3.438*	18.8*	253
2-chloro	4.348*	22.2	253
4,6-dichloro-2-methyl	4.194	13.7	253

TABLE 3.2—*Continued*

Compound	e^2qQ/h (in MHz)	η_N %	Ref.
2,4-dimethoxy-5-methyl:			
N_1	3.824	9.5	253
N_3	3.649	18.6	253
pyrazine	4.857	53.6	31, 32, 206, 240
tetramethylpyrazine	4.672	45	31, 206
phenazine	4.324	37.2	31, 206
s-triazine	4.572	44.3	31, 206
cyanuric acid	3.716	0.5	222
cyanuric chloride	4.083	1.7	207, 208
pyrrole	2.060	26.9	29, 209, 210
2,1,3-benzo-thiadiazole	3.437	14.2	193
(296°K)	3.411	13.7	
2,5-dimethyl-1,3,4-thiadiazole			
(295°K)	4.140	38.3	193
CN Group			
XCN-Derivatives			
X = H	4.018	0.8	211
Cl	3.219	1.6	212
Br	3.35	0.6	192
I (200°K)	3.80	0.0	192
CN	4.269	2.2	191
CH_3	3.738	0.46	213
C_2H_5	3.776	2.08	213
$(CH_3)_2CHCH_2CH_2$	3.830	2.7	36
CH_2Cl	3.894	14.1	202
$CHCl_2$	3.947*	3.4*	36
CCl_3	4.052	0.53	214
$NCCH_2$	3.922	7.6	36
$NC(CH_2)_8$	3.802	4.7	36
$CH_2{=}CH-$	3.800*	4.8*	36
CH_3OCH_2-	3.912	18.2	36
CH_3OCOCH_2-	3.854	8.6	36
$NCCH_2NHCH_2-$	2.785	9.9	36
$(-C(CH_3)_2N{=})_2$	3.974	3.4	36
cyclopropyl	3.700	16.5	36
C_6H_5-	3.885	10.7	214, 36
$C_6H_5-CH_2-$	3.860	1.4	36
$C_6H_5CH_2CH_2-$	3.807	3.05	36
p-XC_6H_4-Derivatives			
X = NO_2	3.960	4.55	37
CF_3	3.935	6.9	37
CN	3.892	5.6	37
F	3.826	11.1	37
OH	3.638	7.54	37
OCH_3	3.766*	20.3*	37

TABLE 3.2—*Continued*

Compound	e^2qQ/h (in MHz)	η_N %	Ref.
p-XC_6H_4-Derivatives			
$N(CH_3)_2$	3.706	23	37
NH_2	3.589	27.3	37
2-cyanopyridine	3.958	7.2	214, 34
3-cyanopyridine	3.885	9.9	32, 36, 34
4-cyanopyridine	3.895	1.44	214, 34
Tetracyanoethylene	4.331*	4.4*	215
7,7'8,8' tetracyano-quinodimethane	4.021*	13.1*	215
CH_3SCN	3.515	47.3	216
C_2H_5SCN	3.590	47.3	216
$NCSC_2H_4SCN$	3.544	46.7	216
Tetracoordinated Nitrogen Sites			
$(H_3N^\oplus-OH)Cl^\ominus$	4.013	4.3	217
$(H_3^\oplus-OH)_2SO_4^{2\ominus}$	4.192; 4.280	7.9; 4.5	217
$(H_3N^\oplus-OH)OCOCH_3^\ominus$	4.361	15.7	217
$(H_3N^\oplus-OCH_2CH_3)Cl^\ominus$	4.081	9.8	217
$(CH_3H_2N^\oplus-OH)Cl^\ominus$	3.895	32.6	217
$(H_3N^\oplus-N^\oplus H_3)2F^\ominus$	4.121	0	217
$(D_3N^\oplus-N^\oplus D_3)2Cl^\ominus$	3.907	0	217
$(H_3N^\oplus-N^\oplus H_3)2Br^\ominus$	3.986	5.6	217
$(H_3N^\oplus-N^\oplus H_3)SO_4^\ominus$	3.889; 4.032	12.1; 5.7	217
$(CH_3H_2N^\oplus-N^\oplus H_2CH_3)2Cl^\ominus$	3.737	45.3	217
$(H_2N-N^\oplus H_3)Br^\ominus$	2.776	12.8	254
Complexes			
$NH_3, \frac{1}{2}H_2O$	3.052	2.7	241
	3.169	1.1	
pyridine–$CHCl_3$	4.535	38.1	41
	4.535	44	42
4-picoline–$CHCl_3$	4.302; 4.295	33.8; 32.9	41
3,5-lutidine–$CHCl_3$	4.432	38.8	41
pyridine–pyrrole			
pyridine	4.287	48	42
pyrrole	2.122	10	42
Triethylamine–$CHCl_3$	4.779	0	41
Triethylenediamine–$CHCl_3$	4.719	0	41
Diethylamine–$CHCl_3$	4.551	10.9	41
Piperidine–$CHCl_3$	4.475	13.2	41
2-methyl pyrimidine, $\frac{1}{2}H_2O$	4.395*	32.6*	253
NH_2-NH_2, H_2O	4.731	83.6	254
Thiourea–cyclohexane	2.974*	41*	218
Miscellaneous Inorganic Compounds			
N_2 4.2°K)	4.648	0	219
N_2 in β-quinol clathrate (4.2°K)	4.78*	0*	220

TABLE 3.2—*Continued*

Compound	e^2qQ/h (in MHz)	η_N %	Ref.
$NO_2^{\ominus}Na^{\oplus}$	5.792	40.5	221
$SCN^{\ominus}K^{\oplus}$	2.432	2.8	223
$SeCN^{\ominus}K^{\oplus}$	2.845	5	223
$Hg(CN)_2$	3.951	3.14	224
$K_3Co(CN)_6$	3.684*	3.0*	224
$K_2Hg(CN)_4$	4.047	0	224
	4.074	1.37	224
$K_2Zn(CN)_4$	4.139	0	224
$K_2Cd(CN_4)$	4.199	0	224
$K_3Cu(CN)_4$	3.964	2.94	224
$K_2Pt(CN)_4, 3H_2O$	3.467*	3.2*	224

[a] All results listed are for liquid nitrogen temperature except when otherwise stated. Most presently available data are given except for compounds presenting very similar χ_N and η_N values, like the series of alkyl nitriles. The values marked * are averaged over several sites.

The determination of the ^{14}N quadrupole coupling on powders is only feasible for couplings which are small compared to the ^{14}N resonance frequency. Large single crystals have been studied and have yielded χ_N values in $NH_4PO_4H_2$ [12], in KN_3 [13] and in glycine [14] (see Table 3.3).

Liquid Phase and Nematic Phase NMR measurements,
    In *liquids* the molecular reorientations modulate the interaction between the nuclear quadrupole and the molecular field gradient and induce very rapid transitions between the quadrupole levels. The lifetime of these levels becomes very short rendering the pure quadrupole resonance line too broad for detection. The quadrupolar interaction then becomes a relaxation mechanism for the nucleus and, if the function describing the molecular motions is known, the quadrupolar coupling may be obtained from relaxation time measurements [see equation (3.1), where A is quadrupolar]. This point will be considered below, together with nitrogen relaxation times. In *nematic phases* the molecular motions are strongly anisotropic. The quadrupolar interaction is not averaged out to zero and may be measured when it is small compared to the Zeeman interaction. The ^{14}N NMR spectra of molecules oriented in the nematic phase of liquid crystals are split by

the quadrupole interaction into two components separated by [15–17]:

$$\Delta_{\text{nematic}} = \frac{3}{2}\left(\frac{e^2qQ}{h}\right)S \qquad (3.29)$$

when $\eta = 0$, where:

$$S = \tfrac{1}{2}\langle 3\cos^2\theta - 1\rangle \qquad (3.30)$$

is the orientation parameter of the molecule in the nematic solution and θ is the angle between the field gradient axis Z and the nematic optic axis. When $\eta \neq 0$ one has:

$$\Delta_{\text{nematic}} = \frac{3}{4}\left(\frac{e^2qQ}{h}\right)(3\cos^2\theta - 1 - \eta\sin\theta) \qquad (3.31)$$

Measurement of the orientation parameter from proton resonance for instance, may provide the information necessary to obtain the quadrupolar coupling when $\eta = 0$ (3.29). In the general case, χ_N and η_N can be obtained separately only when $\theta = 0°$. Such measurements have been performed on CH_3NC [16], and on CH_3CN, CH_3NO_2. $C_6H_5NO_2$ [17] (see Table 3.3).

TABLE 3.3.   Nitrogen-14 Nuclear Quadrupolar Coupling Constants and Asymmetry Parameters Obtained from NMR and EPR Data

Compound	Method	e^2qQ/h (in MHz)	η_N %	Ref.
$NH_4^{\oplus}$ $Cl^{\ominus}$	NMR (solid)	0	0	226
$NH_4^{\oplus}$ $H_2PO_4^{\ominus}$	NMR (solid)	0.0246	0	12
$NO_3^{\ominus}$ $Na^{\oplus}$	NMR (solid)	0.745	0	225
$N_3^{\ominus}$ $K^{\oplus}$	NMR (solid)			
terminal N		1.79	4	13
central N		1.028	3	13
glycine	NMR (solid)	1.200	50	14
	NMR (solid) (140°C)	0.745	61	243
CH_3NC	NMR (nematic)	0.272	0	16
CH_3CN	NMR (nematic)	3.6	0 (ass)	17
CH_3NO_2	NMR (nematic)	1.45	0 (ass)	17
$C_6H_5NO_2$	NMR (nematic)	1.76	0 (ass)	17
NS	EPR (gas)	−3.1	0	18
		−2.86	0	
$(N(SO_3)_2^{2\ominus})$	EPR (solid)	−5.2	23	19
triglycine	NMR (solid) (20°C)	1.008	58.6	242
sulfate		0.983	13.7	
		1.020	13	

Electron Paramagnetic Resonance Measurements

NQCC's may also be obtained from the EPR spectra of radicals. χ_N has been determined in the NS free radical from the spacings of the lines in the hyperfine triplets observed in the gas phase EPR spectrum [18]. The signs and magnitudes of the nitrogen NQCC components in the $(N(SO_3)_2^{2\ominus})$ radical have been calculated from the quadrupolar contribution to the line splittings observed in the EPR spectrum of γ-irradiated single crystals of potassium aminedisulfonate [19].

3.2.3   Theoretical Calculation of Nitrogen Nuclear Quadrupole Coupling Constants

Nitrogen NQCC's χ_N may be calculated by computing the field gradient from molecular wavefunctions and using the quadrupole moment $(eQ)_N$ [2, 18]. Such studies have two important features: they provide a sensitive test of the quality of the wavefunctions and conversely, they allow a detailed description of the origin of the field gradient in molecules. However, the significance of such information depends very much on the quality of the wavefunctions used.

Exact Calculations of Field Gradients

Recently a number of exact calculations of electric field gradients have been performed on several nitrogen containing molecules using Hartree-Fock-SCF-LCAO-MO wavefunctions [5–8, 20, 21, 38]. Using a single-determinental electronic wavefunction Ψ containing n doubly occupied MO's Φ_r, built of M atomic orbitals φ_k, such that

$$\Psi = [(2n!)]^{-1/2} |\Phi_1\alpha(1)\Phi_1\beta(2) \ldots \Phi_n\beta(2n)| \qquad (3.32)$$

$$\Phi_r = \sum_{k=1}^{M} c_{kr}\varphi_k \qquad (3.33)$$

the field gradient tensor $\mathbf{V}_I = eq_I$ at the quadrupolar nucleus I is:

$$eq_I = e\left[\sum_{K \neq 1} \frac{Z_K(3R_{IK}R_{IK} - R_{IK}^2 \mathbf{1})}{R_{IK}^5} \right.$$

$$\left. - 2\sum_{i=1}^{n}\sum_{k,l=1}^{M} c_{ki}c_{li}\left\langle \varphi_k \left| \frac{3r_I r_I - r_I^2 \mathbf{1}}{r_I^5} \right| \varphi_l \right\rangle \right] \qquad (3.34)$$

with Z_K = charge on nucleus K;

   R_{IK} = vector from nucleus I to nucleus K;

   1   = unit dyadic;

   r_I = vector from nucleus I to electron.

The first and second term on the righ-hand side of equation (3.34) represent respectively the nuclear and the electronic contribution to the field gradient. The nuclear term is obtained from the molecular geometry and nuclear charges. Once the eq_{ii} components in the coordinate axis system have been calculated, diagonalization of eq_I gives the principal components and the orientation of the principal axis system. It is also possible to transform to the inertial axis system so as to compare computed and microwave values of the NQCC components. Without going into the details of the calculations, quantitatively satisfactory results may be obtained with extended basis sets of atomic functions; minimal basis sets yield generally only qualitative or semi-quantitative results. The inclusion of polarization functions may be important as shown for O_3 and NO_2 [22]. For our present purpose, the important feature of these calculations is that they permit a complete, non-empirical, interpretation of the χ_N values, thus allowing the analysis of the molecular structural effects on χ_N. These effects are discussed in Section 3.2.4, together with the experimental data.

Approximate Calculations of Field Gradients

Semi-empirical methods of various degrees of sophistication have been used for computing electric field gradients, giving more or less accurate results; these results are generally poorer for nitrogen sites than for halogens, although they may be no worse than the values obtained from low quality non-empirical wavefunctions [23–27]. However, as no allowance is made for core electron polarization, a 'Sternheimer factor', has to be introduced to take this effect into account.

We shall only describe briefly the simple and most widely applied *Townes and Dailey method* [23] (for more details see refs 2, 18 and 23). This method tries to account for the field gradients by considering only the valence p electrons and neglecting the contributions from inner shells, from valence s electrons (because of spherical symmetry) and from the charges external to the atom considered (assuming cancellation of the contributions from the external nuclei and electrons). If the field gradient produced by a single p electron is eq_p, then the field gradients produced by n_x, n_y, n_z electrons in the p_x, p_y, p_z orbitals along the $z \equiv Z$ axis are [2, 3, 28]:—

$$eq_{zz}(p_z) = n_z eq_p \qquad (3.35)$$

$$eq_{zz}(p_x) = -\tfrac{1}{2} n_x eq_p \qquad (3.36)$$

$$eq_{zz}(p_y) = -\tfrac{1}{2} n_y eq_p \qquad (3.37)$$

Similar expressions may be written for eq_{xx} and for eq_{yy}. The total field gradient along z is then:

$$eq_{zz} = [n_z - \tfrac{1}{2}(n_x + n_y)]eq_p \qquad (3.38)$$

and

$$\eta = \tfrac{3}{2}(n_x - n_z)/[n_z - \tfrac{1}{2}(n_x + n_y)] \qquad (3.39)$$

eq_p is not known for the nitrogen atom, however $e^2 q_p Q/h$ may be estimated to be *ca.* 10 ± 2 MHz. In fact, because of the approximations in the method, the $e^2 q_p Q/h$ value depends on the procedure used for calculating the orbital populations.

Despite the more or less drastic assumptions made, as shown by the exact calculations [2, 8] and by interpretative difficulties [29, 30], the Townes and Dailey method may nevertheless provide a quick, simple but necessarily rough, idea about the electron distribution around the nitrogen nucleus. It has been used for this purpose in the interpretation of the nitrogen NQCC's in many compounds, especially nitrogen heterocycles and nitriles (see 28, 31–37 and references in 2); substituent effects have been described in terms of electronic population changes.

3.2.4   Effects of Molecular Structure on Nitrogen Nuclear Quadrupole Coupling Constants

We shall now discuss the effects of molecular structure on the value of χ_N and η_N using the results of exact calculations [6–8, 21] as the main framework, supplemented by results from approximate methods and by consideration of trends in experimental data. Tables 3.1, 3.2 and 3.3 give a list of representative nitrogen NQCC's and asymmetry parameters.

The present discussion will be limited to the more general effects which may also be of predictive value.

Structural Effects

(1) The principal component of the electric field gradient lies approximately along the direction of the *nitrogen lone pair* (except in special cases like diazirine [21]), for both tricoordinated (XYZN) and dicoordinated (X = NY) nitrogen sites as well as of course for XCN groups.

(2) Structural effects leading to partial or complete disappearance of the nitrogen lone pair decrease the field gradient. Conjugation leads to such a decrease as is found for pyrrole or other heterocycles and amides. For instance in the aminopyridines, conjugation with the ring C=N system decreases the coupling constant in the 2-amino and 4-amino derivatives with respect to aniline more than in 3-amino pyridine. The effect is also found at the ring nitrogen atom where the field gradient is decreased in the $2\text{-}NH_2$ and $4\text{-}NH_2$ pyridines, this is probably because dilution of the NH_2 lone pair into the ring π system decreases the electronic unbalance due to the lone pair by enhancing the electronic population in the direction of the π system.

Protonation or quaternization ($NH_4^{\oplus}$, $NR_4^{\oplus}$) and complex formation (H_3N-BH_3) generally lead also to smaller field gradients. Ammonium ions possess very small coupling constants. In general, compounds for which the valence bond formulae place a positive charge on nitrogen or for which mesomeric formulae with positively charged nitrogen may be written, display small coupling constants (nitro groups, isonitriles, nitrates, central N in azide ion etc.).

(3) Heteroatoms directly linked to the nitrogen site produce a marked increase in field gradient, the effect being generally larger for more electronegative heteroatoms (e.g. NF_3, hydrazine, pyrazole, isoxazole, etc.). Large coupling constants may even be found in tetracoordinated nitrogen sites, for instance in $H_3N^{\oplus}-OH$ and $H_3N^{\oplus}-NH_3^{\oplus}$ (see Table 3.2), where the adjacent O and $N^{\oplus}$ atoms distort appreciably the electronic symmetry at the $NH_3^{\oplus}$ groups.

(4) Characteristic field gradient components are found for tri- and dicoordinated nitrogen sites in five-membered heterocycles.

(5) The quadrupolar coupling constants are very sensitive to changes in geometry at the nitrogen site. Distortion towards the inversion transition states of pyramidal and planar nitrogen, by opening the XNY angle in pyramidal XYZN or the XNY angle in bent X = NY, leads to a marked increase in

coupling constant, the main effect arising from the nitrogen lone pair which evolves towards a pure p orbital

(6) Large η_N values are found in $-NH_2$, $-NH-$ and $=N-$ groups. Small values are generally found in nitriles, tertiary amines and quaternary ammonium salts.

Nuclear and Electronic Contributions to the Field Gradient

(1) The most important *nuclear* contributions are by far those of the directly linked atoms; they increase with the nuclear charge of the atom considered and with the shortening of the corresponding bond (contributions: $O > N > C > H$; N in $N=N > N$ in $N-N$).

(2) The largest *electronic* contributions from individual *atoms* are the monocentric ones, on the nitrogen site itself. Two centre and especially three-centre contributions are generally weak. Adjacent heteroatoms (like O, N) contribute however much more than C and H atoms. This may in part be a lone-pair effect, since the electronic charge extends less far away from the atom in a lone pair than in a bond.

(3) The largest *molecular orbital* contribution is due to the nitrogen lone pair MO, followed by the π MO, in dico-ordinated $X = NY$ nitrogen sites, and then by the $X-N\sigma$ bond MO's.

Intermolecular Effects on Quadrupolar Coupling Constants

Nitrogen lone pairs may participate in intermolecular interactions. Since the quadrupolar coupling constant is very sensitive to changes at the lone pair, appreciable medium effects may be expected.

Gas phase coupling constants are generally about 0.3–0.5 MHz, larger than solid state values [pyridine: 4.88 MHz (gas), 4.60 MHz (solid)]. As discussed in Section 3.3, liquid phase values are difficult to determine accurately. But a study of both chlorine and nitrogen quadrupolar coupling and relaxation in ClCN indicates that the liquid and solid phase values of the nitrogen coupling constant are the same [39]. It thus seems valid to use the solid state coupling constants for studies in liquids. A further indication is obtained from the comparison with coupling constants measured in a nematic phase (Table 3.3). For CH_3CN one has: $\chi_N = 4.21$ (gas), 3.60 (nematic) and 3.74 (solid) MHz, for CH_3NC: $\chi_N = 483$ (gas), 272 (nematic) kHz (see Tables 3.1–3.3). The nematic phase values are 0.6 and 0.2 MHz below the gas phase values and are probably comparable to solid state values in both cases. Hydrogen bonding to the nitrogen lone pair should decrease the coupling constant as shown by a theoretical study of the $H_3N\cdots H^\oplus$ system [40]. Some complexes of acyclic and heterocyclic amines with hydrogen bonding compounds (chloroform, methanol, pyrrole) have been studied in the solid state [41, 42]. The measured coupling constants are up to 5% lower than in the parent compound in the case of tertiary amines. Secondary amines however display an increase in coupling, probably because the N—H bond is also affected by intermolecular interactions. The especially large solid state shifts measured for the NQCC in NH_3 [40, 43] (−0.6 MHz) and in formamide [44] (−1.3 MHz) have also been attributed mainly to intermolecular interactions. Changes in nitrogen quadrupolar coupling may occur for various other intermolecular associations such as amine-halogen complexes, pyridinium-$X^\ominus$, ion pairs, etc. Such data should be of great interest to the study of the nature of intermolecular bonding.

3.3   Molecular Motions and Correlation Functions

We now come to the description of molecular motions in liquids, which, by inducing rapid nuclear quadrupole transitions, also induce rapid transitions among the nuclear Zeeman levels, since the nuclear electric quadrupole and magnetic dipole moments are colinear, and thus lead to fast nuclear relaxation of quadrupolar origin (for more detailed treatments see references 1 and 45). The discussion of experimental results is deferred to section 3.4.4 as knowledge of the nuclear relaxation data is needed for study correlation times.

3.3.1   Theoretical Models

The motions of a molecule in a liquid are of two types:

(1) *Intramolecular* motions, i.e. overall rotations of the molecule itself and motions due to internal degrees of freedom (rotations about bonds, inversions).

(2) *Intermolecular* motions, i.e translations of the molecule with respect to the others in the liquid.

Only intramolecular processes reorient the molecular field gradient; intermolecular processes do not contribute to quadrupolar relaxation except in special cases of specific intermolecular interaction as discussed in section 3.3.2. There are then two main steps in the study of molecular motions. First the determination of molecular correlation times and second their interpretation and calculation in terms of molecular structural parameters (size, shape, etc.) and solution properties.

The *autocorrelation function* $G(t_1, t_2)$ representing the correlation between the values $f(t_1)$ and $f(t_2)$ taken by a certain time dependent function of the molecular positions at times t_1 and t_2 is defined as the average:

$$G(t_1 t_2) = \langle f(t_1) f^*(t_2) \rangle$$

or   (3.40)

$$G(\tau) = \langle f(t) f^*(t + \tau) \rangle$$

The *correlation time* τ_c is a time such that $G(\tau)$ is very small when $|\tau| \gg \tau_c$, where τ is the time interval $\tau = t_2 - t_1$. The *spectral densities* $J(\omega)$ are the Fourier transforms of $G(\tau)$:

$$J(\omega) = \int_{-\infty}^{+\infty} G(\tau)\, e^{-i\omega\tau}\, d\tau$$

$$G(\tau) = \frac{1}{2\pi} \int_{-\infty}^{+\infty} J(\omega)\, e^{i\omega\tau}\, d\omega$$

(3.41)

The form of $G(\tau)$ depends on the physical model chosen for describing the molecular motions. The two extreme models used for describing molecular rotations in liquids are:

(1) *The rotational diffusion model*, which describes the reorientation by random small angle jumps (Debye model; Brownian motion).

(2) *The inertial rotation model* where large angle jump, gas phase like rotation is assumed.

These models may be considered as the limiting cases of the Langevin equation (3.42) describing the time dependence of the angular rotational velocity ω_i of the molecule about an axis i [47]:

$$\frac{\partial \omega_i}{\partial t} = [-\xi_i \omega_i + F_i(t)]/I_i$$

(3.42)

where ξ_i is the rotational friction coefficient and I_i the moment of inertia about axis i, $F_i(t)$ is a randomly fluctuating torque arising from the fluctuations of the environment of the rotating molecule. When ξ_i is large, frictional effects pre-dominate and the motions are of the rotational diffusion type.

For small ξ_i, the motions are of the inertial type. At these two limits, the correlation times are given by equations (3.49–3.50) and (3.51) below. It is seen that the temperature dependence is quite different for the two cases, being larger for rotational diffusion $(f(T))$ than for inertial rotation $(f(\sqrt{T}))$. A *"quasi lattice random flight model"* has also been developed [48].

Rotational Diffusion Model

In the case of *isotropic reorientation* the motions of the molecule are represented by a single correlation time τ_c. The diffusion equation leads to a *reduced* correlation function $g(\tau)$ of exponential form:

$$g(\tau) = G(\tau)/G(0) = e^{-|\tau|/\tau_c} \qquad (3.43)$$

The reduced spectral density from equation (3.41), becomes:

$$j(\omega) = J(\omega)/J(0) = \frac{2\tau_c}{1 + \omega^2 \tau_c^2} \qquad (3.44)$$

In the *extreme narrowing conditions* defined by

$$\omega^2 \tau_c^2 \ll 1 \qquad (3.45)$$

(very fast molecular motions with respect to the resonance frequency) one has:

$$j(\omega) = 2\tau_c \qquad (3.46)$$

This is the case in most mobile liquids. One may then try to relate τ_c to molecular parameters. In the hydrodynamic description of a fluid composed of spheres of radius a (molar volume V_m) in a medium of viscosity η, τ_c is related to the Stokes friction coefficient ξ in the following way:

$$\xi = 8\pi\eta a^3 \qquad (3.47)$$

$$\tau_c = \frac{\xi}{6kT} = \frac{4\pi\eta a^3}{3kT} = \frac{V_m\eta}{kT} \qquad (3.48)$$

For real molecules this equation is of course very approximate, leading generally to correlation times which are too long. Corrections have been introduced in the form of mutual viscosity effects ($\tau = \tau_c/2.85$) [49a] and of a microviscosity factor ($\tau = \tau_c/6$) [49b]. In the general case of *anisotropic reorientation* the correlation function is no longer an exponential of a single correlation time of the type given by equation (3.43). The molecular motions may be described in terms of the three principal *diffusion constants* D_a, D_b, D_c of the rotational diffusion tensor in the diffusion principal axis system a, b, c. One has:

$$\tau_i = \frac{1}{6D_i} \tag{3.49}$$

$$D_i = \frac{kT}{\xi_i} \tag{3.50}$$

where $i = a$, b, c.

Thus, in this model, the motions are determined by frictional forces as opposed to inertial forces in the inertial model. The detailed expressions for anisotropic motions have been given [46] and are discussed in Section 3.4.1.

Inertial Rotation Models [47, 50, 51]

When the rotation is assumed to be free as for a molecule in a gas, the friction coefficients ξ_i vanish and the correlation time τ_c may be related to the moments of inertia I_i. This is expected to be valid for small quasispherical molecules having weak intermolecular interactions. The *unconditional inertial rotation* model leads to a correlation time:

$$\tau = \frac{1}{2}\left(\frac{\pi I}{3kT}\right)^{1/2} \tag{3.51}$$

A *conditional inertial rotation* model has also been considered recently [51]. This model, based on a non-markovian treatment of correlation functions [52], makes the occurence of molecular rotation dependent on the formation of an expanded lattice or lattice defect. The attractive feature of this model is that it separates the intrisic properties of the free molecule from those of the lattice, attributing the departure from pure inertial behaviour to lattice dynamics effects. However, up to now only the 'single defect' model has been treated. This may be satisfactory for spheroidal molecules but defects of different sizes and shapes have to be envisaged for markedly anisotropic molecules, each allowing one type of motion. In contrast with this picture, the diffusion constants are statistical quantities incorporating both the properties of the molecule and its interactions with the environment.

Tests for the Different Models

The rotational diffusion model has been the most widely applied to rotation in liquids. A simple test of its applicability [46, 53, 54] consists in calculating the ratio:

$$\chi_i = \left(\frac{\tau_c}{\tau_f}\right)_i = \left(\frac{1}{6D_i}\right)\left(\frac{5}{3}\frac{kT}{I_i}\right)^{1/2} \tag{3.52}$$

where $(\tau_f)_i = (3/5)(I_i/kT)^{1/2}$ is the reorientation time of the free rotor in the gas. If $\chi_i \gg 1$, the motions are small-angle Brownian diffusion at faster rates that gas-like rotations and the model applies. When χ_i is ~1 it does not, and inertial effects become important. χ_i is thus a measure of the importance of intermolecular interactions on molecular rotation in the liquid.

As already pointed out, the two limiting models lead to different temperature dependences of the correlation times [equations (3.49), (3.50) and (3.51)]. The experimentally observed temperature dependence may be used as a criterion. However, in the conditional inertial rotation model [51] the temperature dependence has been modified by introducing an exponential temperature dependence for the formation of the lattice defects. Another test consists in comparing correlation times from dielectric relaxation data and from NMR. For a rotational diffusion process with small angle jumps one should have:

$$\tau_{\text{dielectric}} = 3\tau_{\text{NMR}} \tag{3.53}$$

whereas for rotation through large angle jumps,

$$\tau_{\text{dielectric}} = \tau_{\text{NMR}} \tag{3.54}$$

The proportionality factor between $\tau_{\text{dielectric}}$ and τ_{NMR} varies from 3 to 1 as the jump angle varies from 0° to about 120°. Infrared band shapes may provide another test [55].

Internal Motions.

When internal motions (rotations, inversions) are present which also reorient the atom studied, both overall and internal motions have to be taken into account in the correlation function [45, 46, 56–59]. Assuming isotropic motion, the *effective local correlation time* τ_q of a quadrupolar nucleus in a bond which undergoes internal rotation at a rate τ_i^{-1} around an axis whose reorientation time is described by a correlation time τ_M, is given by equation (3.55) or (3.56) in the following two cases [56–58] [in extreme narrowing conditions, equation (3.45)].

(a) Diffusional internal rotation among a very large number of equilibrium positions:

$$\tau_q = A\tau_M + B\left(\frac{1}{\tau_M} + \frac{1}{\tau_i}\right)^{-1} + C\left(\frac{1}{\tau_M} + \frac{1}{\tau_i}\right)^{-1} \qquad (3.55)$$

(b) Three-fold rotation barrier, with randoms jumps between the three equilibrium positions at an average rate $(3\tau_i)^{-1}$:

$$\tau_q = A\tau_M + (B + C)\left(\frac{1}{\tau_M} - \frac{1}{\tau_i}\right)^{-1} \qquad (3.56)$$

with

$$A = \tfrac{1}{4}(3\cos^2\theta - 1)$$

$$B = \tfrac{3}{4}\sin^2\theta \qquad (3.57)$$

$$C = \tfrac{3}{4}\sin^4\theta$$

(θ: angle between the rotation axis and the main field gradient axis). If $\theta = 109°28'$ one has:

$$A = 0.11; \qquad B = 0.30 \qquad \text{and } c = 0.59. \qquad (3.58)$$

Extension of equations (3.55) and (3.56) to the case of anisotropic motions of the molecule may be found in ref. 57c. The equations outside the extreme narrowing conditions are given in ref. 57a. Equations (3.55) and (3.56) may be written in terms of diffusion constants by the relations:

$$\tau_M = (6D_M)^{-1} \qquad \tau_i = D_i^{-1} \qquad (3.59)$$

If the quadrupolar nucleus is several bonds removed from the largest rigid portion of the molecule, all internal motions have to be taken into account, *each* shortening τ_q by a maximum factor A (= 0.11 for $\theta = 109°28'$) [59]. Equations (3.55) and (3.56) are easily extended to this case. Thus the local motions are decoupled from the slower framework motions.

3.3.2   Intermolecular Effects

Molecular translations have no effect on quadrupolar relaxation. However, modulation of the field gradient may occur through time dependent intermolecular interactions. Such effects may be important, since hydrogen bonding to or protonation of the nitrogen lone pair, for instance, may lead to a pronounced change in field gradient. In this case the time-dependent functions contain exchange terms in addition to reorientation terms. A treatment of such a situation is discussed in Section 3.4.1.

3.4   Nitrogen-14 Nuclear Quadrupolar Relaxation

3.4.1   Theoretical Results

Nuclear Quadrupolar Relaxation (QR) Time Equations

In a liquid, the quadrupolar interaction Hamiltonian equation (3.15) is time dependent because of the rotational motions of the molecule and becomes a relaxation mechanism of the quadrupolar nucleus. One has:

$$\mathcal{H}_Q = \sum_m Q^m F^{(-m)}(t) \qquad (3.60)$$

where the components Q^m of the nuclear quadrupole moment are given by equations (3.8–3.10) and $F^m(t)$ are the lattice functions in a space fixed coordinate system, representing random fluctuations of the field gradient orientation at the nuclear site. These lattice functions may be decomposed into time independent field gradient components in a molecule linked axis system V^m and time dependent second order rotation matrices $\mathcal{D}^{(2)}(t)$, where the time dependence represents the angular fluctuations between the space-fixed and the molecule linked system, i.e. the molecular motions:

$$F^m(t) = \sum_{m'} \mathcal{D}^{(2)*}_{mm'}(t) V_D^{m'} \qquad (3.61)$$

The $V_D^{m'}$ are the field gradient components in the principal axis system of the diffusion tensor. They may themselves be related by a rotation matrix to the $V_{XYZ}^{m'}$ components in the field gradient principal axis system equations (3.20–3.22). The reduced autocorrelation function of $F^m(t)$ is:

$$g(\tau) = G(\tau)/G(0) = \frac{\langle F^m(t) F^{-m}(t+\tau) \rangle}{\langle |F^m(t)|^2 \rangle} \qquad (3.62)$$

where the numerator and denominator are averaged over t.

For isotropic diffusional rotation $g(\tau)$ has the exponential form $e^{-|\tau|/\tau_q}$, where τ_q is the *quadrupolar correlation time*; the Fourier transform [equation (3.41)] of $g(\tau)$ gives the reduced spectral density as:

$$j(\omega) = \frac{2\tau_q}{1 + \omega^2\,\tau_q^2} \qquad (3.63)$$

The master equation of the quadrupolar relaxation process of spin I is [1]:

$$\frac{d\langle I \rangle}{dt} = -\frac{1}{2} \sum_{m=-2}^{+2} \{J_m(m\omega_0)[Q^m,[Q^{-m}, I]](\sigma - \sigma_0)\} \qquad (3.64)$$

where $J_m(m\omega_0)$ is the Fourier transform of $G(\tau)$ and σ_0 is the thermal equilibrium value of the density matrix of the spin system σ. The relaxation equations of the longitudinal I_z and transversal $I_{x,y}$ components of the magnetization are then derived in the form:

$$\frac{d\langle I_z \rangle}{dt} = -\frac{1}{T_1}(\langle I_z \rangle - \langle I_z \rangle_0) \qquad (3.65)$$

$$\frac{d\langle I_{x,y} \rangle}{dt} = -\frac{1}{T_2}\langle I_{x,y} \rangle \qquad (3.66)$$

($\langle I_z \rangle_0$ equilibrium value), where for a nucleus of spin 1 the spin-lattice T_1 and spin–spin T_2 relaxation times are given by:

$$\frac{1}{T_1} = \frac{3}{80}\left(\frac{e^2 qQ}{\hbar}\right)^2 \left(1 + \frac{\eta^2}{3}\right)\{j(\omega_0) + 4j(2\omega_0)\} \qquad (3.67)$$

$$\frac{1}{T_2} = \frac{1}{160}\left(\frac{e^2 qQ}{\hbar}\right)^2 \left(1 + \frac{\eta^2}{3}\right)\{9j(0) + 15j(\omega_0) + 6j(2\omega_0)\} \qquad (3.68)$$

In the extreme narrowing conditions [equation (3.45)], one has:

$$j(0) = j(\omega_0) = j(2\omega_0)$$

Then the quadrupolar relaxation (QR) time T_{qN} is:

$$\frac{1}{T_q} = \frac{1}{T_1} = \frac{1}{T_2} = \frac{3}{16}\left(\frac{e^2 qQ}{\hbar}\right)^2 \left(1 + \frac{\eta^2}{3}\right)j(0) \qquad (3.69)$$

The form of $j(0)$ depends on the form of the reduced correlation function, i.e. on the model used for describing the molecular motion. With the exponential form [equation (3.43)] $j(0)$ is given by equation (3.46) and one has

$$\frac{1}{T_q} = \frac{3}{8}\left(\frac{e^2qQ}{\hbar}\right)^2 \left(1 + \frac{\eta^2}{3}\right)\tau_q \qquad (3.70)$$

Equation (3.70) assumes isotropic reorientation and thus describes the molecular motions by a single correlation time. When internal motions are also reorienting the nitrogen nucleus, τ_q in equation (3.70) is given by equation (3.55) or (3.56) (with $\eta = 0$). In the general case of *anisotropic motions* and within the rotational diffusion model the relaxation equation may be written in terms of the three principal diffusion constants D_a, D_b, D_c [46]. For symmetric or asymmetric rotors with sufficient symmetry (e.g. pyridine) the diffusion principal axis system coincides with the inertial system. It is then sufficient to study three differently placed nuclei in the molecule in order to determine the three diffusion constants, provided the orientation of the field gradient axes is known. For low symmetry asymmetric rotors the orientation of the diffusion axes has also be be determined. It can be shown [46] that this is only possible when the asymmetry parameter of the field gradient $\eta \neq 0$. Thus, since ^{14}N quadrupolar interactions present particularly large η values they are in principle well suited for such studies. For instance, in the case of a ^{14}N nucleus in a planar asymmetric rotor (e.g. pyridine) one has [46]:

$$\frac{1}{T_q} = \frac{3}{32}\left(\frac{e^2qQ}{\hbar}\right)^2 \frac{1}{D_R}\left\{[4D_a + (\eta-1)^2D_b + (\eta+1)^2D_c]\cos^2\varphi\right.$$

$$+ [4D_b + (\eta-1)^2D_a + (\eta+1)^2D_c]\sin^2\varphi - \frac{(\eta-3)^2}{3}\left[\frac{(D_a-D_b)^2}{(D_c+D_s)}\right]$$

$$\left. \cos^2\varphi\sin^2\varphi\right\} \qquad (3.71)$$

$$D_R = 3(D_aD_b + D_bD_c + D_cD_a); \qquad D_s = (1/3)(D_a + D_b + D_c)$$

where a and b are the in-plane diffusion axes and φ is the angle between the field gradient in-plane principal axes and the a and b axes. The equations have been derived for several cases [46, 53, 60]: asymmetric rotor, planar asymmetric rotor without or with internal rotation, symmetric rotor, spherical rotor. In particular, in this last case τ_q in equation (3.70) becomes $(6D)^{-1}$. The relaxation equations for a spin 1 nucleus in the case of spherical and symmetrical rotors have also been derived within the unconditional and conditional inertial rotation models [47, 50, 51]. For instance for unconditional rotation of a spherical rotor one has:

$$\frac{1}{T_q} = \frac{3}{16} \left(\frac{e^2 qQ}{\hbar}\right)^2 \left(\frac{\pi I}{3kT}\right)^{1/2} \qquad (3.72)$$

Where I is the moment of inertia of the rotor.

Analysis of nitrogen quadrupolar relaxation data using detailed equations of the type (3.71) has only been performed for four molecules: acetonitrile [61–63], trichloro-aceto-nitrile [54], dimethyl formamide [53] and pyridine [64] and the conditional inertial model has been applied to ammonia[51]. The internal rotation process in benzyl cyanides [65] has been studied using equations (3.70) ($\eta = 0$) and (3.55). However, most studies have assumed isotropic motions using equation (3.70). The results obtained are of course approximate, but because of uncertainties in the validity of the theoretical models and difficulties in treating complicated molecules in a detailed way, they nevertheless have provided a wealth of very useful data on molecular properties in liquids and on liquid structure.

Quadrupolar Relaxation Time in the Presence of Chemical Exchange

As already noted above, nitrogen sites in molecules are particularly sensitive to intermolecular interactions. The relaxation equation must then contain the effects of chemical exchange between the different nitrogen sites [66, 67]. When exchange occurs between two sites A and B at rates defined by the equations:

$$A \underset{k_{BA}}{\overset{k_{AB}}{\rightleftharpoons}} B; \qquad A \xrightleftharpoons{k_{AA}} A; \qquad B \xrightleftharpoons{k_{BB}} B$$

the relaxation time of a spin 1 nucleus in the extreme narrowing limit, equation (3.45), may be obtained from the conditional probability treatment of the coupled effects of molecular rotation and exchange [66]:†

$$\frac{1}{T_q} = \frac{3}{8} f_A \left(\frac{e^2 q_A Q}{\hbar}\right)^2 \left(1 + \frac{\eta_A^2}{3}\right) \tau_A + \frac{3}{8} f_B \left(\frac{e^2 q_B Q}{\hbar}\right)^2 \left(1 + \frac{\eta_B^2}{3}\right) \tau_B$$

$$(3.74)$$

where

$$\tau_A^{-1} = (\tau_q^A)^{-1} + (\tau_{ex}^B)^{-1}$$
$$\tau_B^{-1} = (\tau_q^B)^{-1} + (\tau_{ex}^B)^{-1}$$

$$(3.75)$$

and

$$(\tau_{ex}^A)^{-1} = k_{AA} + k_{AB}$$
$$(\tau_{ex}^B)^{-1} = k_{BB} + k_{BA}$$

$$(3.76)$$

In this treatment isotropic reorientation at sites A and B is assumed with diffusion constants D_A and D_B. The rotational correlation times are then:

$$(\tau_q^A)^{-1} = 6D_A; \qquad (\tau_q^B)^{-1} = 6D_B \qquad (3.77)$$

f_A, f_B and q_A, q_B are respectively the equilibrium fractions of nuclei at sites A and B and the main principal field gradient components in sites A and B, with asymmetry parameters η_A and η_B in their respective principal axis systems. Thus one allows for changes *both* in field-gradient magnitude and direction between sites A and B. Using equations (3.67) and (3.68) the relaxation equations outside extreme narrowing may also be obtained from the correlation function given in ref. 66. A treatment of the case $\tau_{ex} \gg \tau_q$ had been given previously, with special application to the case where one site is a paramagnetic ion [67]. *Aqueous solutions* represent a very complex case where multiple exchange processes may occur between sites corresponding to various mobile solvation shells or molecular aggregates, as shown by the study of aqueous mixtures of acetonitrile and of pyridine [82]. Many other intermolecular processes are amenable to a detailed study using such treatments of nitrogen relaxation data.

† Exchange is assumed to be fast with respect to the nuclear relaxation rates at sites A and B: $\tau_{ex} \ll T_q(i)$ ($i = A$ or B); the results are appropriate in the limit where the NMR signals for nuclei at different sites have coalesced into a single line.

Quadrupolar Relaxation in Ions

As seen above, intermolecular interactions may perturb the field gradient at the nitrogen site, but the effect is nevertheless relatively weak in comparison to the magnitude of the quadrupolar coupling constant in most molecules (Section 3.2.4). However, when the NQCC is small and especially in ionic substances, quadrupolar relaxation may be induced largely by the fluctuating field gradients produced by the diffusional motions of ions and dipolar solvent molecules in the neighbourhood of the nitrogen nucleus. For instance, ion pair formation and dissociation is expected to alter markedly the field gradient and to lead to efficient relaxation. In aqueous solution, both stable and rapidly exchanging hydration shells of ions may create fluctuating field gradients.

In the case of nitrogen containing ions, intermolecularly induced relaxation is expected to be the dominant relaxation process in symmetrical ammonium ions $NR_4^\oplus$, where the internal field gradient is zero. The intermolecular contribution is expected to be less important in less symmetric ions ($NO_3^\ominus$, $N_3^\ominus$, $NR_3X^\oplus$). Various theories and models describing the relaxation of quadrupolar ions have been developed [45, 68—70).

In extreme narrowing conditions, the quadrupolar relaxation time of a spin 1 nucleus, due to fluctuating field gradients produced by neighbouring charges (ions) and electric dipoles (solvent molecules) may take the following form [69,70]:

$$T_q^{-1} = 8\pi \left(\frac{\beta e Q}{\hbar}\right)^2 \left[\frac{\mu^2 c_s}{r_0^5}\tau_s + \sum_j \frac{z_j^2}{3a_j^3} c_j\tau_j\right] \qquad (3.78)$$

where the first and second term in the brackets represent respectively the effect of the solvent molecules with dipole moment μ and of the j types of ions of charge z_je; the summation is taken over all ions. c_s and c_j are the concentrations of the solvent and of the ions. τ_s and τ_j are the correlation times for the reorientation of the solvent molecules and for the translational diffusion of the ions. r_0 and a_j are the minimum distances of approach of a solvent molecule or of an ion. The factor $\beta = P(1 - \gamma_\infty)$, where $P = (2\epsilon + 3)/5\epsilon$ (ϵ: dielectric constant of the solvent) and γ_∞ is the Sternheimer anti-shielding factor, allows for polarization of the ion by the solvent (P) and for the amplification of external field gradients by distortion of the electronic shells surrounding the nucleus (γ_∞).

Deformation of the electronic shells by collisions also produces field gradients and leads to relaxation [68]. The real microdynamic behaviour of ions in solution is much more complex than assumed in the above model; for instance, the rates of exchange between several different environments and the reorientation time of the ions in these environments both contribute to relaxation in a generalized form of the two sites treatment given above. The complexity of such processes may be appreciated from the analysis of the relaxation of the water protons in diamagnetic electrolyte solutions [45, 71].

Temperature Dependence of Quadrupolar Relaxation Times. Activation Parameters

Considering the molecular motions represented by the quadrupolar correlation time τ_q, equation (3.70), as a thermally activated process and using the Eyring rate equation one has:

$$\tau_q^{-1} = KT \exp(-\Delta G^{\ddagger}/RT) \tag{3.79}$$

where $K = k_B f/h$ (k_B: Boltzman's constant; f: frequency factor taken equal to unity) and $\Delta G^{\ddagger}$ is the free energy of activation of the correlation time for molecular reorientations. Equation (3.79) may be written as:

$$\log (\tau_q T)^{-1} = \log K - \frac{\Delta G^{\ddagger}}{4.57T} = \log K + \frac{\Delta S^{\ddagger}}{4.57} - \frac{\Delta H^{\ddagger}}{4.57T} \tag{3.80}$$

Using equation (3.70) one obtains:

$$\log (\tau_q T)^{-1} = \log (T_q/T) + \log \frac{3}{8}\left(\frac{e^2 qQ}{\hbar}\right)^2 \left(1 + \frac{\eta^2}{3}\right) \tag{3.81}$$

and from equations (3.80) and (3.81)

$$\log (T_q/T) = K' - \frac{\Delta H^{\ddagger}}{4.57T} \tag{3.82}$$

with

$$K' = \log K - \log \frac{3}{8}\left(\frac{e^2 qQ}{\hbar}\right)^2 \left(1 + \frac{\eta^2}{3}\right) + \frac{\Delta S^{\ddagger}}{4.57}$$

If $(e^2 qQ/h)$ and η are known, a plot of $\log(T_q/T) = f(1/T)$ should give a straight line whose slope and intercept respectively yield the enthalpy of activation $\Delta H^{\ddagger}$ and the entropy of activation $\Delta S^{\ddagger}$ for the reorientation process. One may also write an Arrhenius activation equation in the form:

$$\tau_q = \tau_q^{\circ} \exp\left(E_a/RT\right) \qquad (3.83)$$

and obtain the activation energy E_a in a similar way.

3.4.2   Measurement of Nitrogen Quadrupolar Relaxation Times, T_{qN}

We shall only mention briefly the experimental methods available for measuring nitrogen quadrupolar relaxation times T_{qN}. More details may be found in the literature [1, 72].

Despite the low sensitivity of ^{14}N NMR (sensitivity of the ^{14}N nucleus relative to the proton at constant field $\sim 10^{-3}$) nitrogen-14 nuclear relaxation times have been measured for a wide variety of compounds. The method used depends on the range of relaxation rates to be covered. The *spin-echo method* [72, 73] which consists in the observation of the decay of the magnetization resulting from the application of a certain RF pulse sequence, may be used to determine either T_1 or T_2 depending of the type of pulse sequence. The method may be used for a very broad range of relaxation times (from microseconds to minutes). Using another pulse technique, both the nitrogen relaxation time and the $J(\text{N}-\text{X})$ spin-spin coupling constant may be obtained by observing spin-echo trains of the dipolar X nucleus and determining their dependence on pulse-repetition rate [74]. *The adiabatic fast passage* method [51, 72, 75] may be used to determine T_1 from the observed amplitudes of the signal when sweeping rapidly through the resonance in forward and reverse directions using either two different sweep frequencies or different sweep rates in the forward and return directions. This method is suitable for relaxation times longer than *ca.* 0.05 sec. NMR resonance lines generally have a Lorentzian shape. T_2 is related to the full *linewidth* at half-height Δ (in Hz) by:

$$\Delta^* = 1/\pi T_2 \qquad (3.84)$$

If the derivative of the Lorentzian line is recorded, one has:

$$L = 1/\sqrt{3}\pi T_2 \qquad (3.85)$$

where L is the separation of the two peaks of the line. In the extreme narrowing conditions $T_1 = T_2$, and measurement of the linewidth represents a simple method for determining T_q. However, several effects (magnetic field homogeneity, saturation, modulation broadening†) render the measured linewidth Δ^* larger than the quadrupolar width Δ_q. Assuming a Lorentzian distribution for the field inhomogeneity and applying correction factors† one may subtract the non-quadrupolar contributions Δ' from the total linewidth Δ^*:

$$\Delta_q = \Delta^* - \Delta' = 1/\pi T_q \qquad (3.86)$$

The method is suited for lines which are markedly broader than the correction factors. In the *saturation-recovery* method one observes the recovery of the signal after rapid switching from a saturating RF level to a level below saturation. The relaxation time T_1 is obtained from the integral form of equation (3.65) as the time constant in the exponential function:

$$\langle I_z \rangle = \langle I_z \rangle_0 \, [\, 1 - \exp\,(-t/T_1) \qquad (3.87)$$

Progressive saturation [72, 77] relies on the observation of the signal amplitude while increasing progressively the RF field intensity. This method yields the product $T_1 T_2$; either $T_1 = T_2$ may be used in extreme narrowing conditions or T_2 is determined from the linewidth using equation (3.86). The accuracy of the measurements may be markedly improved by repetitive scanning and accumulation into a digital memory system. Spin-echo, adiabatic fast passage and linewidth measurements may be performed with a 5% accuracy.

When the relaxation time is long (narrow ^{14}N resonance), and if the nitrogen nucleus is spin–spin coupled to a dipolar nucleus, T_{qN} may also be obtained by lineshape analysis of the resonance signal of the dipolar nucleus; this method is discussed in Section 3.5.2.

In a number of cases, different methods have been used for determining the nitrogen relaxation time in the same molecule. In the case of the linear symmetrical $N_3^{\ominus}$ ion the ratio of the linewidths of central and terminal nitrogens [78] (1/3.3) agrees with the ratio of the corresponding quadrupolar coupling constants [13] (1/3.05); but the ratio of the relaxation times measured by the saturation recovery method is only 1.6 [79] probably because of partial saturation of the central nitrogen resonance. The nitrogen relaxation time in

---

† For modulation broadening corrections see ref. 76.

acetonitrile has been obtained at about 25°C by spin-echo measurements [61, 62, 80], by progressive saturation [77], from the linewidth [81, 82] and from the proton T_2 data [74c] (Table 3.5). The results agree with the average value of 4.1 msec except for one spin-echo measurement [80], where a 50% error was estimated, and for the values obtained from T_2 of the protons [74c] and from progressive saturation data [77].

3.4.3   Quadrupolar Effects in ^{14}N NMR Spectra

We have already seen in Section 3.2.2 that nitrogen-14 NMR spectra of solid samples or of samples dissolved in a nematic solvent show a splitting of the resonance signal by the quadrupolar interaction, providing a means of measuring NQCC's which are small in comparison to the ^{14}N resonance frequency.

In liquids, molecular rotations render the lifetime of the discrete quadrupolar levels so short that no quadrupolar splittings may be observed. The quadrupolar interaction serves then as a relaxation mechanism which is generally far more efficient than any other, except in some highly symmetrical ammonium salts discussed in Section 3.4.4. Consequently, the linewidth of a given resonance is determined by the quadrupolar relaxation rate, equations (3.84) and (3.86). Inspection of a nitrogen-14 NMR spectrum thus allows an immediate appreciation of relaxation effects. On the other hand, since the quadrupolar coupling constants are often quite large, the ^{14}N relaxation time is short; the signals are broad and may overlap when several different nitrogen atoms are present in the molecule.

In a magnetic field of 23,487 kG the nitrogen resonance frequency is 7.2238 MHz (proton: 100.00 MHz) and the chemical shifts cover ca. 900 p.p.m., i.e. 6500 Hz. However, amines, amides, nitriles, nitro derivatives cover only a total range of ca. 3000 Hz (400 p.p.m.) in which each type of functional group is found in a range of about 500–700 Hz. Now, from the results given in Table 3.5 it is seen that the linewidths of amino nitrogen signals for instance, amount to several hundreds of Hz even in small molecules and increase rapidly with molecular size, as the reorientation time τ_q increases [equation (3.70)]. (Δ^* ~200 Hz in pyridine and 960 Hz in quinoline at 25°C.) In nitriles and nitro derivatives, narrower lines are obtained, but they still become very broad when the molecular size increases (Δ^* ~85 Hz in acetonitrile and 390 Hz in benzonitrile; Δ^* ~20 Hz in nitromethane and 100 Hz in nitrobenzene). It is seen that the efficiency of the quadrupolar relaxation mechanism may cause the nitrogen resonance lines to overlap strongly, for instance in the amino-pyridines. This difficulty may be avoided in some cases by using low viscosity solvents or by working at

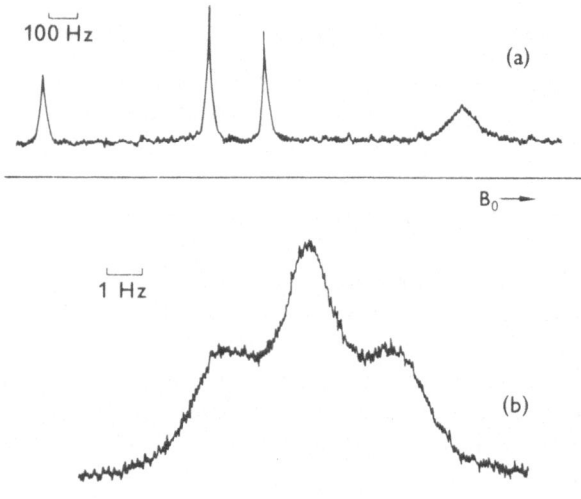

Fig. 3.1. (a) ^{14}N NMR spectrum of CH_3N_3 with CH_3NO_2 as internal reference, at 4.33 MHz and at room temperature; from left to right: CH_3NO_2, $=\overset{\oplus}{N}=$, $=N^{\ominus}$ and $=N-CH_3$ resonances (after Witanowski [83]). (b) ^1H NMR spectrum of CH_3N_3 in benzene at 60 MHz and at 75°C, showing the quadrupole broadened CH_3 triplet due to spin–spin coupling between the CH_3 protons and the central $CH_3-N=\overset{\oplus}{N}=N^{\ominus}$ nitrogen nucleus [145].

temperatures above room temperature. There are nevertheless a number of cases where individual resonances may be observed for different nitrogen atoms, especially for functional groups associated with symmetrical nitrogen sites, i.e. low NQCC's, for instance isonitriles, ammonium salts, azides and nitro compounds (Tables 3.1-3.3). Figure 3.1 shows the spectrum of methyl azide $CH_3-\overset{\ominus}{N}=\overset{\oplus}{N}=\overset{\ominus}{N}$ and of CH_3-NO_2 [83]. Four well-separated signals are observed one of them, the $CH_3-N=$ nitrogen, being much larger (100 Hz) than the three others (CH_3N_3 = 17 Hz (central) = 19 Hz (terminal); CH_3NO_2 = 20 Hz). A particularly striking case of the co-existence of a sharp and a very broad signal is the spectrum of Et_2N-NO_2 (Fig. 3.2) where the sharp NO_2 signal ($\sim$12 Hz) is sitting on a very broad Et_2N signal (1000 Hz) with a chemical shift difference of about 175 p.p.m. (*ca.* 770 Hz at 14.1 kG) [84].

Spin–spin coupling constants between nitrogen-14 and protons or fluorine are weak unless they are directly bonded and it is generally not possible to observe nitrogen multiplets. Liquid NH_3 displays a strongly broadened quartet [85a]

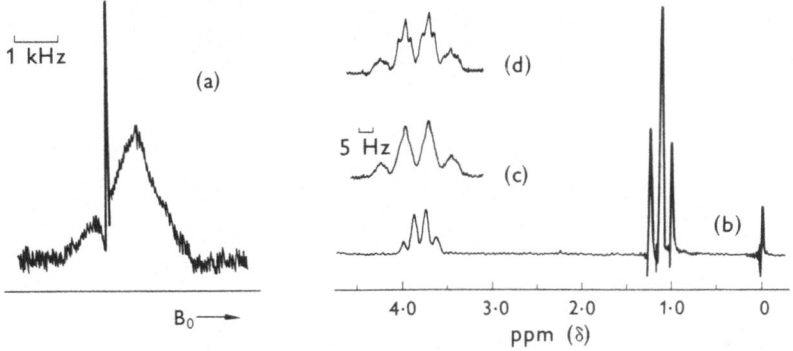

Fig. 3.2 (a) ^{14}N NMR spectrum of neat $(CH_3CH_2)_2N-NO_2$ at 4.33 MHz and at room temperature; the sharp and broad signals are respectively due to the $-NO_2$ and the $(CH_2CH_2)_2N-$ group (after Kintzinger, Lehn and Williams [84]). (b) ^1H NMR spectrum of $(CH_3CH_2)_2N-NO_2$ in pyridine at 60 MHz and at 33°C. (c) Enlarged CH_2 signal in same conditions as (b). (d) Enlarged CH_2 signal at 70°C showing better resolved splittings due to coupling with the NO_2 nitrogen nucleus.

$(JN-H = 46$ Hz)†whereas the symmetrical $NH_4^{\oplus}$ ion shows a sharp quintet ^{14}N signal ($J(N-H) = 51$ Hz) (Fig. 3.3). Splittings due to coupling with ^{19}F have also been observed (ONF_3, O_2NF, NF_3) [86].

In cases of nitrogen sites with very small NQCC's it is even possible to observe small long-range couplings, for instance in CH_3NC ($J(H-C-N) = 2.3$ Hz) [87] (Fig. 3.4) and in $(CH_3)_3\overset{\oplus}{N}-CH=CH_2\ Br^{\ominus}$ [88]. The quadrupolar width of each component of the nitrogen multiplet is $1/\pi T_{qN}$ [1, 89] (see also Section 3.5.1).

† It may be noted that the nitrogen relaxation time in liquid NH_3 is 38 msec [51] giving a linewidth of 8.3 Hz. The difference between this value and the $\sim$40 Hz width of Fig. 3.3 is probably due to proton exchange broadening catalyzed by traces of water [85c].

3.4.4   Nitrogen Quadrupolar Relaxation Data, Structural and Medium Effects

Table 3.5 presents a list of nitrogen-14 relaxation times. They generally lie in the 0.5 to 5 msec range. To discuss these values in terms of molecular structure and of medium effects we must remember that T_{qN} depends both on the ^{14}N NQCC and on the correlation time for molecular rotations, equation

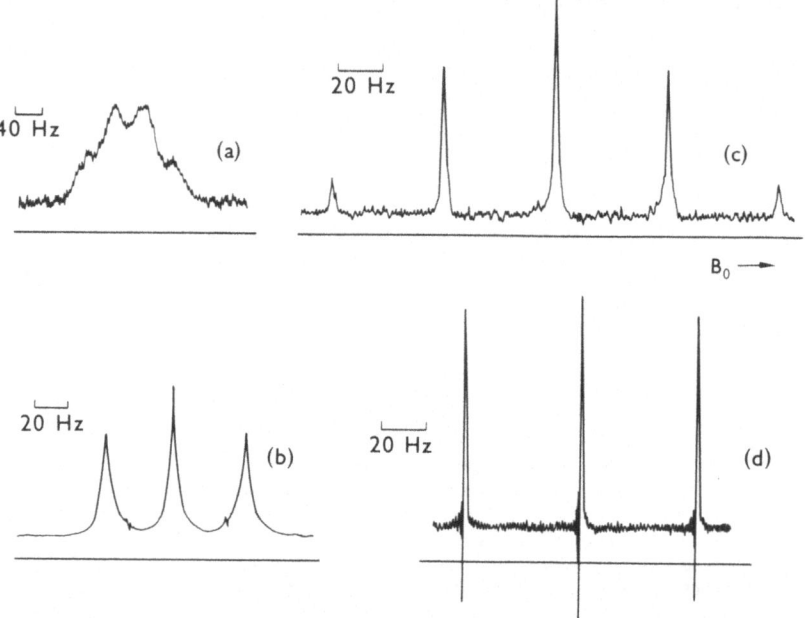

Fig. 3.3.   (a) ^{14}N NMR spectrum of liquid NH_3 at 3 MHz (after Ogg and Ray [85a]). (b) ^{1}H NMR spectrum of liquid NH_3 at 30 MHz showing a 43.8 Hz N–H coupling (after Ogg and Ray [85b]). (c) ^{14}N NMR spectrum of a saturated and acidified solution of NH_4Cl in H_2O at 6.50 MHz and at room temperature. (d) ^{1}H NMR spectrum of a saturated and acidified solution of NH_4Cl in H_2O at 60 MHz and at room temperature (N–H coupling constant = 51.5 Hz).

(3.70), and that given structural changes may either influence the NQCC or the motions or both. These effects will now be discussed.

Main Features of Nitrogen Quadrupolar Relaxation Data

Before proceeding further, we will summarize the main features of T_{qN} data, as they result from the preceding sections, with respect to their use in studying NQCC values and quadrupolar correlation times.

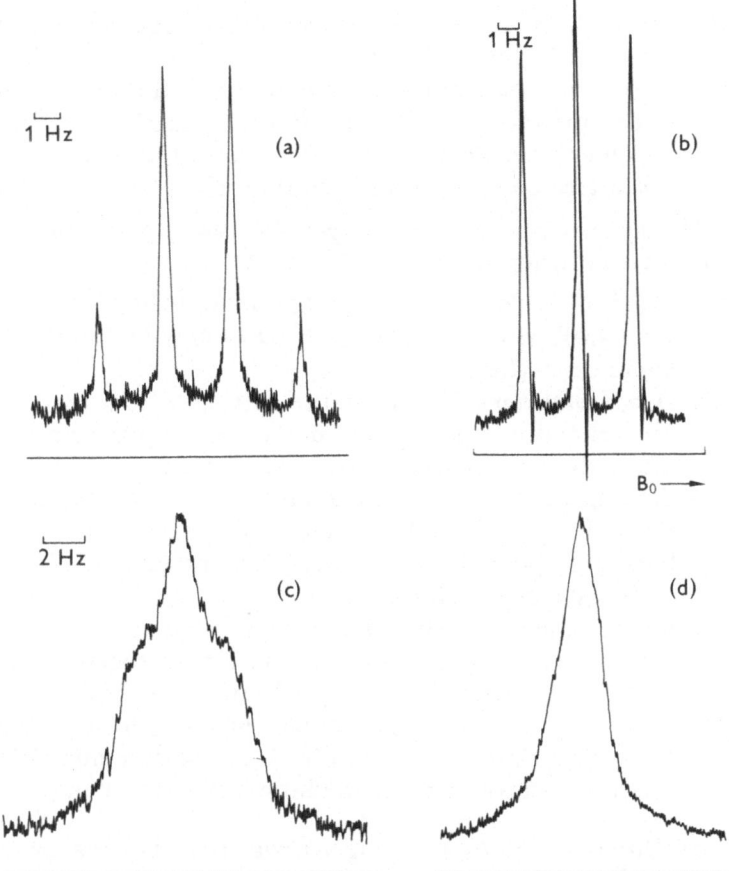

Fig. 3.4.   (a) ^{14}N NMR spectrum of neat CH_3NC at 7.22 MHz and at room temperature (after Moniz and Poranski [87]). (b) ^1H NMR spectrum of CH_3NC in pyridine at 60 MHz and at room temperature (N–H coupling constant = 2.7 Hz). (c) ^1H NMR spectrum of the complex of CH_3NC with $ZnCl_2$ in D_2O at 60 MHz and at room temperature [145]. (d) Same as in (c), but $CdCl_2$ complex of CH_3NC [145]; (c) and (d) display the disappearance of the N–H splitting reflecting the increase in nitrogen-14 relaxation rate on complexation.

(1) Nitrogen-14 relaxation has the general advantages of quadrupolar relaxation:

(a) Quadrupolar relaxation is generally by far the *dominant* relaxation mechanism; the quadrupolar interaction term in equation (3.70) is in the MHz range (Tables 3.1–3.3) (except in some special cases like $NR_4^{\oplus}$ ions: see Section 3.5) whereas the dipole–dipole interaction with a proton at a distance of *ca.* 1 Å is very small, of the order of 10^{-2} MHz.

(b) Since the quadrupolar interaction is large, T_q is very sensitive to motional changes.

(c) Only *rotational* motions contribute to the relaxation; in combination with dipole–dipole relaxation studies it becomes possible to separate the contributions of intra and intermolecular (translation) motions.

(2) The special case of the *nitrogen-14* quadrupolar nucleus calls for the following remarks:

(a) The ^{14}N nucleus has a very high natural abundance but low NMR sensitivity; this difficulty may be avoided in some cases (Section 5.2).

(b) *Intramolecular* effects on the nitrogen field gradient are generally largely predominant over intermolecular ones; this is, however, not always so. If an independent measure of molecular motions is then available ('double quadrupolar labelling', see below) it becomes possible to study the nature of intermolecular interactions.

(c) Very numerous organic, inorganic and biological substances of great interest contain nitrogen atoms and thus present nitrogen quadrupolar effects.

(d) ^{14}N NQCC's lie in a range where the resulting relaxation times and linewidths may still be measured relatively easily in most medium-size molecules.

Study of Molecular Motions Using Nitrogen Relaxation Data: Detailed Analysis of Molecular Rotation

We have seen previously that it is possible to analyse in detail the dynamical behaviour of molecules within a given physical model using quadrupolar relaxation data (Section 3.4.1). The rotational diffusion model has been applied to CD_3CN [61–63], CCl_3CN [54], $(CD_3)_2N$–$CD^{17}O$ [53] and α, β and γ deuteropyridines [64], the principal diffusion constants have been determined by analyzing the relaxation data for the different quadrupolar nuclei. For instance in CD_3–CN, the diffusion constant for CD_3 reorientation ($D_\parallel = 15 \times 10^{11}$ sec^{-1}) indicates a much faster process with a much smaller activation energy than does the diffusion constant for reorientation of the CCN axis ($D_\perp = 1.5 \times 10^{11}$ sec^{-1}) (see Table 3.4).† Increasing the pressure [63]

---

† Anisotropic rotation of the CH_3–CN molecule in the liquid has been suggested in an earlier paper [80].

TABLE 3.4. Diffusion Constants and Activation Energies for Anisotropic Molecular Reorientation[a]

Compound	$10^{-11} \times D_{\parallel}$ (sec^{-1})	$10^{-11} \times D_{\perp}$ (sec^{-1})	T (°C)	$E_{a\parallel}$ (kcal/ mole)	$E_{a\perp}$ (kcal/ mole)	Ref.
CD_3CN	15	1.5	25°	0.7	1.8	61, 6:
CCl_3CN	0.9	0.4	25°	1.9	2.7	54
pyridine	1.6	1.9	80°	1.0	3.1	64
	1.3	0.8	30°			
	1.0	0.3	−20°			
$3\text{-}CH_3\text{-}C_6H_4\text{-}CH_2CN$	0.4[b]	0.2[c]	65°	1.5[b]	3.8[c]	65

[a] The $\parallel$ axis is either the CCN axis or, in pyridine, the axis perpendicular to the molecular plane.
[b] Internal rotation of the $-CH_2CN$ group.
[c] Overall molecular rotation.

does not affect the rate of CD_3 reorientation, $D_{\parallel}$, up to 2000 bar, but markedly decreases $D_{\perp}$. The reported values are $D_{\perp} = 0.7 \times 10^{11}$ sec^{-1} at 2000 bar, $D_{\parallel}/D_{\perp} = 10$ at 1 bar and $= 22$ at 2000 bar. The corresponding activation volumes are $\Delta V^{\ddagger}$ $(D_{\perp}) > \Delta V^{\ddagger}(D_{\parallel}) \sim 0$. The CD_3 reorientation is of 'gas-like' type, whereas the reorientation of the CCN axis may be described by the rotational diffusion equations. In the case of pyridine, anisotropic motion has been found, with a change from solid like behaviour $(D_{\perp} < D_{\parallel})$ to gas-like behaviour $(D_{\perp} > D_{\parallel})$ as the temperature of the liquid increases from above the melting point to below the boiling point [64]. For $(CD_3)_2N-CDO$ the three diffusion constants (30, 6 and 0.4×10^{10} sec^{-1}) and the orientation of the diffusion principal axis system have been obtained [53]. A less accurate treatment has been performed on the 2-, 3- and 4-methyl benzyl-cyanides for which the diffusion constants for overall and internal rotation have been obtained (see Table 3.5) [65]. Anisotropic molecular reorientation has also been studied in liquid cyano-pyridines and the diagonal elements of the diffusion tensor have been estimated from the nitrogen QR data [90, 91]. An approximate analysis of anisotropic rotation has been performed for a number of organic liquids e.g. pyridine, aniline and nitrobenzene derivatives [91b].

Application of the conditional inertial rotation model to liquid NH_3 and ND_3 leads to a description of the molecular motion in terms of inertial rotation of the molecule in a lattice having an activation energy of 1.88 kcal/mole for the formation of the defects in which the molecule may rotate [51]. The quasilattice model has also been applied to NH_3 [48].

An attempt to calculate the $D_\perp$ and $D_\parallel$ rotational diffusion constants of acetonitrile from molecular parameters and viscosity has shown that a modified Hill method [49a], using a friction coefficient given by equation (3.88), leads to reasonable agreement between calculated and experimental values [92]. This was not at all the case when either equation (3.47) or a microviscosity factor [49b] were used.

$$\xi_i = 12 \frac{\eta \sigma}{m} \frac{I_{BB} I_i}{I_{BB} + I_i} \qquad (3.88)$$

with $I_{BB} = 4r^2 m$ (η = viscosity of the liquid; σ, m, and r = diameter, mass and radius of the molecule respectively, I_i = moment of inertia about axis i).

Approximate Studies of Molecular Rotation in Liquids

In most studies molecular rotation was assumed to be isotropic. Then the correlation time τ_q for molecular re-orientation may be obtained from the nitrogen relaxation time using equation (3.70) together with χ_N and η_N values measured by other methods. In this way it is possible to obtain in a relatively simple and straightforward fashion, an approximate but very useful piece of information about molecular motions. Table 3.5 gives a list of nitrogen QR times measured by various methods together with correlation times τ_q calculated from equation (3.70) and data in Tables (3.2–3.3). The asymmetry parameter may generally be neglected, since even for its largest values $\eta^2/3$ amounts only to about 5%.

The QR times obtained from ^{14}N linewidths may be quite different from one report to another. These differences and inaccuracies may arise from several factors: modulation broadening, saturation, use of an internal reference which changes the viscosity, temperature not specified, concentration of solutions unspecified or different from one experiment to another. For this reason we do not list in Table 3.5 all the linewidth data reported in the literature but only a selection of those which seemed well suited for the present discussion. [14]N QR data have been used for studying molecular motions in a series of molecules [80, 81, 91]. Correlation times have been obtained and attempts have been made to calculate them from molecular parameters and liquid properties. *Structural effects* on correlation times may be analyzed by means of the data in Table 3.5. Some main features emerge. Molecular motions are slowed down when the molecular size and weight increases. This may be seen by comparing for instance $N(CH_3)_3$ and $N(CH_2CH_3)_3$, pyridine and quinoline, ClCN and BrCN, CH_3CN and C_6H_5CN, CH_3NO_2 and $C_6H_5NO_2$; in all cases τ_q increases by at least a factor of 3. An interesting *symmetry effect* is also found when comparing pyridine, the methylpyridines and quinoline. Indeed, the addition of just one CH_3 group has quite a pronounced effect on τ_q. This can be taken as indicating that the relatively easy orientation of pyridine which may result from its approximate six-fold symmetry [64], is strongly hindered as soon as a substituent group is introduced. A further indication is found when considering the dimethylpyridines, where the more symmetrical 2,6 compound reorients faster than the less symmetrical ones [91, 93]. This type of symmetry is reminiscent of the fact that high symmetry potential barriers to internal rotation are often lower than low symmetry ones; a similar effect may contribute to lowering the barriers to rotation of symmetrical molecules in the liquid. It is indeed known that molecules like benzene or methane undergo rapid reorientation even in the solid state. Nitrogen QR time measurements on NH_3 [94] and hexamethylenetetramine [95] have also shown that these symmetric molecules undergo fast reorientation in the solid state. As discussed in Section 3.3.1, *internal motions* decrease τ_q and increase T_{qN}. This is illustrated by the case of $C_6H_5CH_2CN$ for which one finds a shorter τ_q (4.3 psec) than for C_6H_5CN (5.5 psec), despite its larger molecular size. A longer τ_q is also found for 2-methylbenzylcyanide where the *ortho* CH_3 group hinders internal rotation as compared to the 3-methyl and 4-methyl isomers [65]. The same correlation time is found for NO_2 in $O_2N(CH_2)_6NO_2$, where several internal rotation degrees of freedom decouple the local motion from the overall motion and for the appreciably smaller $O_2NCH_2CH_2NO_2$ molecule where fewer internal motions are possible. Increasing the *temperature* shortens τ_q. Activation parameters for molecular

TABLE 3.5.   Nitrogen-14 Quadrupolar Relaxation Times and Correlation times[a]

Compound	Method[b]	Δ^*(Hz)	or T_1 (msec)	T_2 (msec)	$\tau_q{}^c$ (psec)	Ref.
N$_2$	SE (106°K)		15		0.21	227
NH$_3$	SE; AFP		38		0.18	51, 94a
ND$_3$	AFP		27		0.23	51
N(CH$_3$)$_3$	LW	77		4.1	0.6	228
N(C$_2$H$_5$)$_3$	LW	255		1.25	2.1	145
pyridine	SE		1.4			80
	LW	197		1.62		81
	LW	187		1.70		93
	LW	215		1.48		82
	LW	213		1.49		108
	SE		1.65		1.95	64
	SE		2.3			80
pyridine-I$_2$	LW	266		1.2		93
2-methylpyridine	LW	325		0.98	3.3	258
3-methylpyridine	LW	330		0.98	3.3	258
4-methylpyridine	LW	480		0.66	4.9	258
2,4-dimethylpyridine	LW (388°K)	230		1.39	2.6	91b
2,6-dimethylpyridine	LW (388°K)	150		2.12	1.7	91b
2,4,6-trimethylpyridine	LW (371°K)	345		0.92	4	91b
2-cyanopyridine	LW (427°K)					90, 91a
cyano		243		1.31	3.3	
ring		210		1.52	2.0	
3-cyanopyridine	LW (427°K)					90, 91a
cyano		285		1.12	4	
ring		210		1.52	2	

Compound	Method					Ref.
4-cyanopyridine: cyano	LW (427°K)	275		1.16	3.85	90, 91a
ring		360		0.88	3.20	
quinoline	LW	960		0.33	9.7	259
isoquinoline	LW	650*		0.49	6.6	255
	LW	1670		0.19	17	259
	LW	680*		0.47	6.9	255
pyrrole	LSA		2		8	140
pyrrolidine	LW	265		1.2		145
N-methylpyrrole	LW	118		2.8	2.95	145
N-methylpyrrolidine	LW	168		1.9	1.42	145
aniline	LW	1300		0.24	18.2	93
o-toluidine	LW (370°K)	375		0.85		91b
m-toluidine	LW (370°K)	360		0.88		91b
p-toluidine	LW (370°K)	405		0.79		91b
o-nitroaniline: NH$_2$	LW (463°K)	275		1.16		91b, 90
NO$_2$		36		8.8		
m-nitroaniline: NH$_2$	LW (463°K)	360		0.88		91b
NO$_2$		45		7.1		
p-nitroaniline: NH$_2$	LW (463°K)	440		0.72		91b
NO$_2$		72		4.4		
cyclohexylamine	SE	400	1.5		2.8	80
dimethylcyclohexylamine	LW	160		0.80		145
dimethylformamide	LW			2.0		81
	SE		1.4			80
	SE		5			80
CH$_3$CN	LW	88		3.9†		229
	LW	89		4.0†		81

TABLE 3.5—Continued

Compound	Method[b]	Δ^*(Hz) or T_1(msec)		$T_2^{\oplus}$(msec)	$\tau_q^{\,c}$(psec)	Ref.
		Δ^*(Hz)	T_1(msec)			
CH_3CN	LW	78		4.1†	1.2	82
	SE		3.9			61
	SE		4.35			62
	LW	81		4.0†		102
	SED		4.9			74c
	PS		7			77
CH_3CH_2CN	LW	122*		2.8†	1.7	229
Me_3CCN	LW	190*		1.7	2.7	229
CH_2ClCN	LW	175*		1.9	2.4	229
CH_2Cl-CH_2CN	SE		0.8		5.7	80
CCl_3CN	SE		1.5			
	LW	220*			2.7	229
$ClCN$	SE		5.9		1.1	39
$BrCN$	LW	165*		2	3	229
$NCCH_2CN$	LW	280*		1.2	3	229
	SE		0.8		5.5	80
$NC(CH_2)_2CN$	LW	370*		0.86	5.2	229
$NC(CH_2)_3CN$	LW	400*		0.80	5.8	229
$CH_2=CH-CN$	LW	110;*108		3.1; 3.2	1.5	81, 229
	SE		3.5		1.3	80
$CH_2=CH-CH_2CN$	LW	166*		2.0	2.3	229
C_6H_5CN	LW	390*		0.82	5.5	229
$C_6H_5CH_2CN$	LW	300*		1.05	4.3	229
2-methylbenzylcyanide	LW (313°K)	450		0.71	6.4	65
3-methylbenzylcyanide	LW (313°K)	1000		0.32	14.2	65
4-methylbenzylcyanide	LW (313°K)	600		0.53	8.5	65
	LW (313°K)	600		0.53	8.5	65
CH_3NC	LW	0.26		1200	0.8	87

Compound	Method					Ref.
CH_3CH_2NC	LW	0.31		1000	0.9	87
CH_3NO_2	LW	21; 18		26;† 26†	1.3	81, 230
	SE		22			80
$CH_3CH_2NO_2$	LW	30; 28*		17;† 15†	1.8	230, 231
	SE		15.8			80
$(CH_3)_2C(NO_2)_2$	LW	34*		12†		231
$CH_3CH(NO_2)_2$	LW	11		64†		230
$C_6H_5NO_2$	LW	108*		3.2†	6.8	81
	SE		3.5		6.2	80
$O_2N(CH_2)_2NO_2$	LW	60*		6†	5.3	231
$O_2N(CH_2)_6NO_2$	LW	60*		6†	5.3	231
o-nitrotoluene	LW	155		2.1		93
	SE		3.8			80
m-nitrotoluene	LW (364°K)	42		7.6		91b
p-nitrotoluene	LW (364°K)	61		5.2		91b
	LW (364°K)	74		4.3		91b
CH_3ONO_2	LSA		65			84
$C_2H_5ONO_2$	LW	12		53		230
	SE		45			80
C_2H_5ONO	SE		5.2			80
CH_3SCN	SE		2.0		2.7	80
CH_3NCS	SE		5.8			80
C_6H_5NCS	SE		1.3			80
HNCO	LW	20		16		232
CH_3NCO	LW	35		9		232
C_6H_5NCO	SE		1.5			80
	LW	50		6.4		232
$(C_2H_5)_2NNO_2$ NO_2	LSA		80			84
N	LW	1100		0.29		84

TABLE 3.5—Continued

Compound	Method[b]	Δ^*(Hz) or	T_1 (msec)	$T_2^{\oplus}$ (msec)	$\tau_q{}^c$ (psec)	Ref.
CF$_3$NO$_2$	LSA (190°K)		33.1			153
CF$_3$N=NOCF$_3$	LSA (190°K)		5.55			153
CH$_3$N$_3$						
central N	LW	17*		26.5†		83
terminal N	LSA, LW	19*	100*	22.7†		145, 83
R—N	LW	101*		3.3†		83
C$_2$H$_5$N$_3$						
central N	LW, PS	22*	4.5	18.7†		83, 79
terminal N	LW, PS	28*	2.2	13.8†		83, 79
R—N=	LW, PS	122*	1.1	2.7†		83, 79
thiazole	LW	160		2.0		108
isothiazole	LW	118		2.8		108
isoxazole	LW	200		1.6		108
2-bromothiazole	LW	515		0.62		109
2-fluoropyridine	SED		1.24			74c, d
3,4,5-trichloro-2,6-	LSA			0.233		111, 152
difluoropyridine	LW	1360		0.237		111, 152
3,5-dichloro-2,4,6-	LSA (337°K)		0.879	0.742		152
trifluoropyridine	LW (333°K)	453				152
α-picoline-N-oxyde	LW (351°K)	70		4.5		91b
β-picoline-N-oxyde	LW (351°K)	110		2.3		91b
Ionic Compounds^d						
XNO$_3$^{f}	LW	18^e				233
NH$_4$X^g	LW	11^e				228
N(CH$_3$)$_4$X^h	LW	-12^e				228
NaNO$_2$	LW	360		0.9		233
	LW	520				83

Compound	Method				References
$N(CH_3)_4I$	LSA		>2000		143a
$X\text{-pyridinium-}\overset{\oplus}{N}-C_2H_5$	LSA				129, 145
$\quad X = CH_3$			40		
$\quad C_2H_5$			40		
$\quad H$			100		
$\quad COOH$			140		
$\quad CONH_2$			140		
$\quad COOEt$			140		
$\quad CF_3$			310		
$Et_3\overset{\oplus}{N}(CH_2)_n\overset{\oplus}{N}Et_3, 2Br^{\ominus}$	LSA, LW				142, 118
$\quad n = 2$		105		3†	
$\quad 3$		27.8		17†	
$\quad 4$		13	60	80†	
$\quad 5$		10.8	220	200†	
$\quad 6$		9.2			
[structure] Br	LW	500		0.6	118
[structure] bBr$^{\ominus}$	LSA, LW	14.5	60	57†	118
NaN_3					83, 78, 79
$\quad$ central N	LW, PS	22; 20	5.8	22†	
$\quad$ terminal N	LW, PS	73; 60	3.6	6†	
$(CN)^{\ominus}K^{\oplus}$	LW	52		7†	229, 99
$(SCN)^{\ominus}K^{\oplus}$	LW	147		2.2	83
$(OCN)^{\ominus}K^{\oplus}$	LW	13		40†	83

TABLE 3.5—Continued

Compound	Method[b]	Δ^* (Hz)	T_1 (msec) or T_{1N}	$T_2^\oplus$ (msec)	τ_q^c (psec)	Ref.
$(CH_2NO_2)^\ominus Na^\oplus$	LW	450		0.7		234
$(HC(NO_2)_2)^\ominus Na^\oplus$	LW	46		7.8†		234
$(C(NO_2)_3)^\ominus K^\oplus$	LW	25		16†		234

[a] The nitrogen-14 QR times listed are either measured T_{1N} values or T_{2N} values calculated from measured linewidth after correction for field inhomogeneity equations (3.84) and (3.86). One expects $T_{1N} = T_{2N} = T_{qN}$. The T_{1N} and T_{2N} values listed are for pure liquids, except for those marked*, (which are for liquids containing an internal reference) at about +25°C (except when otherwise stated). T_2 values marked† are calculated from linewidth after inhomogeneity correction.

[b] SE: spin-echo; AFP: adiabatic fast passage; LW: nitrogen-14 NMR linewidths Δ^*; LSA: lineshape analysis of the resonance signal of coupled protons or fluorines; SED: spin-echo measurement of a coupled dipolar nucleus; PS: progressive saturation.

[c] Correlation times calculated using equation (3.70), taking χ_N from Tables 3.2 and 3.3 and neglecting η_N.

[d] All values listed for ionic compounds have been measured on aqueous solutions at the same concentration within a series of compounds. Correlation times could not be calculated since the χ_N values are generally not available.

[e] These linewidths have been measured in a 15-mm sample tube. Using a 7-mm tube decreases the linewidth of NO_3X and NH_4X to 9 and 4.5 Hz respectively.

[f] $X = NH_4^\oplus$, $Li^\oplus$, $Na^\oplus$, $K^\oplus$, $Ag^\oplus$, $Ag(NH_3)_2^\oplus$, $Zn^\oplus_{2}$, $Pb^\oplus_{2}$, $Cd^\oplus_{2}$, $UO_2^\oplus_{2}$, $Al^\oplus_{3}$.

[g] $X = NO_3^\ominus$, $Cl^\ominus$, $Br^\ominus$, $I^\ominus$, $H_2PO_4^\ominus$, $C_2H_3O_2^\ominus$, $SO_4^\ominus_{2}$, $S^\ominus_{2}$.

[h] $X = Cl^\ominus$, $Br^\ominus$.

rotation have been obtained from the temperature dependence of τ_q as discussed in Section 3.4.1. Using equation (3.83) Arrhenius activation energies have been calculated [80]. The E_a values lie in the 2–3 kcal/mole region and increase with molecular size as expected. Low-viscosity solvents shorten τ_q of the nitrogen containing solute. Dilution studies at constant temperature have been used to estimate the molecular volume [81]. Intermolecular associations are also expected to have a strong effect on molecular correlation times as will be seen later. The *calculation of correlation times* from molecular parameters and liquid properties represents the next step of interpretation of τ_q values obtained from relaxation data. It is found that equation (3.48) leads to calculated correlation times which are a factor of 6–10 longer than the experimental ones. Better agreement is obtained using mutual viscosity [49a] or microviscosity correction factors [49b, 92] discussed in Section 3.3.1. It seems that in the Hill formulation [49a] the important factor is the inclusion of the moment of inertia of the molecule in equation (3.88) [92, 96]. A method for calculating an 'apparent molecular volume', defined as the volume traced out by a molecule undergoing rotation, has been proposed [77]. The quasilattice model applied to ammonia leads to a correlation time which is too long [48]. The inertial model does not account for the temperature dependence of T_{qN} of NH_3 [50] and the conditional inertial model ascribes the departure of liquid NH_3 from the free rotor behaviour to the lattice dynamics [51].

Nitrogen NQCC's from Quadrupolar Relaxation Data

The converse of the procedure used for studying molecular motions has been applied to the determination of NQCC's by introducing a calculated correlation time into equation (3.70) and thus obtaining e^2qQ/h from the measured relaxation time [77, 80, 91]. This procedure is both inaccurate and unreliable because of the difficulties involved in calculating τ_q, as we have seen above. However, estimates of the NQCC may be obtained which may be of interest when no accurate values are available. Comparative studies along a series of molecules of similar size and shape may provide reliable data about the variations in NQCC along the series [91]. Correlation times may also be transferred among similar molecules [80]. However, since obtaining reliable NQCC's in liquids is of great interest for studying specific effects characteristic of the liquid state (associations, exchange, etc.), a better method is needed. A way around the problem is to use a *double quadrupolar labelling procedure*. This method consists of independently determining τ_q from the QR time T_{qA} of a second quadrupolar nucleus A in the same molecule, whose NQCC is known. This τ_q value may then be used together with the measured nitrogen QR time for obtaining the value of χ_N. The procedure is most accurate when the second quadrupolar nucleus is placed in such a way in the molecule that it experiences the same motions as the nitrogen nucleus; this is especially important when the molecular rotation is strongly anisotropic, unless of course the three diffusion constants may be determined using several A nuclei. In the extreme narrowing conditions one has:†

$$\chi_N^2 = \frac{2I_A + 3}{5I_A^2(2I_A - 1)} \chi_A^2 \left[\left(1 + \frac{\eta_A^2}{3}\right) \Big/ \left(1 + \frac{\eta_N^2}{3}\right) \right] \frac{T_{qA}}{T_{qN}} \qquad (3.89)$$

where I_A, χ_A and η_A are respectively the nuclear spin, the NQCC and the asymmetry parameter of nucleus A. The χ_N values obtained in this way may be expected to be both much more accurate and more reliable than those determined with calculated correlation times. They should still be satisfactory when isotropic rotation is assumed for not too anisotropic molecules.

This method, although very promising for the study of χ_N values in liquids, has not been much used up to now. It has been applied to the determination of χ_N in liquid ClCN [39] and in water solutions of $Et_3N^{\oplus}CD_2CH_2CH_2CD_2\overset{\oplus}{N}Et_3$, $2Br^{\ominus}$ (Section 3.5.3) taking respectively Cl and D as the second quadrupolar nucleus. If it is difficult to dispose of a second quadrupolar nucleus one may use the correlation time for the dipole–dipole relaxation of a dipolar nucleus (^1H, ^{19}F, ^{13}C) in the molecules as a measure of τ_{qN}. In addition to the problem of motional anisotropy, this procedure is less valid on the grounds that in the general case dipole–dipole and quadrupolar correlation times represent different motions, but we feel that it should still be notably more accurate than using calculated correlation times. ^1H dipole–dipole correlation times have been used to determine χ_D in the corresponding deuterated molecules [59, 97, 98]. The method

---

† Equation (3.89) is valid outside extreme narrowing only when $I_A = 1$[1].

should also be a powerful tool for studying intermolecular effects.

Intermolecular Effects on Nitrogen Quadrupolar Relaxation

Intermolecular interactions and medium effects are expected to influence both χ_N and τ_q, so that T_{qN} measurements alone do not allow us to separate the two contributions. For this reason, the double or multiple quadrupolar labelling method outlined above, should be of particular interest to many fundamental studies of intermolecular and medium effects using χ_N as the *electronic structure label* and for instance τ_q (deuterium) as the *microdynamic label*. It is of course necessary that χ_A of the second nucleus A be largely unaffected by the intermolecular interactions; using deuterium in a C–D bond as the A nucleus is particularly favourable in view of the weakness of the intermolecular interactions of such bonds. When exchange is present and exchange rates have to be taken into account, treatments of the type discussed in Section 3.4.1. may be applied using, for example, the deuterium nucleus for describing τ_A^{-1} and τ_B^{-1} and the nitrogen nucleus for q_A, q_B, η_A and η_B (equation 3.74). Many problems may be studied in this fashion: field gradients, electronic structure and exchange rates in N $\cdots$ H hydrogen bonded systems, in acceptor–donor complexes, in $NR_4^{\oplus} X^{\ominus}$ ion pairs, etc. Intermolecular associations will slow down the rate of molecular reorientation and increase the corresponding activation energies. The relaxation times of aqueous mixtures of pyridine and acetonitrile have been analysed; correlation times and solution structure have been discussed; the results suggest the formation of different types of molecular aggregates for different compositions of the mixtures [82, 97]. Dissolution of ionic salts in CD_3CN decreases T_{qN} and T_{qD} [61]. Since complexation would be expected to decrease χ_N, the effect is of motional nature. The results indicate that whereas $D_{\parallel}$ is nearly unchanged (10×10^{11} sec^{-1}) $D_{\perp}$ is strongly slowed down (0.13×10^{11} sec^{-1}) with respect to the pure liquid, in agreement with the formation of a slowly reorienting complex. However, exchange rates have not been taken into account in this study. Strong solvation of silver cations by acetonitrile has also been observed [99].

Hydrogen bonding and association have been studied in a number of systems using nitrogen-14 data [100–102]. Enthalpies of hydrogen-bonded complex formation between

CH_3CN or CH_3NC and methanol have been measured [102]. Rate constants and activation parameters have been obtained for the exchange of pyridine in the $CoPy_4Cl_2$ complex [103] and of acetonitrile in the coordination spheres of Mn(II) [104] and Ni(II) [105] ions by analysing the nitrogen-14 relaxation time according to the treatment of Swift and Connick [67]. ^{14}N relaxation times have been measured in metal cyanide complexes [106] and electron exchange rates between ferri- and ferrocyanide ions have been determined from ^{14}N linewidths [107].

3.5 Nitrogen-14 Quadrupolar Effects on the NMR Spectra of Spin $\frac{1}{2}$ Nuclei

3.5.1 Theoretical Results

A nitrogen-14 nucleus may affect the NMR signal of a neighbouring dipolar (spin $\frac{1}{2}$) nucleus X by direct dipole-dipole interaction or by indirect scalar spin–spin coupling. We shall see that in the second case interesting quadrupolar effects arising from the nitrogen relaxation may be found in the X spectrum.

Dipole–dipole Interaction

In liquids, the dipole–dipole interaction serves as a relaxation mechanism and in extreme narrowing conditions its contributions T_{1D} and T_{2D} to T_1 and to T_2 are equal. Because of the small magnetic moment of the ^{14}N nucleus, the contribution $T_D^X(X,N)$ of the N, X dipole–dipole interaction to the relaxation of the X nucleus is very small compared to the contribution $T_D^X(X,Y)$ of (X,Y) dipole–dipole interactions when X and Y are ^1H or ^{19}F nuclei, which have a comparatively large magnetic moment [108].

One may show that [1]:

$$\frac{[T_D^X(X,Y)]^{-1}}{[T_D^X(X,N)]^{-1}} = C\,\frac{\gamma_Y^2}{\gamma_N^2}\,\frac{I_Y(I_Y+1)}{I_N(I_N+1)} = C\,\frac{3}{8}\,\frac{\gamma_Y^2}{\gamma_N^2} \qquad (3.90)$$

where X and Y are dipolar nuclei and $C = \frac{3}{2}$ if $X = Y$ and $C = 1$ if $X \neq Y$. Equation (3.90) is equal to 108, 95, 72 and 63; for $X \equiv Y = H$, $X \equiv Y = F$, $X = F$, $Y = H$, and $X = H$, $Y = F$ respectively. Thus, unless the dipolar nucleus is completely isolated magnetically we can ignore the X, N dipole–dipole contribution to the X relaxation.

Lineshape of the spin $\frac{1}{2}$ resonance signal

When the X and ^{14}N nuclei are spin–spin coupled, the lineshape of the X resonance signal depends on the spin lattice relaxation time $T_{1N} = T_{qN}$ of the nitrogen nucleus. If T_{qN} is long compared to the inverse of the spin–spin coupling constant, $1/J$, a triplet is obtained. As T_{qN} shortens (intermediate range) the triplet structure progressively disappears and a single line is obtained. At very short T_{qN} the interaction serves as a scalar contribution to the relaxation of X [1, 108, 109]. The X spectrum is described over the entire range of T_{qN} values by the following lineshape equation [1, 110]:

$$I(\omega) \propto Re\{W . A^{-1} . 1\} \qquad (3.91)$$

where Re denotes 'real part of', and W is a line matrix of dimension $2I_N + 1 = 3$ whose components are proportional to the relative probabilities of the nitrogen spin states. In the present case we take W = (1 1 1) since the X, N spin system may be considered to give a first-order spectrum. 1 is a column matrix:

$$1 = \begin{bmatrix} 1 \\ 1 \\ 1 \end{bmatrix}$$

A^{-1} is the inverse of the lineshape matrix A:

$$A_{mm'} = \left[i(\omega_0 - \omega + 2\pi mJ) - \frac{1}{\tau_m} - \frac{1}{T'_{2X}} \right] \delta_{mm'} + P_{mm'} \qquad (3.92)$$

where ω_0 is the frequency at the centre of the multiplet, J is the X,N coupling constant in Hertz, τ_m is the lifetime of spin state m ($m = +1, 0, -1$) and $P_{mm'}$, is the transition probability between states m and m'. The latter two parameters are related by:

$$\frac{1}{\tau_m} = \sum_{m \neq m'} P_{mm'} \qquad (3.93)$$

$$P_{m, m \pm 1} = \frac{1}{2T_{qN}} \frac{(2m \pm 1)^2 (I_N \pm m + 1)(I_N \mp m)}{(2I_N - 1)(2I_N + 3)} =$$

$$= \frac{(2m \pm 1)^2 (2 \pm m)(1 \pm m)}{10T_{qN}} \qquad (3.94)$$

$$P_{m, m\pm2} = \frac{1}{2T_{qN}} \frac{(I_N \mp m)(I_N \mp m - 1)(I_N \pm m + 1)(I_N \pm m + 2)}{(2I_N - 1)(2I_N + 3)}$$

$$= \frac{(1 \mp m)(\mp m)(2 \pm m)(3 \pm m)}{10T_{qN}} \tag{3.95}$$

In equation (3.92) T'_{2X} is a phenomenological relaxation time representing the 'natural' linewidth at halfheight Δ'_X of the X resonances in the absence of any quadrupolar contrubution:

$$1/T'_{2X} = \pi\Delta'_X \tag{3.96}$$

The lineshape equation (3.91) now becomes:

$$I(\omega) \propto \mathrm{Re} \left\{ (1, 1, 1) \begin{bmatrix} i(\Delta\omega + 2\pi J) - \dfrac{3}{5T_{qN}} - \dfrac{1}{T'_{2X}} & \dfrac{1}{5T_{qN}} & \dfrac{2}{5T_{qN}} \\[2mm] \dfrac{1}{5T_{qN}} & i\Delta\omega - \dfrac{2}{5T_{qN}} - \dfrac{1}{T'_{2X}} & \dfrac{1}{5T_{qN}} \\[2mm] \dfrac{2}{5T_{qN}} & \dfrac{1}{5T_{qN}} & i(\Delta\omega - 2\pi J) - \dfrac{3}{5T_{qN}} - \dfrac{1}{T'_{2X}} \end{bmatrix}^{-1} \begin{bmatrix} 1 \\ 1 \\ 1 \end{bmatrix} \right\}$$

$$\tag{3.97}$$

where $\Delta\omega = \omega_0 - \omega$ is the distance from the centre of the multiplet in rad sec^{-1} and T_{qN} is the nitrogen quadrupolar relaxation time in seconds. Neglecting the term $1/T'_{2X}$, the lineshape equation may be written in analytical form [110, 111]:

$$I(x) = \frac{2\bar{\eta}}{\pi J} \frac{45 + \bar{\eta}^2(5x^2 + 1)}{225x^2 + \bar{\eta}^2(34x^4 - 2x^2 + 4) + \bar{\eta}^4(x^6 - 2x^4 + x^2)} \tag{3.98}$$

where $x = \Delta\nu/J$, $\Delta\nu = \Delta\omega/2\pi$, and $\bar{\eta} = 10\pi T_{qN}J$. The appearance of the X spectrum depends solely on the dimensionless parameter $\bar{\eta}$. Figure 3.5 shows the lineshape changes as a function of $\bar{\eta}$ when the relaxation time T_{qN} changes from small to large values with respect to J, for two cases:

    (a) zero natural linewidth in the absence of quadrupolar relaxation broadening, $T'_{2X} = \infty$.
    (b) a natural linewidth of 0.3 Hz, $T'_{2X} = 1.06$ sec.

It is seen that marked differences occur especially in the two limits of large and small $\bar{\eta}$ values.

Large $\bar{\eta}$ Values:
    (a) and T'_{2X} much longer than T_{qN}; the lineshape is governed by the quadrupolar effect and the X triplet shows

three lines of equal area but of relative width $\frac{3}{5}$, $\frac{2}{5}$, $\frac{3}{5}$. T_{qN} may be obtained from the quadrupolar contribution Δ_{qX} to the total linewidth Δ_X^* of a given X component:

$$\Delta_{qX} = \Delta_X^* - \Delta_X' = \frac{d}{\pi T_{qN}} \qquad (3.99)$$

(with $d = \frac{3}{5}$ for the outer lines and $\frac{2}{5}$ for the central line).

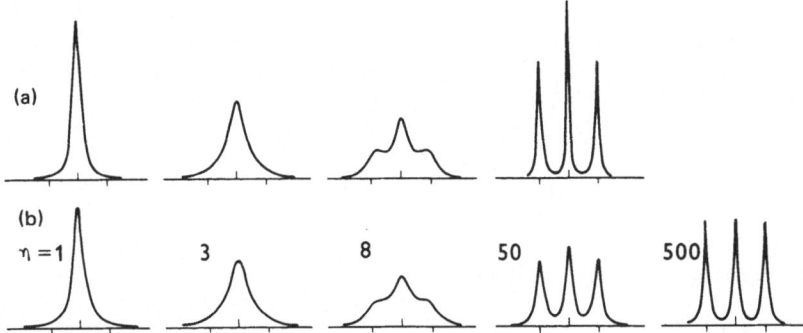

Fig. 3.5. Calculated NMR lineshapes using equation (3.97), of a dipolar nucleus X coupled to a nitrogen-14 nucleus as a function of the ^{14}N relaxation time T_{qN} represented by the parameter $\bar{\eta} = 10\pi J T_{qN}$ [equation (3.99)]; two cases are shown: (a) zero 'natural' linewidth ($T_2' X = \infty$), and (b) a 'natural' linewidth of 0.3 Nz ($T_2' X = 1.06$ sec) for the X resonances (abscissa: $x = \Delta\nu/J$).

(b) and T_{qN} much longer than $T_2' X$; equation (3.97) (with $T_{qN} = \infty$) now becomes a sum of three Lorentzian functions having equal heights and widths $1/\pi T_2' X$:

$$\mathbf{I}(\nu) = \frac{T_{2X}'}{1 + 4\pi^2 T_{2X}'^2 (\Delta\nu + J)^2} + \frac{T_{2X}'}{1 + 4\pi^2 T_{2X}'^2 \Delta\nu^2} +$$

$$+ \frac{T_{2X}'}{1 + 4\pi^2 T_{2X}'^2 (\Delta\nu - J)^2} \qquad (3.100)$$

In this case the lineshape of the X spectrum is governed by other contributions than that due to the nitrogen quadrupole. Therefore the quadrupolar broadening is negligeable with respect to the linewidth of the X signals arising from field inhomogeneity and other relaxation processes of the X nucleus. This kind of lineshape is observed for instance in the proton NMR spectrum of the $N(CH_2 CH_3)_4^{\oplus}$ ion and of CH_3-NC where the nitrogen quadrupolar relaxation time is long and proton–proton dipole–dipole relaxation as well as inhomogeneity determine the proton NMR lineshape (Figs. 3.3, 3.4, and 3.6). The relaxation times of the individual X lines have been studied [112, 113]. When χ_N is small but not zero zero ($\geqslant 0.2$ MHz) the nitrogen QR averages out the spin-lattice relaxation rates of the three X resonances, the value of this average being independent of χ_N. However, on selective saturation of a single component of the X triplet, the initial relaxation rate depends on the QR rate.

Very Small $\bar{\eta}$ Values:

When $\bar{\eta}^2 \ll 1$, equation (3.98) reduces to equation (3.101) [111]:

$$I(x) = \frac{2\bar{\eta}}{\pi J}\frac{45}{225x^2 + 4\bar{\eta}^2} \qquad (3.101)$$

which may also be written as:

$$I(\nu) = \frac{6T_{2X}}{1 + 4\pi^2 T_{2X}^2 \Delta\nu^2} \qquad (3.102)$$

One observes for X a single Lorentzian line of relative intensity 3 centred at ν_0 and having a linewidth at halfheight given by:

$$\Delta_X = 1/\pi T_{2X} \qquad (3.103)$$

with

$$T_{2X} = 15/4\pi\bar{\eta}J = 3/8\pi^2 J^2 T_{qN} \qquad (3.104)$$

Incorporating the T_{2X}' contribution due to effects other than the quadrupolar one, the X resonance lineshape is determined by the relaxation parameter T_{2X}^*:

$$1/T_{2X}^* = 1/T_{2X} + 1/T_{2X}' \qquad (3.105)$$

T_{2X}' can be obtained from lines in the spectrum which are not broadened by quadrupolar effects and for which inhomogeneity broadening is dominant. In the limit of very short values of T_{qN} the scalar coupling J may be considered to contribute to the X relaxation through a scalar relaxation mechanism, the nitrogen nucleus being treated as part of the lattice. The correlation time of the scalar contributions to X relaxation T_{1X}^S and T_{2X}^S is T_{qN}. One has [1]:†

$$\frac{1}{T_{1X}^S} = \frac{8\pi^2 J^2 I_N (I_N + 1)}{3} \left(\frac{T_{qN}}{1 + (\omega_X - \omega_N)^2 T_{qN}^2} \right) \quad (3.106)$$

$$\frac{1}{T_{2X}^S} = \frac{4\pi^2 J^2 I_N (I_N + 1)}{3} \left(T_{qN} + \frac{T_{qN}}{1 + (\omega_X - \omega_N)^2 T_{qN}^2} \right) \quad (3.107)$$

It is seen that $T_{1X}^S \neq T_{2X}^S$. Since T_{qN} is generally not shorter than $10^{-4} - 10^{-5}$ sec and $(\omega_X - \omega_N)$ is of the order of $10^7 - 10^8$ Hz, $(\omega_X - \omega_N)^2 T_{qN}^2 \gg 1$ and equation (3.107) becomes:

$$1/T_{2X}^S = 8\pi^2 J^2 T_{qN}/3 \quad (3.108)$$

which is the same as T_{2X}, (equation (3.104)) of the Lorentzian lineshape equation (3.102) obtained from the general expression (3.98), in the $\bar{\eta}^2 \ll 1$ limit.

The above lineshape equation (3.97) was derived for a single X nucleus coupled to a single ^{14}N nucleus. *Multispin systems* of the type $NX_m Y_n$ containing respectively m and n dipolar nuclei X and Y spin–spin coupled to ^{14}N, and systems containing several quadrupolar nuclei have been treated recently [89]. For $NX_m Y_n$ it is found that the overall dipolar spectrum is a sum of lineshapes of the type given by equation (3.98) having the same relative intensities and centred at the same positions as the components of the X and Y multiplets in the absence of the ^{14}N nucleus (obtained for instance in a ^{14}N decoupling experiment). Furthermore, inclusion of a phenomenological parameter T_{2X}' (equation (3.96)) as is done in equation (3.97) for simulating the inhomogeneity broadening of the X, Y signal, is shown to be valid over the whole range of lineshapes if the inhomogeneity is Lorentzian. The *lineshape of the nitrogen spectrum* is the summation of Lorentzian lines of total width $\Delta_N^* = \Delta_{qN} + \Delta'$ with $\Delta_{qN} = 1/\pi T_{qN}$ being the quadrupolar contribution and Δ' the 'natural' linewidth contribution used in equation (3.86). In the treatments of these complex multispin systems the general relaxation matrix method has been used and second order effects in the dipolar spectrum have been studied [89].

† If the spin of the quadrupolar nucleus is >1, these relations are only valid in extreme narrowing conditions.

Lineshapes in the Presence of Chemical Exchange

When the dipolar X nuclei in an NX_m system undergo chemical exchange, the X spectrum is affected by both T_{qN} and by the exchange rate [114–116]. We shall first consider the case of exchange between *two identical sites*. If τ_e is the lifetime of a nucleus X in an environment characterized by a certain spin state of the ^{14}N nucleus, the probability of X experiencing a change in ^{14}N spin state from m to m' through chemical exchange is:

$$R_{mm'} = \tau_e^{-1} \qquad (3.109)$$

This probability is independent of m and m'. The total probability of a given spin X experiencing an m to m' change in ^{14}N spin state becomes:

$$\mathscr{P}_{mm'} = P_{mm'} + R_{mm'} \qquad (3.110)$$

The lineshape equation is then (3.91) where in the $A_{mm'}$ and $1/\tau_m$ expressions, equations (3.92) and (3.93), $P_{mm'}$ is replaced by $\mathscr{P}_{mm'}$. Since there are three nitrogen spin states only two-thirds of the exchange events will bring a given nucleus into a site with a different ^{14}N spin state than the original one. Thus, the probability that a nucleus X undergoes chemical exchange is:

$$k_e = \frac{3}{2} \frac{1}{\tau_e} \qquad (3.111)$$

In order to relate k_e to the chemical exchange rates, the number of equivalent X nuclei on each site and the exchange mechanism have to be taken into account. If exchange of the X nuclei occurs between *two different sites* A and B of populations p_a and p_b, chemically shifted by Ω rad/sec (= $\omega_a - \omega_b$), with forward and reverse rate constants k_{ab} and k_{ba} the problem is more complex.

Suppose the X nuclei are coupled to ^{14}N in site B with a coupling constant of J Hz, the ^{14}N relaxation rate being $R_q = T_{qN}^{-1} = T_{1N}^{-1}$ and that they can be scrambled with equal probability among the 3 subsites of ^{14}N by a reaction with specific rate k_s. One can show that the lineshape of the X resonance becomes [117]:

$$I(\omega) = -\mathbf{Re}\left\{[p_b\alpha + p_a(\beta + \gamma) - 4\bar{k}p_ap_b]/[\alpha(\beta + \gamma) - 4\bar{k}^2p_ap_b]\right\}$$

$$(3.112)$$

with $\alpha = -R'_{2X} - k_{ab} + i(\omega_a - \omega)$;

$\qquad \beta = -R'_{2X} - k_{ba} + i(\omega_b - \omega)$;

$\qquad \gamma = \frac{2}{3}(2\pi J)^2 (\beta - 0.6R_q - k_s)/[(\beta - 0.6R_q - k_s)$

$\qquad\qquad (\beta - R_q - k_s) + (2\pi J)^2/3]$;

$\qquad \bar{k} = (k_{ab} + k_{ba})/2$; and

$R'_{2X} = T'^{-1}_{2X}$ from equation (3.96).

For fast quadrupolar relaxation and fast chemical exchange one obtains one line of width Δ^*_X:

$$\pi\Delta^*_X = \frac{1}{T^*_{2X}} = \frac{1}{T'_{2X}} + \frac{\Omega^2 p_a p_b}{2\bar{k}} + \frac{8\pi^2}{3} \frac{p_b J^2}{R_q + k_{ab} + k_s} \quad (3.113)$$

For slow exchange and fast relaxation one has two lines of width Δ^*_{aX} and Δ^*_{bX}:

$$\pi\Delta^*_{aX} = \frac{1}{T'_{2X}} + k_{ab} \quad (3.114)$$

$$\pi\Delta^*_{bX} = \frac{1}{T'_{2X}} + k_{ba} + \frac{8\pi^2}{3} \frac{J^2}{(R_q + k_{ba} + k_s)} \quad (3.115)$$

Thus, the linewidth of the resonance at site A depends only on the exchange rate but this is not the case for site B. For slow exchange–slow relaxation one obtains one line for site A and three lines for site B. The two sites exchange case is represented for instance by the proton exchange of ammonia, amines, amides and ammonium salts catalyzed by water. Exchange between free and ion-paired sites in ammonium salts has also been considered [143b].

3.5.2 Lineshape Analysis of the Dipolar Resonances

From the previous discussion it is clear that computer fitting of theoretical lineshapes $I(\omega)$, equation (3.97), to experimental spectra of nuclei coupled to ^{14}N allows the determination of the ^{14}N relaxation time [84, 118, 119]. The lineshape analysis (LSA) method for determining T_{qN} has advantages and limitations:

(1) LSA of the resonance of protons or fluorines coupled to a ^{14}N nucleus allows T_{qN} to be determined with proton or fluorine magnetic resonance *sensitivity* (the gain is a factor of about 10^3 with respect to nitrogen NMR). This advantage

is essential for the use of T_{qN} measurements in biophysical studies.

(2) The lineshape changes are appreciable and easily detected.

(3) The temperature dependence of T_{qN} may be studied over temperature ranges of *ca.* 100° or more.

(4) From a practical stand point, regular NMR spectrometers may be used to perform the measurements.

(5) The main limitation is that in many compounds ^{14}N relaxation is too fast for appreciable lineshape changes to be observed, except in compounds with relatively symmetrical nitrogen sites or when the N—X coupling constant is large as in directly linked N—X groups. This point is considered in Section 3.5.3 together with specific cases.

The lineshape of the dipolar nuclei is of course affected by any change in T_{qN} and thus by any structural, medium or temperature effect on T_{qN} as discussed in Section 3.4.4. It may also be pointed out that the lineshape changes due to quadrupolar effects have a characteristic *inverse temperature dependence* as compared to chemical exchange effects. In the quadrupolar case a single line is observed at low temperature (short T_{qN}, slow molecular motion) which broadens and splits into a triplet at high temperature (long T_{qN}, fast motions). In the case of chemical exchange separate signals are observed at low temperature (slow exchange) and coalesced lines at high temperature (fast exchange).

3.5.3   Nitrogen-14 Quadrupolar Effects on Proton NMR Lineshapes

^{14}N—H Spin–Spin Coupling Constants

Nitrogen-14-proton spin–spin coupling constants are only observed in compounds where the ^{14}N QR time is long enough to avoid washing out the N—H splittings in the proton spectrum. The nitrogen 15 isotope has a nuclear spin of $\frac{1}{2}$. ^{15}N—H couplings have been measured in a number of isotopically enriched compounds and ^{14}N—H coupling constants may be calculated from the ^{15}N—H coupling using the respective gyromagnetic ratios $\gamma(^{15}$N$)$ and $\gamma(^{14}$N$)$. One has:

$$J(^{14}\text{N}-\text{H}) = \frac{\gamma(^{15}\text{N})}{\gamma(^{14}\text{N})} J(^{15}\text{N}-\text{H}) = -0.7129 J(^{15}\text{N}-\text{N}) \quad (3.116)$$

Nitrogen proton coupling constants are discussed in Chapters 1 and 5. We shall limit ourselves to briefly considering the cases where ^{14}N–H couplings may be observed. Protons directly linked to nitrogen show large N–H couplings of the order of 50–60 Hz [85, 120]; broad triplets are generally observed. Their lineshape is also influenced by proton exchange rates. Geminal N–C–H and vicinal N–C–C–H couplings have been observed in compounds where the field gradient at nitrogen is zero by symmetry or very weak. This is the case in quaternary ammonium salts [88, 118, 121-128, 238], heterocyclic ammonium salts [129, 130] nitrilium salts [131], isonitriles [87, 132-134, 138], nitramines [84, 135, 136], alkyl nitrates [84] and methyl azide [118]. Some spectra are shown in Fig. 3.1-3.4, 3.6 and 3.7. Geminal N–H couplings are generally very small ($\leqslant 0.5$ Hz) whereas vicinal couplings are of the order of 2-3 Hz. Some selected couplings are listed in Table 3.6. The angular dependence of vicinal couplings [127, 134] and the signs of N–C–H (generally negative) and N–C–C–H (positive) couplings [88, 128], have been studied. It may be noted that the separation of the triplet components of the proton resonance is only equal to J(N–H) when the quadrupolar broadening is very small.

Nitrogen-14 QR Data from Proton LSA (Table 3.5)

The two factors in equation (3.70), ^{14}N NQCC and correlation time, which affect T_{qN} also influence the proton lineshapes. The temperature effects on proton lineshape in directly bonded N–H groups have been observed for a number of coupounds [120]. A detailed LSA of the NH$_3$ proton triplet as a function of temperature has yielded the temperature dependence of the nitrogen relaxation time [139]. The temperature dependence of T_{qN} in pyrrole has also been obtained from a LSA of the N–H resonance [140].

Ammonium Salts

Structure, solvent and temperature effects on the lineshape of N–C–CH$_3$ groups in ammonium salts have been reported for a number of compounds [88, 122, 124, 129, 141-144], but an accurate LSA has generally not been performed.

At high temperatures, as the molecular motions become faster, T_{qN} becomes longer and better resolved triplets are observed. The disappearance of triplet splittings in non-aqueous solutions of ammonium salts [141, 143a] may arise

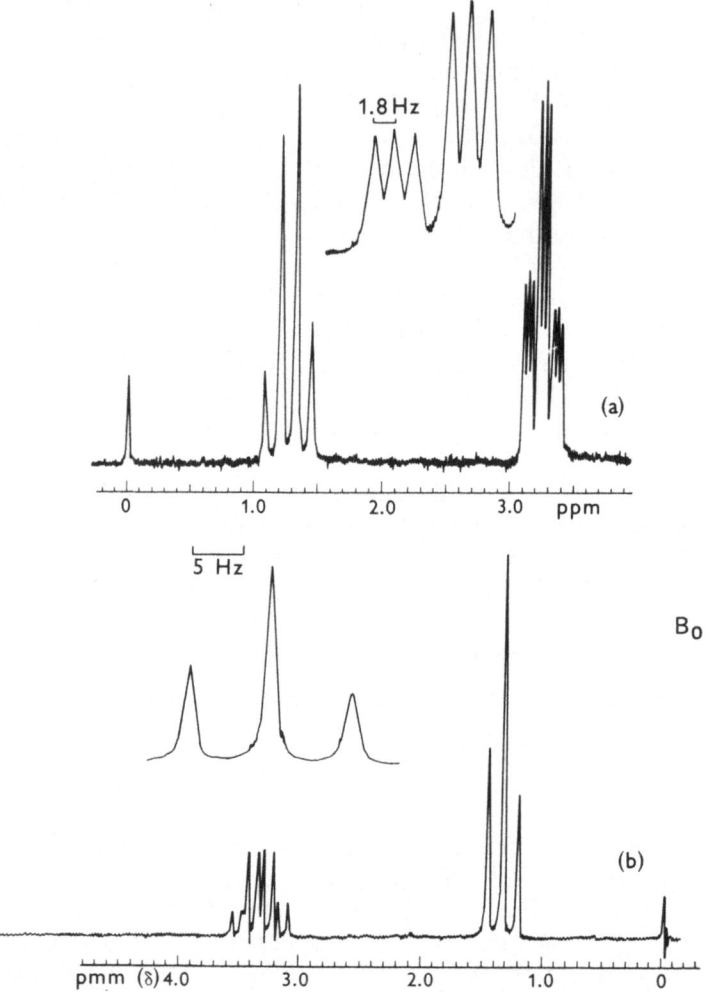

Fig. 3.6.   (a) ^{1}H NMR spectrum of a molar solution of $(CH_3CH_2)_4N^{\oplus}Br^{\ominus}$ in D_2O at 60 MHz and at room temperature (internal reference: HDO signal); the expanded trace shows part of the CH_3 resonance displaying a 1.8 Hz N–H coupling. (b) ^{1}H NMR spectrum of a molar solution of $Et_3NH^{\oplus}Cl^{\ominus}$ in 6N HCl at 60 MHz and at room temperature [internal reference: $(CH_3)_3SiCH_2CH_2CH_2-SO_3Na$]; the expanded trace shows the CH_3 resonance.

from the effect of ion-pairing both on the field gradient at nitrogen and on the molecular correlation time. Distortion of the electric symmetry at nitrogen occurs in ammonium salts with different nitrogen substituents; this is clearly apparent in the spectra of $\overset{\oplus}{N}Et_4$ and $H\overset{\oplus}{N}Et_3$, where the correlation time change would be expected to give the reverse effect, $\overset{\oplus}{N}HEt_3$ being smaller than $\overset{\oplus}{N}Et_4$ (Fig. 3.6). LSA of $(CH_3)_3N^{\oplus}R$ compounds have been performed, taking into account exchange between free and ion-paired sites; micelle formation, leading to low activation energies for reorientation [143b]. The effect of

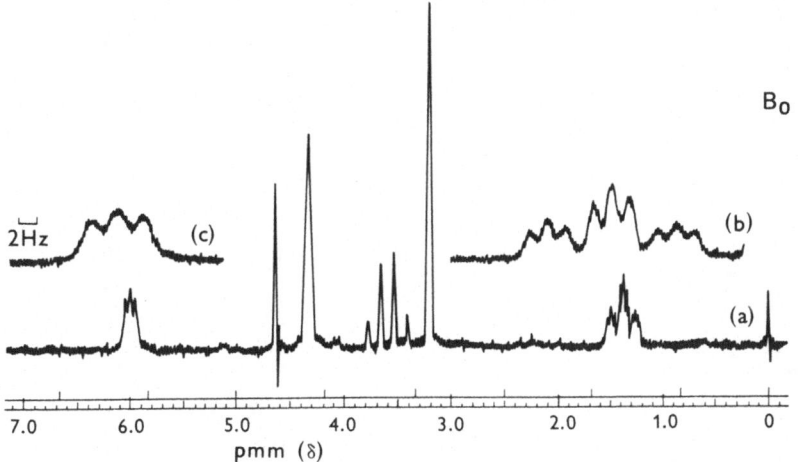

Fig. 3.7. (a) ^1H NMR spectrum of a molar solution of ⌷N⌷ $Br^{\ominus}$ in D_2O at 60 MHz and at room temperature [internal reference: $(CH_3)_3SiCH_2CH_2CH_2SO_3Na$]. (b) and (c) are expanded traces of the signals due to the CH_3CH_2 and to the vinylic protons showing coupling with the nitrogen nucleus (see Table 3.6) [145].

para-substituents in N-methyl-pyridinium salts [120] may be interpreted in terms of changes in electric symmetry at the nitrogen site. Accurate LSA of the N–C–CH$_3$ signal has also been performed on these compounds [145] (Table 3.5). Assuming similar correlation times for these compounds of similar molecular size, it is found that electron attracting para-substituents diminish the NQCC since T_{qN} increases on going from para-CH$_3$ to para-CF$_3$. This result is opposite to the substituent effects on NQCC's in para-substituted pyridines (Table 3.2). It may be explained by considering that electron attraction away from the nitrogen site along the σ and π bonds in substituted pyridines increases the electron unbalance with respect to the lone pair, whereas in the pyridinium salts the disappear-

TABLE 3.6.　Selected Nitrogen-14, Proton Spin–Spin Coupling Constants (in Hertz)

Compound	$J_{\text{N–H}}$	Ref.
NH_3	43.8	139
$(NH_4)^\oplus$	52.5	235
H–C(=O)NH_2	60	120
HNCO	64, 69	236, 237
pyrrole N–H	69.5	140
pyridinium N–H	68	83
$CH_3\overset{\oplus}{N}H_3$	49	120
$(CH_3)_2\overset{\oplus}{N}H_2$	53	120
CH_3NC	2.7	132
CH_3CH_2NC	2.0	137
CH_3CH_2NC	2.5	137
$(CH_3)_4N^\oplus$	0.56	143
$(CH_3\text{-}CH_2)_4N^\oplus$	1.8	121, 122
$(CH_3\text{–}CH_2)N^\oplus(CH_3)_3$	2.2	121
H_A H_K C=C H_B $\overset{\oplus}{N}(CH_3)_3$	5.6 (A)	88
	2.6 (B)	88
	3.6 (K)	88
H_A H_A ... $\overset{\oplus}{N}$ CH_3 CH_2 CH_3	3.0 (A)	118
	2.1 (CH_3)	118
H_A CF_3 ... $\overset{\oplus}{N}$ CH_3	3.0 (A)	129
	2.4 (CH_3)	129
$CH_3 N N N$	2.9	118
$(CH_3)_2 N NO_2$	1.7	135, 136, 84
$CH_3 ONO_2$	2.75	84
$CH_3 C{\equiv}\overset{\oplus}{N}{-}CH_3$	2.7	131
$CH_3 C{\equiv}\overset{\oplus}{N}{-}CH_3$	1.7	131
$CH_3{-}C{\equiv}\overset{\oplus}{N}{-}CH_2 CH_3$	1.5	131
$(CH_3)_3 C{-}NCB(CH_3)_3$	2.0	146
$(CH_3)_2 N^\oplus{=}CH_2, SbCl_6^\ominus$	1.7 (CH_3)	244
	2.0 (CH_2)	

ance of the lone pair transfers the origin of the field gradient to the electron unbalance between the σ and π electrons at the nitrogen site. Electron-attracting substituents may then lower the field gradient by diminishing the difference in σ and π electron populations. This result is in agreement with the common picture of substituents effects in aromatic systems and provides a more quantitative appreciation of the effect in terms of electron population transfer. Inclusion of the nitrogen site in a ring [118, 125, 144, 145] shows that distortion of the C—N—C angles in strained systems leads to disappearance of the N—C—CH_3 triplet splittings. LSA allows the estimation of NQCC's. The nitrogen atom site is symmetrical in piperidinium and pyrrolidinium salts but is distorted in azetidinium and especially in aziridinium salts for which NQCC's of about 0.44 and 3.1 MHz may be estimated. These values may be interpreted in terms of the difference in electronic population between endo- and exocyclic bonds, with the result that $[p(C-N)_{exo}] - [p(C-N)_{endo}] = 0.3$ and 0.04 for the aziridinium and azetidinium ions respectively [145].

Distortion of the electric symmetry at the nitrogen site in ammonium salts has also been brought about by shortening the chain length in $Et_3 \overset{\oplus}{N}-(CH_2)_n-\overset{\oplus}{N}Et_3 \, 2Br^{\ominus}$ salts [118, 142, 145]. As n decreases from 6 to 2, one observes a progressive disappearance of the N—CH_2—CH_3 triplet splittings and a progressive broadening of the ^{14}N NMR signal. Again in these compounds the effect is of an electric nature, correlation time changes would predict T_{qN} to increase when the chain becomes shorter, whereas the reverse is observed. Interionic and solvent effects appear also to be important [equation (3.78)]. Indeed, the changes in NQCC observed when the $N^{\oplus}$ groups come nearer to each other are much larger than expected from a single point charge at the appropriate distance. An idea of the complex processes involved may be gained from a study of the microdynamic structure of aqueous NEt_4Br solutions [45]. In all of the cases discussed above, it is only possible to compare relative ^{14}N NQCC's in series of compounds of similar size so as to keep the correlation time approximately constant along the series.

The *double quadrupolar labelling* method (Section 3.4.4) has been applied to $Et_3 NCD_2 CH_2 CH_2 CD_2 \overset{\oplus}{N}Et_3, \, 2Br^{\ominus}$; and the correlation time has been obtained from the deuterium relaxation time. Introducing this correlation time in the ^{14}N relaxation equation leads to a NQCC of about 150 kHz for a molar aqueous solution.† Such deuterium–nitrogen double labelling should give very interesting results by allowing much more accurate determinations of NQCC's than those obtained using correlation times estimated from molecular size and viscosity as discussed in Section 3.4.4. Studies of nitrogen field gradients and molecular motions in ammonium salts are also of broad interest because of the importance of such compounds in biological systems, especially in the field of neurochemistry (choline derivatives, phospholipids in membrances, ganglion-blocking agents, etc.).

Nitro Derivatives X—NO$_2$

T_{qN} data obtained by LSA of the CH$_3$ resonance in $(CH_3)_2N-NO_2$ and in CH_3ONO_2 have been used in conjunction with data on $(CH_3)CH-NO_2$ for analyzing the electronic structure of nitro compounds, especially the low nitrogen χ_N values found in nitramines and in nitrates as compared to nitro-alkanes [84].

Isonitriles

No detailed LSA of proton signals of isonitriles has been published. The persistance of the sharp CH$_3$ triplet in CH_3-NCBH_3 has been taken to indicate that the adduct is linear [146].

In complexes of CH_3NC with cadmium(II) and zinc(II) ions, the triplet splitting has partially coalesced the more so for the heavier Cd(II) than for Zn(II). This suggests that it is an effect due more to an increase in correlation time than to an increase in the field gradient at nitrogen (Fig. 3.4) [145].

Nitrogen-Containing Heterocycles

The signals due to the protons in the α position to the nitrogen atom in heterocyclic molecules (pyridine, thiazole, isothiazole, oxazole, etc.) are markedly broadened by an incomplete washing out of an 8-10 Hz J(N–C–H) coupling [108]. By lowering the temperature, the rate of nitrogen relaxation increases, bringing about more effective N–H de-coupling. The broad lines sharpen and previously hidden H–H

---

† Rotation about the C–N bond is expected to be slow and not to shorten appreciably the correlation time of deuterium with respect to the nitrogen correlation time.

splittings may become resolved. The quadrupolar broadening may be calculated using an independent measure of T_{qN} (from ^{14}N linewidths) and of J(N—H) (from ^{15}N—H data). A double resonance study has also been performed on 2-bromo-thiazole [109].

Combined Quadrupolar Relaxation and Exchange Effects

Combined effects of quadrupolar and chemical exchange effects on proton lineshapes have been studied in aqueous urea [147] and thiourea [117] using the LSA treatment discussed in Section 3.5.2. Rates of protolysis have been obtained in this way. Ion-pairing in ammonium salts has been considered [143b]. LSA of the NH_3 signal in the NH_3—KNH_2 system has been used for determining the rate constant $(1.51 \times 10^7$ liter/mole sec at $25°C)$ and the activation parameters for the bimolecular exchange process [139]:

$$NH_3 + NH_2^{\ominus} \rightleftharpoons NH_2^{\ominus} + NH_3$$

LSA of the NH_3 proton resonance of a dilute potassium-liquid ammonia solution shows a large contribution to the ^{14}N relaxation of the dipolar interaction with the unpaired electrons; it leads to the average lifetime of a given ^{14}N in a given electron-ammonia complex $(10^{-12}$ sec) and to the average solvation number of the electron (20–40) [239].

3.5.4   Nitrogen-14 Quadrupolar Effects on Fluorine NMR Lineshapes.

Table 3.7 lists some available N—F coupling constants [148–155]. It is seen that these couplings are appreciably larger than the corresponding N—H couplings; as a consequence ^{19}F NMR lineshape changes may in principle be observed over a wider range of nitrogen relaxation rates, extending to much shorter T_{qN}'s than in the case of proton NMR lineshapes. The lineshape of the ^{19}F NQR signal of NF_3 changes from a single line at $-205°C$ to a broad triplet at $+20°C$ [148]. A gross estimate of T_{qN} from the reported spectra using theoretical lineshapes (Fig. 3.5) gives $T_{qN} = 1$ msec at $-165°C$; then taking $e^2qQ/h = 7.068$ MHz (Table 3.2) one obtains the correlation time $\tau_q(NF_3) = 1.35$ psec.

As expected for a tetrahedral species, $NF_4^{\oplus}$ displays a

TABLE 3.7. ^{14}N, ^{19}F Spin–Spin Coupling Constants and ^{14}N Quadrupolar Relaxation Times in Fluorinated Nitrogen Compounds

Compound	Solvent	J(N–F) (in Hz, at $T°$C)	T_{qN} (in msec, at $T°$C)	Reference
NF$_3$	neat	155 (20°)	1 (−185°)[a]	148, 86
NF$_4^{\oplus}$ AsF$_6^{\ominus}$	HF liq.	234 (25°)	−	149
NF$_3$O	CCl$_3$F	135.5 (−82°)	30 (−82°)[a]	150, 86
N$_2$F$^{\oplus}$AsF$_6^{\ominus}$	HF liq.	328 (25°)	−	151
FNO$_2$	neat	112.5 (−100°?)	−	257, 86
cis-FN=NF	CCl$_3$F	±145; ∓37	−	155
trans-FN=NF	CCl$_3$F	±136; ∓73	−	155
CF$_3$–NO$_2$	neat	15.1 (∼25°)	43 (−57°)	153
CF$_3$N=N(O)CF$_3$	neat	14.5 (∼50°)	10 (−58°)	153
3,4,5-trichloro, 2,6-difluoropyridine	neat	37.2 (∼25°)	0.23 (25°)	111, 152
3,5-dichloro, 2,4,6-trifluoropyridine	neat	37.4 (35°)	1.33 (35°)	152
2-fluoropyridine	neat	42 ± 6 (25°)	1.2 (25°)	74c,d

[a] Estimated by the present authors from reported data (see also text).

sharp triplet (linewidth ⩽ 10 Hz) [149]; the same is true for NF$_3$O [150] for which a sharp triplet [linewidths: 6 Hz (central line); 8 Hz (outer lines)] is observed even at −82°C, indicative of a very low field gradient at the nitrogen nucleus.

A broad ^{19}F triplet is observed for N$_2$F$^{\oplus}$ (linewidth ∼105 Hz) for which a linear structure N≡N̈–F has been proposed [151]. Detailed LSA has been performed over a range of temperature on the ^{19}F resonance signal of 2,6-fluorinated pyridines [111, 152] and on the CF^3 group in CF$_3$NO$_2$ and CF$_3$N=N(O)CF$_3$ [153].† Nitrogen relaxation times have been obtained and activation energies for molecular rotational diffusion have been calculated from their temperature dependence.

The long T_{qN} found for CF$_3$–NO$_2$ agrees very nicely with the study of X–NO$_2$ compounds by proton LSA (Section 3.5.3) [84]; indeed the larger electronegativity of the CF$_3$ group with respect to CH$_3$ lowers the nitrogen field gradient in a way similar to the effect of O and N atoms in nitrates and nitramines. It is also worth noting that in CF$_3$N=N(O)CF$_3$ only the N(O)CF$_3$ group shows a triplet

† In these two cases, however, no 'natural' linewidth correction was included in the treatment.

splitting the NCF_3 group being a single line. This is, of course, to be ascribed to the absence of the nitrogen lone pair in the first case, which renders the electrical environment of the nitrogen nucleus much more symmetrical than in the NCF_3 group.

The ^{19}F resonances in 2-fluoro-quinoline [154] and in fluorinated imines [156] are markedly broadened at room temperature and sharpen on cooling, allowing fine structure to be observed in a way similar to the changes observed in the proton NMR spectra of nitrogen heterocycles [108] discussed in section 3.5.3.

Solvent and temperature effects on T_{qN} and on $J(N-F)$ in 2-fluoropyridine have been studied by spin-echo measurements [74].

3.6   Conclusions

We have seen that quadrupolar effects depend both on electronic structure and motions of molecules. Thus, quadrupolar labelling, and especially double or multiple labelling, represents a particularly attractive method of investigating molecular properties and liquid structure. The nitrogen-14 nucleus, because of its widespread occurence in all types of systems, is of particular interest in studying electron distributions, molecular reorientations and intermolecular time dependent interactions. It is very probable that such studies will acquire more and more importance in the future and will become more and more frequent, especially with the availability of multinuclear spectrometers and new techniques, such as heteronuclear Fourier transform NMR spectroscopy.

References

1. A. ABRAGAM, *Les Principes du Magnétisme Nucléaire*, Presses Universitaires de France, Paris, 1961.
2. E. A. C. LUCKEN, *Nuclear Quadrupole Coupling Constants*, Academic Press, New York, 1969.
3. C. T. O'KONSKI, in *Determination of Organic Structures by Physical Methods*, Vol. 2, (F. C. NACHOD and W. D. PHILLIPS, Eds.), Academic Press, New York 1962, p. 661.
4. C. C. LIN, *Phys. Rev.*, 119, 1027 (1960).
5. P. E. CADE, K. D. SALES, and A. C. WAHL, *J. Chem. Phys.*, 44, 1973 (1966).
6. C. W. KERN and M. KARPLUS, *J. Chem. Phys.*, 42, 1062 (1965);

C. W. KERN, *J. Chem. Phys.*, **46**, 4543 (1967); S. ELETR, T. K. HA and C. T. O'KONSKI, *J. Chem. Phys.*, **51**, 1430 (1969).

7. C. T. O'KONSKI and T. K. HA, *J. Chem. Phys.*, **49**, 5354 (1968).
8. R. BONACCORSI, E. SCROCCO and J. TOMASI, *J. Chem. Phys.*, **50**, 2940 (1969).
9. M. H. COHEN and F. REIF in *Solid State Physics*, Vol. 5 (F. SEITZ and D. TURNBULL, Eds.) Academic Press, New York, 1957.
10. T. P. DAS and E. L. HAHN, in *Nuclear Quadrupole Resonance Spectroscopy, Solid State Physics*, Supplement I (F. SEITZ and D. TURNBULL, Eds.) Academic Press, New York, 1958.
11. P. A. CASABELLA, *J. Chem. Phys.*, **40**, 149 (1964).
12. T. CHIBA, *Bull. Chem. Soc. Japan*, **38**, 490 (1965).
13. R. A. FORMAN, *J. Chem. Phys.*, **45**, 1118 (1966).
14. L. O. ANDERSSON, M. GOURDJI, L. GUIBE, and W. G. PROCTOR, *Compt. Rend. Acad. Sci.*, **267B**, 803 (1968).
15. P. DIEHL and C. L. KHETRAPAL, in *NMR Basic Principles and Progress*, (P. DIEHL, E. FLUCK, and R. KOSFELD, Eds,) Vol. 7 Springer Verlag, Berlin, 1969, p. 1.
16. C. S. YANNONI, *J. Chem. Phys.*, **52**, 2005 (1970).
17. M. J. GERACE and B. M. FUNG, *J. Chem. Phys.*, **53**, 2984 (1970).
18. A. CARRINGTON, B. J. HOWARD, D. H. LEVY, and J. C. ROBERTSON, *Mol. Phys.*, **15**, 187 (1968); H. UEHARA and Y. MORINO, *Mol. Phys.*, **17**, 239 (1969).
19. D. M. CLOSE and H. N. REXROAD, *J. Chem. Phys.*, **50**, 3717 (1969).
20. E. SCROCCO, in *Advances in Chemical Physics*, (I. PRIGOGINE and S. A. RICE, Eds.) Vol. 5 1963, p. 318.
21. E. KOCHANSKI, J. M. LEHN, and B. LEVY, *Chem. Phys. Letts.* **4**, 75 (1969); *Theoret. Chim. Acta* **22**, 111 (1971), and references therein.
22. S. ROTHENBERG and H. F. SCHAEFFER III, *Mol. Phys.*, **21**, 317 (1971).
23. C. H. TOWNES and B. P. DAILEY, *J. Chem. Phys.*, **17**, 782 (1949).
24. F. A. COTTON and C. B. HARRIS, *Proc. Nat. Acad. Sci. U.S.A.*, **56**, 12 (1966).
25. J. M. SICHEL and M. A. WHITEHEAD, *Theoret. Chim. Acta*, **11**, 263 (1968).
26. D. W. DAVIES and W. C. MACKRODT, *Chem. Commun.*, 1226 (1967).
27. W. D. WHITE and R. S. DRAGO, *J. Chem. Phys.*, **52**, 4717 (1970); H. WRÜBEL and J. VOITLÄNDER, *Z. Naturforsch.*, **24a**, 282 (1969); H. BETSUYAKU, *J. Chem. Phys.*, **50**, 3117 (1964).
28. E. A. C. LUCKEN, *Trans. Faraday Soc.*, **57**, 729 (1961).
29. E. SCHEMPP and P. J. BRAY, *J. Chem. Phys.*, **48**, 2381 (1968).
30. P. L. OLYMPIA, Jr. and M. K. OLYMPIA, *J. Chem. Phys.*, **54**, 1421 (1971).

31. L. GUIBE and E. A. C. LUCKEN, *Mol. Phys.*, 10, 273 (1966); 14, 79 (1968).
32. E. SCHEMPP and P. J. BRAY, *J. Chem. Phys.*, 46, 1186 (1967); 49, 3450 (1968).
33. R. A. MARINO, L. GUIBE, and P. J. BRAY, *J. Chem. Phys.*, 49, 5104 (1968).
34. R. IKEDA, S. ONDA, D. NAKAMURA, and M. KUBO, *J. Phys. Chem.*, 72, 2501 (1968).
35. L. GUIBE, P. LINSCHEID, and E. A. C. LUCKEN, *Mol. Phys.*, 19, 317 (1970).
36. A. COLLIGIANI, L. GUIBE, P. J. HAIGH, and E. A. C. LUCKEN, *Mol. Phys.*, 14, 89 (1968); 19, 144 (1970).
37. A. COLLIGIANI, R. AMBROSETTI, and L. GUIBE, *J. Chem. Phys.*, 54, 2105 (1971).
38. P. L. OLYMPIA, Jr., *J. Chem. Phys.*, 54, 745 (1971).
39. K. T. GILLEN and J. H. NOGGLE, *J. Chem. Phys.*, 52, 4905 (1970).
40. T. K. HA and C. T. O'KONSKI, *J. Chem. Phys.*, 51, 460 (1969).
41. J. P. LUCAS and L. GUIBE, *Mol. Phys.*, 19, 85 (1970).
42. T. K. HA and C. T. O'KONSKI, *Z. Naturforsch.*, 25a, 1509 (1970).
43. S. S. LEHRER and C. T. O'KONSKI, *J. Chem. Phys.*, 43, 1941 (1965).
44. L. GUIBE and E. A. C. LUCKEN, *C. R. Acad. Sci. Paris*, 263b, 815 (1966).
45. H. G. HERTZ, in *Progress in NMR Spectroscopy*, Vol. 3 (J. W. EMSLEY, J. FEENEY, and L. H. SUTCLIFFE, Eds.) Pergamon Press, Oxford, 1967, p. 159.
46. W. T. HUNTRESS, Jr., in *Advances in Magnetic Resonance*, Vol. 4 (J. WAUGH, Ed.) Academic Press, New York, 1970, p. 1. *J. Chem. Phys.*, 48, 3524 (1968).
47. W. A. STEELE, *J. Chem. Phys.*, 38, 2404, 2411 (1963).
48. D. E. O'REILLY, *J. Chem. Phys.*, 49, 5416 (1968); 55, 2155 (1971).
49. (a) N. A. HILL, *Proc. Phys. Soc. London*, B67, 149 (1954); (b) A. von GIERER and K. WIRTZ, *Z. Naturforsch.* 8a, 532 (1953).
50. P. W. ATKINS, *Mol. Phys.*, 17, 321 (1969).
51. P. W. ATKINS, A. LOEWENSTEIN and Y. MARGALIT, *Mol. Phys.*, 17, 329 (1969).
52. J. E. ANDERSON, *J. Chem. Phys.*, 47, 4879 (1967).
53. D. WALLACH and W. T. HUNTRESS, Jr., *J. Chem. Phys.*, 50, 1219 (1969).
54. K. T. GILLEN and J. H. NOGGLE, *J. Chem. Phys.*, 53, 801 (1970).
55. H. SHIMIZU, *J. Chem. Phys.*, 43, 2453 (1965).
56. E. O. STEJSKAL and H. S. GUTOWSKI, *J. Chem. Phys.*, 28, 388 (1958).
57. (a) D. E. WOESSNER, *J. Chem. Phys.*, 36, 1 (1962); 42, 1855 (1965); (b) *J. Chem. Phys.*, 37, 647 (1962); (c) D. E. WOESSNER, B. S. SNOWDEN, Jr., and G. H. MEYER, *J. Chem. Phys.*, 50, 719 (1969).

58. M. D. ZEIDLER, *Ber. Bunsenges. Physik. Chem.*, **69**, 659 (1965).
59. D. WALLACH, *J. Chem. Phys.*, **47**, 5258 (1967).
60. H. SHIMIZU, *J. Chem. Phys.*, **40**, 754 (1964).
61. T. J. BOPP, *J. Chem. Phys.*, **47**, 3621 (1967).
62. D. E. WOESSNER, B. S. SNOWDEN, Jr., and E. THOMAS STROM, *Mol. Phys.*, **14**, 265 (1968).
63. T. E. BULL AND J. JONAS, *J. Chem. Phys.*, **53**, 3315 (1970).
64. J. P. KINTZINGER and J. M. LEHN, *Mol. Phys.*, **22**, 273 (1971).
65. D. WALLACH, *J. Phys. Chem.*, **73**, 307 (1969).
66. A. G. MARSHALL, *J. Chem. Phys.*, **52**, 2527 (1970).
67. T. J. SWIFT and R. E. CONNICK, *J. Chem. Phys.*, **37**, 307 (1962).
68. C. DEVERELL, in *Progress in NMR Spectroscopy*, Vol. 4 (J. W. EMSLEY, J. FEENEY, and L. H. SUTCLIFFE, Eds.) Pergamon Press, Oxford, 1969, p. 235.
69. H. G. HERTZ, *Z. Electrochem.*, **65**, 20 (1961).
70. K. A. VALIEV, *Zhurn. Struk. Khim.*, **3**, 630 (1962); **5**, 477 (1964).
71. H. G. HERTZ, *Ber. Bunsenges. Phys. Chem.*, **71**, 979 (1967).
72. J. A. POPLE, W. G. SCHNEIDER, and H. J. BERNSTEIN, *High Resolution Nuclear Magnetic Resonance*, McGraw-Hill Book Co., New York, 1959.
73. H. Y. CARR and E. M. PURCELL, *Phys. Rev.*, **94**, 630 (1954).
74. (a) H. S. GUTOWSKI, R. L. VOLD, and E. J. WELLS, *J. Chem. Phys.*, **43**, 4107 (1965); (b) A. ALLERHAND and E. THIELE, *J. Chem. Phys.*, **45**, 902 (1966); (c) N. BODEN, J. DECK, E. GORE, and H. S. GUTOWSKI, *J. Chem. Phys.*, **45**, 3875 (1966); (d) T. D. ALGER and H. S. GUTOWSKI, *J. Chem. Phys.*, **48**, 4625 (1968).
75. L. E. DRAIN, *Proc. Phys. Soc. (London)*, **62A**, 301 (1949).
76. O. HAWORTH and R. E. RICHARDS, in *Progress in NMR Spectroscopy*, Vol. 1 (J. W. EMSLEY, J. FEENEY, and L. H. SUTCLIFFE, Eds.) Pergamon Press, Oxford, 1966, p. 1.
77. G. J. JENKS, *J. Chem. Phys.*, **54**, 658 (1971).
78. R. A. FORMAN, *J. Chem. Phys.*, **39**, 2393 (1963).
79. T. KANDA, Y. SAITO, and K. KAWAMURA, *Bull. Chem. Soc. Japan*, **35**, 172 (1962).
80. W. B. MONIZ and H. S. GUTOWSKI, *J. Chem. Phys.*, **38**, 1155 (1963).
81. D. HERBISON-EVANS and R. E. RICHARDS, *Mol. Phys.*, **7**, 515 (1964).
82. E. V. GOLDAMMER and H. G. HERTZ, *J. Phys. Chem.*, **74**, 3734 (1970).
83. M. WITANOWSKI, *J. Amer. Chem. Soc.*, **90**, 5683 (1968).
84. J. P. KINTZINGER, J. M. LEHN, and R. L. WILLIAMS, *Mol. Phys.*, **17**, 135 (1969).
85. (a) R. A. OGG, Jr. and J. D. RAY, *J. Chem. Phys.*, **26**, 1339 (1957); (b) R. A. OGG, Jr. and J. D. RAY. *J. Chem. Phys.*, **26**, 1515 (1957); (c) R. A. OGG, Jr., *J. Chem. Phys.*, **22**, 560 (1954).
86. A. M. QURESHI, J. A. RIPMEESTER, and F. AUBKE, *Can. J. Chem.*, **47**, 4247 (1969).
87. W. B. MONIZ and C. F. PORANSKI, Jr., *J. Phys. Chem.*, **73**, 4145 (1969).

88. M. OHTSURU, K. TORI, J. M. LEHN, and R. SEHER, *J. Amer. Chem. Soc.*, **91**, 1187 (1969).

89. N. C. PYPER, *Mol. Phys.*, **19**, 161 (1970); **20**, 449 (1971), **21**, 1 (1971); **21**, 961 (1971); **21**, 977 (1971), R. K. HARRIS and N. C. PYPER, *Mol. Phys.*, **20**, 467 (1971).

90. T. SALUVERE and E. T. LIPPMAA, *Eesti NSV Tead. Akad. Toim. Füüs. Mat.*, **18**, 445 (1969).

91. (a) T. SALUVERE and E. T. LIPPMAA, *Eesti NSV Tead. Akad. Toim., Füüs. Mat.*, **19**, 436 (1970); (b) *Eesti NSV Tead. Akad. Toim., Keem.-Geol.*, **19**, 275 (1970).

92. R. A. ASSINK and J. JONAS, *J. Phys. Chem.*, **73**, 2445 (1969).

93. M. BOSE, N. DAS, and N. CHATTERJEE, *J. Mol. Spectrosc.*, **18**, 32 (1965).

94. (a) J. L. CAROLAN and T. A. SCOTT, *J. Mag. Res.*, **2**, 243 (1970); (b) D. E. O'REILLY, E. M. PETERSON, and S. R. LAMMERT, *J. Chem. Phys.*, **52**, 1700 (1970).

95. S. ALEXANDER and A. TZALMONA, *Phys. Rev. Letts.*, **13**, 546 (1964); *Phys. Rev.*, **138**, A845 (1965); G. W. SMITH, *J. Chem. Phys.*, **42**, 3341 (1965).

96. R. W. MITCHELL and M. EISNER, *J. Chem. Phys.*, **33**, 86 (1960).

97. E. V. GOLDAMMER and M. D. ZEIDLER, *Ber. Bunsenges. Phys. Chem.*, **73**, 4 (1969).

98. J. G. POWLES and M. RHODES, *Mol. Phys.*, **12**, 399 (1967).

99. M. WITANOWSKI and H. JANUSZEWSKI, *J. Chem. Soc.*, B, 1062 (1967).

100. H. SAITÔ, K. NUKADA, H. KATO, T. YONEZAWA, and K. FUKUI, *Tetrahedron Letts.*, 111 (1965).

101. H. SAITÔ, and K. NUKADA, *J. Amer. Chem. Soc.*, **93**, 1072 (1971); H. SAITÔ, Y. TANAKA, and K. NUKADA, *J. Amer. Chem. Soc.*, **93**, 1077 (1971).

102. A. LOEWENSTEIN and Y. MARGALIT, *J. Phys. Chem.*, **69**, 4152 (1965).

103. G. D. HOWARD and R. S. MARIANELLI, *Inorg. Chem.*, **9**, 1738 (1970).

104. W. L. PURCELL and R. S. MARIANELLI, *Inorg. Chem.*, **9**, 1724 (1970).

105. I. D. CAMPBELL, J. P. CARVER, R. A. DWEK, A. J. NUMMELIN, and R. E. RICHARDS, *Mol. Phys.*, **20**, 913 (1971); I. D. CAMPBELL, R. A. DWEK, R. E. RICHARDS, and M. N. WISEMAN, *Mol. Phys.*, **20**, 933 (1971).

106. M. SHPORER, G. RON, A. LOEWENSTEIN, and G. NAVON, *Inorg. Chem.*, **4**, 358 (1965).

107. M. SHPORER, G. RON, A. LOEWENSTEIN, and G. NAVON, *Inorg. Chem.*, **4**, 361 (1965); A. LOEWENSTEIN and G. RON, *Inorg. Chem.*, **6**, 1604 (1967).

108. J. P. KINTZINGER and J. M. LEHN, *Mol. Phys.*, **14**, 133 (1968).

109. A. KUMAR, N. RAMA KRISHNA, and B. D. NAGESWARA RAO, *Mol. Phys.*, **18**, 11 (1970).

110. J. A. POPLE, *Mol. Phys.*, **1**, 168 (1958); M. SUZUKI and M. KUBO, *Mol. Phys.*, **7**, 201 (1964).

111. A. V. CUNLIFFE and R. K. HARRIS, *Mol. Phys.*, **15**, 413 (1968).

112. R. L. VOLD and H. S. GUTOWSKY, *J. Chem. Phys.*, 47, 4782 (1967).
113. J. M. ANDERSON, *Mol. Phys.*, 8, 505 (1964).
114. C. S. JOHNSON, Jr., in *Advances in Magnetic Resonance*, Vol. 1 (J. WAUGH, Ed.) Academic Press, New York, 1965, p. 33.
115. R. LYNDEN-BELL, in *Progress in NMR Spectroscopy*, Vol. 2 (J. W. EMSLEY, J. FEENEY, and L. H. SUTCLIFFE, Eds.) Pergamon Press, Oxford, 1967 p. 163.
116. D. W. AKSNES, S. M. HUTCHINSON, and K. J. PACKER, *Mol. Phys.*, 14, 301 (1968).
117. R. L. VOLD and A. CORREA, *J. Phys. Chem.*, 74, 2674 (1970).
118. J. P. KINTZINGER, Thèse de Doctorat d'Etat, Université de Strasbourg (1970).
119. J. R. YANDLE and J. P. MAHER, *J. Chem. Soc.*, A, 1549 (1969).
120. J. D. ROBERTS, *J. Amer. Chem., Soc.*, 78, 4495 (1956).
121. M. FRANCK-NEUMANN and J. M. LEHN, *Mol. Phys.*, 7, 197 (1963).
122. J. M. ANDERSON, J. D. BALDESCHWIELER, D. C. DITTMER, and W. D. PHILLIPS, *J. Chem. Phys.*, 38, 1260 (1963).
123. E. BULLOCK, D. G. TUCK, and E. J. WOODHOUSE, *J. Chem. Phys.*, 38, 2318 (1963).
124. M. FRANCK-NEUMANN, Thèse 3é cycle, Université de Strasbourg (1964).
125. P. G. GASSMAN and D. C. HECKERT, *J. Org. Chem.*, 30, 2859 (1965).
126. K. TORI, T. IWATA, K. AONO, M. OHTSURU, and T. NAKAGAWA, *Chem. Pharm. Bull. Japan*, 15, 329 (1967).
127. Y. TERUI, K. AONO and K. TORI, *J. Amer. Chem. Soc.*, 90, 1069 (1968).
128. W. Mc. FARLANE and R. R. DEAN, *J. Chem. Soc.*, A, 1187 (1968).
129. J. F. BIELLMANN and H. CALLOT, *Bull. Soc. Chim. Fr.*, 397 (1967).
130. T. GOTO, M. ISOBE, M. OHTSURU, and K. TORI, *Tetrahedron Letts.*, 1511 (1968).
131. G. A. OLAH and T. E. KIOVSKY, *J. Amer. Chem. Soc.*, 90, 4666 (1968).
132. I. D. KUNTZ, P. von R. SCHLEYER, and A. ALLERHAND, *J. Chem. Phys.*, 35, 1533 (1961).
133. D. S. MATTESON and R. A. BAILEY, *J. Amer. Chem. Soc.*, 90, 3761 (1968).
134. A. A. BOTHNER-BY and R. H. COX, *J. Phys. Chem.*, 73, 1830 (1969).
135. A. H. LAMBERTON, I. O. SUTHERLAND, J. E. THORPE, and H. M. YUSUF, *J. Chem. Soc.*, B, 6 (1968).
136. P. HAMPSON and A. MATHIAS, *Chem. Comm.*, 825 (1968).
137. J. P. MAHER, *J. Chem. Soc.*, A, 1855 (1966).
138. W. Mc. FARLANE, *J. Chem. Soc.*, A, 1660 (1967).
139. T. J. SWIFT, S. B. MARKS, and W. G. SAYRE, *J. Chem. Phys.*, 44, 2797 (1966).
140. E. RAHKAMAA, *J. Chem. Phys.*, 48, 531 (1968).

141. Y. KAWAZOE, M. TSUDA, and M. OHNISHI, *Chem. Pharm. Bull. Japan*, **15**, 214 (1967).
142. J. M. LEHN and M. FRANCK-NEUMANN, *J. Chem. Phys.*, **43**, 1421 (1965).
143. (a) D. W. LARSEN, *J. Phys. Chem.*, **74**, 3380 (1970); (b) *J. Phys. Chem.*, **75**, 509 (1971).
144. J. M. LEHN, *Z. Anal. Chem.*, **235**, 10 (1968).
145. J. P. KINTZINGER and J. M. LEHN, unpublished results.
146. J. CASANOVA, Jr. and R. E. SCHUSTER *Tetrahedron Letts.*, 405 (1964).
147. R. L. VOLD, E. S. DANIEL, and S. O. CHAN, *J. Amer. Chem. Soc.*, **92**, 6771 (1970).
148. E. L. MUETTERTIES and W. D. PHILLIPS, *J. Amer. Chem. Soc.*, **81**, 1084 (1959).
149. K. O. CHRISTIE, J. P. GUERTIN, A. E. PAVLATH, and W. SAWODNY, *Inorg. Chem.*, **6**, 533 (1962).
150. N. BARTLETT, J. PASSMORE, and E. J. WELLS, *Chem. Comm.*, 213 (1966).
151. D. MOY and A. R. YOUNG, *J. Amer. Chem. Soc.*, **87**, 1889 (1965).
152. R. K. HARRIS, N. C. PYPER, R. E. RICHARDS, and G. W. SCHULZ, *Mol. Phys.*, **19**, 145 (1970).
153. R. FIELDS, J. LEE, and D. J. MOWTHORPE, *Trans. Faraday Soc.*, **65**, 1 (1969).
154. C. A. FRANZ, R. T. HALL, and C. E. KASLOW, *Tetrahedron Letts.*, 1947 (1967).
155. J. H. NOGGLE, J. D. BALDESCHWIELER, and C. B. COLBURN, *J. Chem. Phys.*, **37**, 182 (1962).
156. L. CAVALLI and P. PICCARDI, *Chem. Commun.*, 1132 (1969).
157. G. R. GUNTHER-MOHR, R. L. WHITE, A. L. WHITE, A. L. HAWLOW, W. E. GOOD, and D. K. COLES, *Phys. Rev.* **94**, 1184 (1954).
158. J. SHERIDAN and W. GORDY, *Phys. Rev.*, **79**, 513 (1950).
159. D. R. LIDE, Jr. and D. E. MANN, *J. Chem. Phys.*, **28**, 572 (1958).
160. J. E. WOLLRAB and V. W. LAURIE, *J. Chem. Phys.*, **48**, 5058 (1968).
161. D. R. LIDE, Jr., *J. Chem. Phys.*, **38**, 456 (1963).
162. W. H. KIRCHHOFF and D. R. LIDE, Jr., *J. Chem. Phys.*, **51**, 467 (1969).
163. H. E. RADFORD, *Phys. Rev.*, **136A**, 1571 (1964).
164. J. W. SIMMONS, W. E. ANDERSON, and W. GORDY, *Phys. Rev.*, **77**, 77 (1950).
165. J. K. TYLER AND J. SHERIDAN, *Trans. Faraday Soc.*, **59**, 2661 (1963).
166a. C. H. TOWNES, A. N. HOLDEN, and F. R. MERRITT, *Phys. Rev.*, **74**, 113 (1948).
166b. A. G. SMITH, M. RING, W. V. SMITH, and W. GORDY, *Phys. Rev.*, **73**, 633 (1948).
167. M. K. KEMP, J. M. POCHAN, and W. H. FLYGARE, *J. Phys. Chem.*, **71**, 765 (1967).

168. Y. S. LI and M. D. HARMONY, *J. Chem. Phys.*, 50, 3674 (1969).
169. A. A. WESTENBERG and E. B. WILSON, *J. Amer. Chem. Soc.*, 72, 199 (1950).
170. T. SHERIDAN and W. GORDY, *J. Chem. Phys.*, 20, 591 (1952).
171. R. VARMA and K. S. BUCKTON, *J. Chem. Phys.*, 46, 1565 (1967).
172. J. SHERIDAN and A. C. TURNER, *Proc. Chem. Soc.*, 21 (1960).
173. L. PIERCE, L. NELSON, and C. THOMAS, *J. Chem. Phys.*, 43, 3423 (1965).
174. P. B. BLACKBURN, R. D. BROWN, F. R. BURDEN, J. G. CROFTS, and I. R. GILLARD, *Chem. Phys. Letts.*, 7, 102 (1970); H. K. BODENSEH and K. MORGENSTERN, *Z. Naturforsch.*, 25a, 150 (1970).
175. M. WINNEWISSER and H. K. BODENSEH, *Z. Naturforsch.*, 22a, 1724 (1970).
176. H. DREIZLER, H. D. RUDOLPH, and H. SCHLESSER, *Z. Naturforsch.*, 25a, 1643 (1970).
177. H. D. HESS, A. BAUDER, and Hs. H. GÜNTHARD, *J. Mol. Spectrosc.*, 22, 208 (1967).
178. R. A. FORMAN and D. R. LIDE, Jr., *J. Chem. Phys.*, 39, 1133 (1963).
179. W. H. KIRCHHOFF and E. B. WILSON, Jr., *J. Amer. Chem. Soc.*, 84, 334 (1962).
180. M. K. KEMP and W. H. FLYGARE, *J. Amer. Chem. Soc.*, 90, 6267 (1968).
181. Y. S. LI, M. D. HARMONY, D. HAYES, and E. L. BEESON, Jr., *J. Chem. Phys.*, 47, 4514 (1967).
182. B. BAK, J. L. MAHLER, L. NYGAARD, and G. O. SORENSEN, cited in G. O. SORENSEN, *J. Mol. Spectrosc.*, 22, 325 (1967).
183. H. D. RUDOLPH, H. DREIZLER, and H. SEILER, *Z. Naturforsch.*, 22a, 1738 (1967).
184. L. NYGAARD, J. T. NIELSEN, J. KIRCHHEIMER, G. MALTESEN, J. RASTRUP-ANDERSEN, and G. O. SORENSEN, *J. Mol. Struct.* 3, 491 (1969).
185. H. DREIZLER private communication cited in ref. 184; W. ARNOLD, H. DREIZLER and H. D. RUDOLPH, *Z. Naturforsch.*, 23a, 301 (1968).
186. J. M. MACDONALD, D. TAYLOR, and J. K. TYLER, *J. Mol. Spectrosc.*, 26, 285 (1968).
187. C. T. O'KONSKI and T. J. FLAUTT, *J. Chem. Phys.*, 27, 815 (1957).
188. G. A. MATZKANIN, T. A. SCOTT, and P. J. HAIGH, *J. Chem. Phys.*, 42, 1646 (1965).
189. P. J. HAIGH, P. C. CANEPA, G. A. MATZKANIN, and T. A. SCOTT, *J. Chem. Phys.*, 48, 4234 (1968).
190. A. COLLIGIANI, R. AMBROSETTI, and R. ANGELONE, *J. Chem. Phys.*, 52, 5022 (1970).
191. P. J. HAIGH and L. GUIBE, *Compt. Rend. Acad. Sci. Paris*, 261, 2328 (1965).
192. G. D. WATKINS and R. V. POUND, *Phys. Rev.*, 85, 1062 (1952).
193. L. KRAUSE and M. A. WHITEHEAD, *J. Chem. Phys.*, 52, 2787 (1970).

194. D. Y. OSOKIN, I. N. SAFIN, and I. A. NURETDINOV, *Dokl. Akad. Nauk. SSSR*, **190**, 357 (1970).
195. S. KOJIMA, M. MINEMATSU, and M. TANAKA, *J. Chem. Phys.*, **31**, 271 (1959).
196. L. GUIBE, *Compt. Rend. Acad. Sci. Paris*, **250**, 1635 (1960).
197. C. T. O'KONSKI and K. TORIZUKA, *J. Chem. Phys.*, **51**, 461 (1969).
198. D. H. SMITH and R. M. COTTS, *J. Chem. Phys.*, **41**, 2403 (1964).
199. C. T. YIM, M. A. WHITHEAD, and D. H. LO, *Can. J. Chem.*, **46**, 3595 (1968).
200. Y. ABE, Y. KAMISHIWA, and S. KOJIMA, *J. Phys. Soc. Japan*, **21**, 2083 (1966).
201. Y. ABE, *J. Phys. Soc. Japan*, **18**, 1804 (1963).
202. L. GUIBE, Thèse, *Ann. Phys.*, **7**, 177 (1962).
203. T. K. HA, C. T. O'KONSKI, *Z. Naturforsch.*, **25a**, 1155 (1970).
204. A. TZALOMONA, *Phys. Letts.*, **20**, 478 (1966).
205. L. GUIBE, *Compt. Rend. Acad. Sci. Paris*, **250**, 3014 (1960).
206. L. GUIBE, E. A. C. LUCKEN, *Proc. XIIIth Ampère Conf. Louvain*, North Holland Pub. Co., Amsterdam, 1964, p. 241.
207. S. KOJIMA, and M. MINEMATSU, *J. Phys. Soc. Japan*, **15**, 355 (1960).
208. Y. MORINO, T. CHIBA, T. SHIMOZAWA, and M. J. TOYANA, *J. Phys. Soc. Japan*, **13**, 869 (1958).
209. L. GUIBE and E. A. C. LUCKEN, *Mol. Phys.*, **14**, 73 (1968).
210. E. SCHEMPP, and P. J. BRAY, *Physics Letts.*, **A25**, 414 (1967).
211. H. NEGITA, P. A. CASABELLA, and P. J. BRAY, *J. Chem. Phys.*, **32**, 314 (1960).
212. P. A. CASABELLA and P. J. BRAY, *J. Chem. Phys.*, **28**, 1182 (1958).
213. P. A. CASABELLA and P. J. BRAY, *J. Chem. Phys.*, **29**, 1105 (1958).
214. H. NEGITA and P. J. BRAY, *J. Chem. Phys.*, **33**, 1876 (1960).
215. S. ONDA, R. IKEDA, D. NAKAMURA, and M. KUBO, *Bull. Chem. Soc. Japan*, **42**, 2740 (1969).
216. R. IKEDA, D. NAKAMURA, and M. KUBO, *J. Phys. Chem.*, **70**, 3626 (1966).
217. R. A. MARINO and T. OJA, *Chem. Phys. Letts.*, **4**, 489 (1970).
218. R. CLEMENT, M. GOURDJI, and L. GUIBE, *Mol. Phys.*, **21**, 247 (1971).
219. T. A. SCOTT, *J. Chem. Phys.*, **36**, 1459 (1962).
220. H. MEYER and T. A. SCOTT, *J. Phys. and Chem. of Solids*, **11**, 215 (1959).
221. T. OJA, R. A. MARINO, and P. J. BRAY, *Phys. Letts.*, **A26**, 11 (1967); R. IKEDA, M. MIKAMI, D. NAKAMURA and M. KUBO, *J. Mag. Res.*, **1**, 211 (1969).
222. R. H. WIDMAN, *J. Chem. Phys.*, **43**, 2922 (1965).
223. R. IKEDA, D. NAKAMURA, and M. KUBO, *Bull. Chem. Soc. Japan*, **40**, 701 (1967).
224. R. IKEDA, D. NAKAMURA, and M. KUBO, *J. Phys. Chem.*, **72**, 2982 (1968).
225. M. GOURDJI and L. GUIBE, *Compt. Rend. Acad. Sci. Paris*, **260**, 1131 (1965).

226. M. LINZER and R. A. FORMAN, *J. Chem. Phys.*, **46**, 4690 (1967).
227. R. L. ARMSTRONG and P. A. SPEIGHT, *J. Mag. Res.*, **2**, 141 (1970).
228. B. M. SCHMIDT, L. C. BROWN, and D. H. WILLIAMS, *J. Mol. Spectrosc.*, **2**, 539 (1958).
229. M. WITANOWSKI, *Tetrahedron*, **23**, 4299 (1967).
230. C. F. PORANSKI, Jr. and W. B. MONIZ, *J. Phys. Chem.*, **71**, 1142 (1967).
231. M. WITANOWSKI and L. STEFANIAK, *J. Chem. Soc.*, *B*, 1061 (1967).
232. K. F. CHEW, W. DERBYSHIRE, N. LOGAN, A. H. NORBURY, and A. I. P. SINHA, *Chem. Comm.*, 1708 (1970).
233. B. M. SCHMIDT, L. C. BROWN, and D. WILLIAMS, *J. Mol. Spectrosc.*, **2**, 551 (1958).
234. M. WITANOWSKI and S. A. SHEVELEV, *J. Mol. Spectrosc.*, **33**, 19 (1970).
235. R. J. S. BROWN and D. D. THOMPSON, *J. Chem. Phys.*, **34**, 1580 (1961).
236. J. NELSON, R. SPRATT, and S. M. NELSON, *J. Chem. Soc.*, *(A)*, 583 (1970).
237. K. M. MACKAY and S. R. STOBART, *Spectrochim. Acta*, in ref. 232.
238. H. G. HERTZ and W. SPALTHOFF, *Z. Elektrochem.*, **63**, 1105 (1959).
239. R. A. PINKOWITZ and T. J. SWIFT, *J. Chem. Phys.*, **54**, 2858 (1971).
240. B. L. BARTON, *J. Chem. Phys.*, **55**, 1984 (1971).
241. S. ELETR and C. T. O'KONSKI, *J. Chem. Phys.*, **54**, 4312 (1971).
242. R. BLINC, M. MALI, R. OSREDKAR, A. PRELESNIK, I. ZUPANČIČ, and L. EHRENBERG, *J. Chem. Phys.*, **55**, 4843 (1971).
243. R. BLINC, M. MALI, R. OSREDKAR, A. PRELESNIK, I. ZUPANČIČ, and L. EHRENBERG, *Chem. Phys. Letts.*, **9**, 85 (1971).
244. F. KNOLL and U. KRUMM, *Chem. Ber.*, **104**, 31 (1971).
245. C. L. NORRIS and W. H. FLYGARE, *J. Mol. Spectrosc.*, **40**, 40 (1971).
246. J. CASADO, L. NYGAARD, and G. O. SORENSEN, *J. Mol. Struct.*, **8**, 211 (1971).
247. W. WERNER, H. DREIZLER, and H. D. RUDOLPH, *Z. Naturforsch.*, **22a**, 531 (1967).
248. L. NYGAARD, E. ASMUSSEN, J. H. HOG, R. C. MAHESHWARI, C. H. NIELSEN, I. B. PETERSEN, J. RASTRUP-ANDERSEN, and G. O. SORENSEN, *J. Mol. Struct.*, **8**, 225 (1971).
249. L. NYGAARD, R. L. HANSEN, and G. O. SORENSEN, *J. Mol. Struct.*, **9**, 163 (1971).
250. S. G. KUKOLICH, *Chem. Phys. Letts.*, **10**, 52 (1971).
251. E. SCHEMPP, *Chem. Phys. Letts.*, **8**, 562 (1971).
252. J. KOO and Y. N. HSIEH, *Chem. Phys. Letts.*, **9**, 238 (1971).
253. E. SCHEMPP and P. J. BRAY, *J. Mag. Res.*, **5**, 78 (1971).

254. R. IKEDA, S. NODA, D. NAKAMURA, and M. KUBO, *J. Mag. Res.*, **5**, 54 (1971).
255. M. WITANOWSKI, L. STEFANIAK, H. JANUSZEWSKI, and G. A. WEBB, *Tetrahedron*, **27**, 3129 (1971).
256. G. L. BLACKMAN, R. D. BROWN, F. R. BURDEN, and A. MISHRA, *J. Mol. Struct.*, **9**, 465 (1971).
257. R. A. OGG, Jr. and J. D. RAY, *J. Chem. Phys.*, **25**, 797 (1956).
258. M. WITANOWSKI, T. SALUVERE, L. STEFANIAK, H. JANUSZEWSKI and G. A. WEBB, *Mol. Phys.*, **23**, 1071 (1972).
259. R. ZINZIUS, Thèse de Doctorat de ze cycle, Université Louis Pasteur, Strasbourg (1972).

CHAPTER 4

Nitrogen Chemical Shifts in Organic Compounds

M. Witanowski, L. Stefaniak, and H. Januszewski

Institute of Organic Chemistry, Polish Academy of Sciences, Warsaw, Poland

4.1   Introduction

The range of nitrogen chemical shifts in organic compounds is about 800 p.p.m. and may be even larger for paramagnetic complexes. However, the usefulness of the nuclear magnetic resonance spectra of any nucleus for structural investigations depends very much not only on the magnitude of the chemical shifts but also on the possibility of finding reasonable correlations of the shifts with molecular structure. Another important factor is the possible differentiation of closely related structures, such as those of isomeric compounds or molecules occurring in various chemical equilibria. Fortunately, a great deal has already been done by means of systematic investigation of organic compounds by nitrogen nuclear magnetic resonance, and rather simple empirical relationships between nitrogen chemical shift and molecular environment have been established for many important classes of such compounds. Theoretical explanations of observed chemical shifts usually provide a tool for probing deeply into the electronic structure of organic molecules. Such explanations have been reported and seem to work successfully for nitrogen chemical shifts of groups of structurally related compounds, but the situation is less favourable in a general theoretical interpretation of the shifts for very different structures. However, at least some of the most decisive sources of nitrogen chemical shift changes have been recognized (Chapter 1). Nitrogen chemical shifts are particularly valuable as a means of distinguishing between isomeric molecular structures which are very common in organic chemistry. This also includes isomeric equilibria such as tautomerism. In most of the typical isomeric or tautomeric pairs of molecules the corresponding relative shifts in nitrogen NMR are large, even in comparison with the entire range of nitrogen NMR spectra.

Table 4.1 may serve as a general guide to nitrogen chemical shifts in organic compounds. Only approximate ranges are given there and, in practice, most of them may be further divided into sub-sections according to the finer details of molecular structure. This makes spectral assignments even simpler (Section 4.4). So far, systematic investigations of nitrogen chemical shifts have been carried out mainly in ^{14}N NMR spectroscopy but since the advent of various modern experimental techniques (Chapter 2) the share of ^{15}N chemical shift measurements has been increasing.

4.2   Equivalence of ^{14}N and ^{15}N Shifts

It is mentioned in Chapter 1 (Section 1.1.3) that ^{14}N and ^{15}N chemical shifts may be used interchangeably. This means that there are no intrinsic differences of any importance between them other than those resulting from errors, different calibration methods and experimental conditions [131]. However, the situation has not been clear until recently. Precise double-resonance studies [101] of the ^{14}N and ^{15}N resonance frequencies of tetramethylammonium iodide, benzyl isocyanide, methyl nitrate, and dimethylnitrosamine, carried out with very pure substances and in the same experimental conditions within each pair of ^{14}N and ^{15}N isotopomers, indicate that the ratio (R) of the corresponding resonance frequencies (Table 4.2) is constant to better than 1 in 10^{-7} and so is the ratio

$$C = \frac{1 - \sigma_i(^{15}\text{N})}{1 - \sigma_i(^{14}\text{N})} \qquad (4.1)$$

where σ_i is the corresponding screening constant of nucleus i. This seems to exclude any significant difference between the ^{14}N and ^{15}N screening constants for a given compound since otherwise the difference would have to be constant to 0.1 p.p.m. over the observed 300 p.p.m. range of chemical shifts. It also positively excludes any practical differences in the *relative* chemical shifts in ^{14}N and ^{15}N NMR spectra, of any two substances. It seems [101] that the earlier values of ratio R, such as $1.40275486 \pm 20 \times 10^{-8}$ reported [16] for $NH_4^{\oplus}$, and $1.40275800 \pm 16 \times 10^{-8}$ found for MeNC (R. Price, Ph.D. thesis, quoted in ref. 101), which are appreciably different from the average value of R in Table 4.2, contain a much larger experimental error than specified. The latest value of R for $NH_4^{\oplus}$ (R. Price, quoted in ref. 101), $1.40275710 \pm 7 \times 10^{-8}$, is in agreement with the data in Table 4.2.

There have been some remarkable discrepancies between the ^{14}N and ^{15}N shifts of a number of organic compounds but they have recently [131] been explained as due to experimental errors. The apparently opposite trends in the direction of the ^{14}N shifts [41] and ^{15}N shifts [33] in *para*-substituted nitrobenzenes have been shown to be actually parallel by reinvestigation [67] of the ^{15}N data. The anomalous differences [22] reported between the ^{14}N and ^{15}N shifts of ethyl nitrate, dimethylformamide, and hydrazine, respectively, ascribed to an undefined effect of hindered internal rotation,

TABLE 4.1.   Approximate Range of Nitrogen Chemical Shifts in Organic Molecules

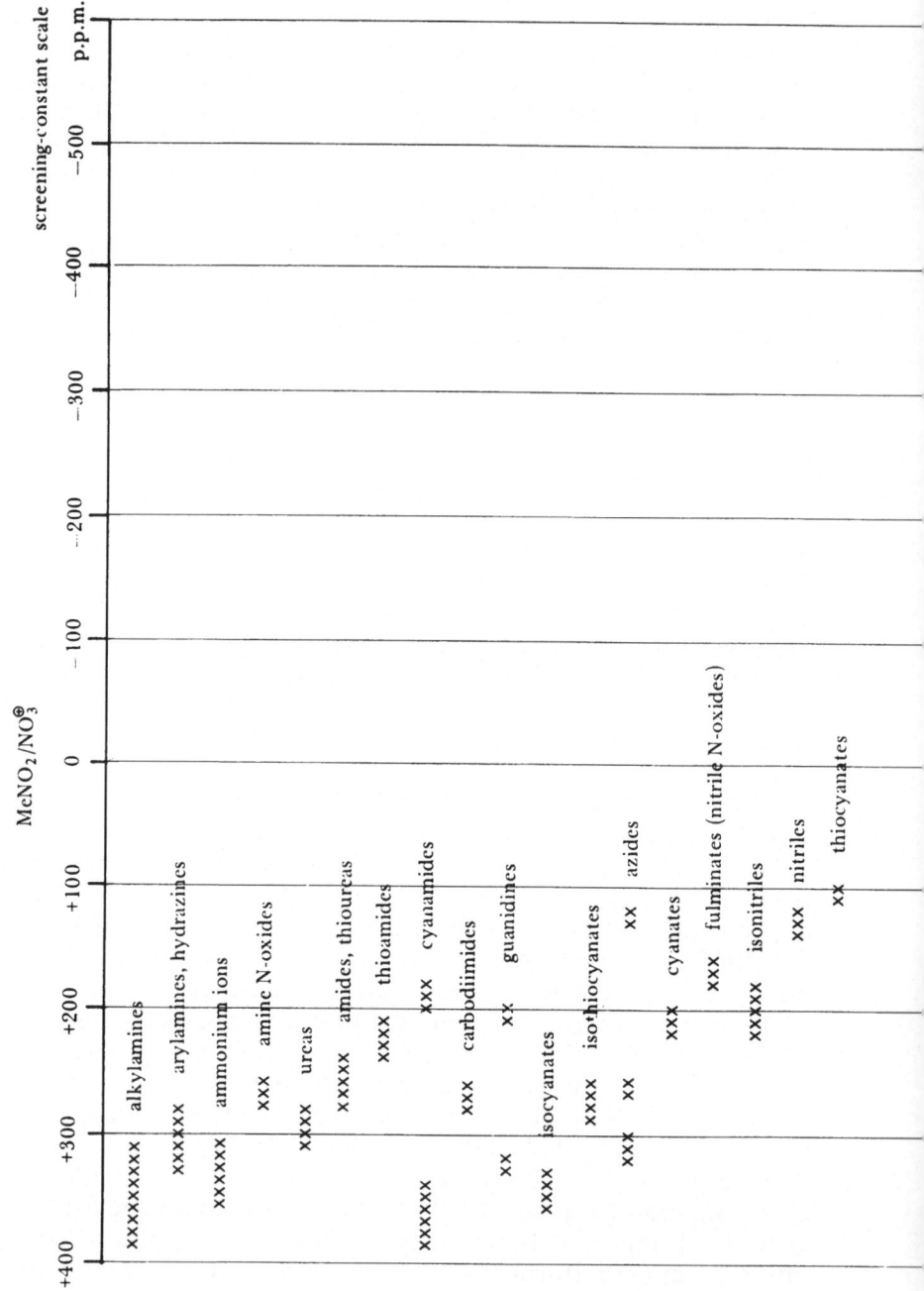

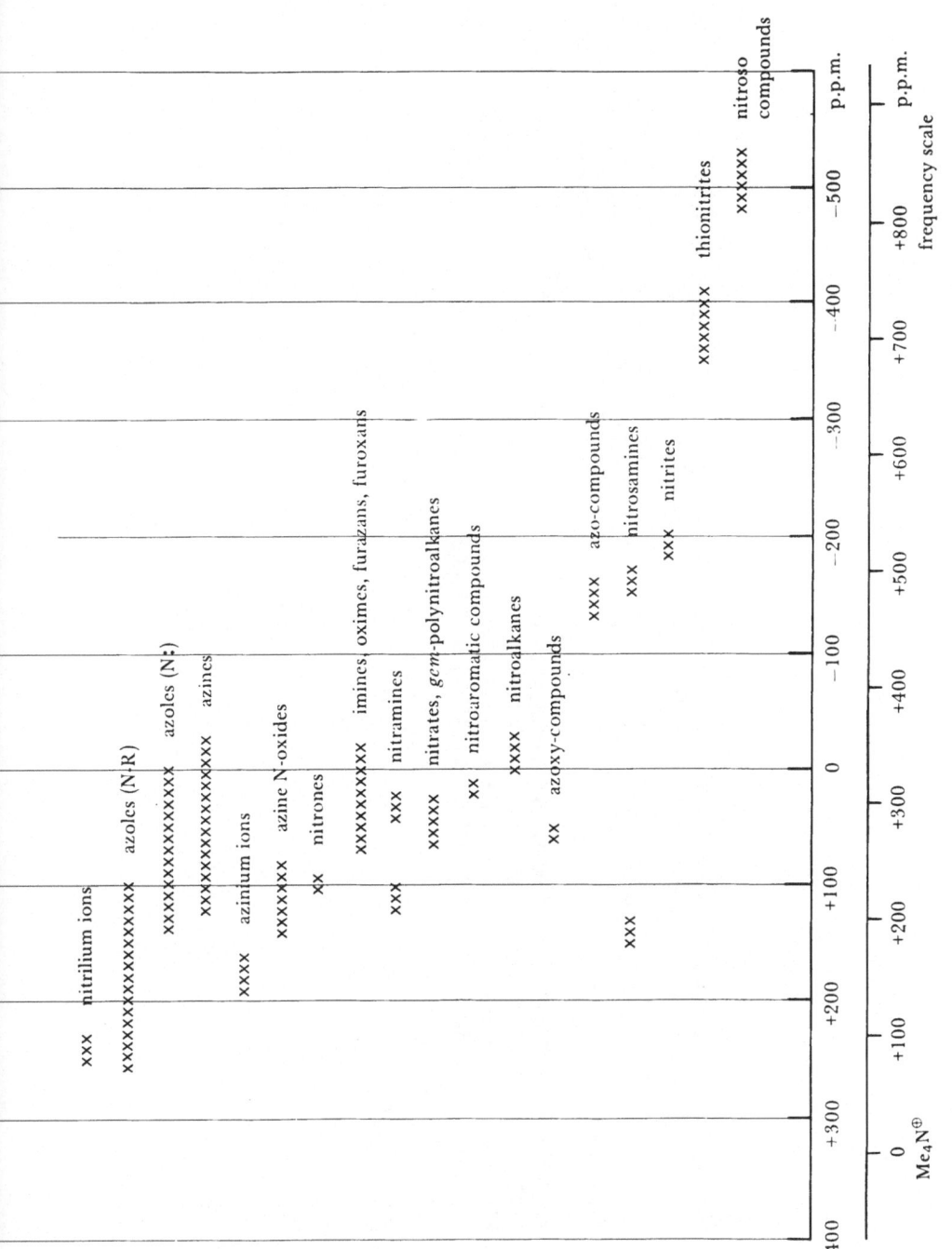

TABLE 4.2.   Comparison of ^{14}N and ^{15}N Resonance Frequencies at the Field for Which the Proton Resonance Frequency of Internal Tetramethylsilane (TMS) is Exactly 60 MHz (Data from ref. 101)

Compound	Solvent	^{14}N frequency (Hz)	^{15}N frequency (Hz)	$R = \dfrac{^{15}\text{N frequency}}{^{14}\text{N frequency}}$	Chemical shift (frequency scale) (p.p.m)
Me$_4$N$^\oplus$I$^\ominus$	DMSO-d$_6$	6080010.6 ± 0.1	4334329.0 ± 0.1	1.40275706 ± 6 × 10^{-8}	0
PhCH$_2$NC	CDCl$_3$	6080773.7 ± 0.1	4334873.7 ± 0.2	1.40275683 ± 9 × 10^{-8}	125.5
MeONO$_2$	Acetone	6081826.0 ± 0.1	4335623.4 ± 0.1	1.40275698 ± 6 × 10^{-8}	298.6
Me$_2$NNO$_2$ (nitro group)	CDCl$_3$	6081905.5 ± 0.2	4335680.32 ± 0.02	1.40275691 ± 6 × 10^{-8}	311.6
			average value	1.40275695 ± 8 × 10^{-8}	

do not exist according to later work [112, 131]. The ^{14}N chemical shift of ethyl nitrate is about 37 p.p.m. [60] to high field from $MeNO_2$ instead of a few p.p.m., as found in the early work [22], and the recent data [112] on the ^{15}N shifts of methylformamide and dimethylformamide show a difference of 15 p.p.m. rather than about 200 p.p.m. found earlier [22]. Thus, the interchangeability of ^{14}N and ^{15}N chemical shifts seems now to be well-proven. The possible discrepancies arise from non-uniformity of experimental conditions or simply from experimental errors.

4.3 Scale of Nitrogen Chemical Shifts

To compare nitrogen chemical shifts of nuclei in various molecular environments, one must assume a common reference point and set up a calibration scale. Ideally, the resonance of bare nitrogen nuclei should serve as the reference but this is practically unfeasible, and other standards must be chosen. The most convenient scale of chemical shifts is expressed in terms of changes in the dimensionless screening constant σ [equation (1.8)]. The numerical values of the latter are so small that it is customary to express them in multiplies of 10^{-6} (parts per million, p.p.m.). A natural system is to assign the positive direction of the shifts to an increase in the screening constant. This is equivalent to assigning a plus sign to shifts to higher magnetic fields at a constant resonance frequency or to lower frequencies at a constant magnetic field. This system will be called here the *screening-constant scale*. It is theoretically sound since the screening constant is a quantity which enters directly into the equations of the general theory of chemical shifts (Chapter 1). However, there is another method of approach to the problem of the direction of chemical shifts, based on preferences drawn from other branches of spectroscopy. This denotes as positive the direction of the increasing energy of observed transitions, and in the case of NMR spectra leads to a reverse in sign of chemical shifts with respect to the screening-constant scale by rendering positive all shifts to higher resonance frequencies at a constant external field or to lower magnetic fields at a constant frequency. Such a system will be called here the *frequency scale* of nitrogen chemical shifts. The frequency scale has some advantage from the experimental point of view since in many modern NMR spectrometers where the frequency sweep and double-resonance techniques are used, transition frequencies are actually measured. It is also common to express chemical shifts measured at either a constant frequency or a constant magnetic field in frequency units (Hertz), and this is more compatible with the frequency scale system.

An important question is the choice of suitable standards in order to set up a zero reference point for the comparison of nitrogen chemical shifts. Because of the experimental inaccessibility of the resonance of a bare nitrogen nucleus, other resonance signals must be considered which should be as inert as possible to intermolecular effects. It seems that the proton resonance of tetramethylsilane (TMS) is the best choice [75, 102], particularly since heteronuclear field/frequency locking systems have been introduced in modern NMR spectrometers. However, direct and accurate measurements of nitrogen chemical shifts relative to TMS are not always possible or convenient, particularly in ^{14}N NMR spectroscopy, and a reference point within the range of nitrogen chemical shifts should be chosen in order to facilitate the measurements. A substance for which the position of the resonance signal of both ^{14}N and ^{15}N may be measured with high accuracy [102] by direct and double-resonance methods and which does not depend very much on solvent is tetramethylammonium iodide. Other anions accompanying the tetramethylammonium ion may also be used. No appreciable effect on the resonance frequency is found [102] due to the gegenion (Table 4.3) and solvent effects are not larger than 2 p.p.m., an acceptable limit with respect to the entire range of nitrogen chemical shifts. Thus, it seems that $Me_4N^{\oplus}$ should be a good internal reference, with $Et_4N^{\oplus}$ as an auxiliary standard, for relating nitrogen resonance frequencies to that of ^1H in TMS. The nitrogen resonance signal of the tetramethylammonium ion may constitute a convenient zero reference point for the frequency scale of nitrogen chemical shifts since it is placed at the low-frequency (high-field) side of the total range, so that the shifts for most molecules, except amines, ammonium ions and isocyanates, are positive on this scale. We shall use henceforth the resonance signal of $Me_4N^{\oplus}$ as the origin of the frequency scale of nitrogen shifts.

However, $Me_4N^{\oplus}X^{\ominus}$ salts are not very convenient as practical standards to be employed in direct measurements of the shifts, particularly in ^{14}N measurements, because of their

TABLE 4.3. ^{14}N Resonance of some Molecules of the Type $R_4N^{\oplus}X^{\ominus}$
(Data from ref. 102)

Solvent (solute concentration 0.075M)	Anion	Resonance frequency (Hz) at magnetic field corresponding to ^{1}H resonance of TMS at exactly 220 MHz	
		$Me_4N^{\oplus}X^{\ominus}$	$Et_4N^{\oplus}X^{\ominus}$
DMSO-d_6	$Cl^{\ominus}$	15 892 547	15 892 869
	$Br^{\ominus}$	548	870
	$I^{\ominus}$	548	869
	$F^{\ominus}(3 H_2O)$	545	—
	$NO_3^{\ominus}$	545	—
	$MeSO_4^{\ominus}$	545	—
	$PF_6^{\ominus}$	545	—
	$BF_4^{\ominus}$	546	—
D_2O	$Cl^{\ominus}$	526	—
	$I^{\ominus}$	527	—
	$MeSO_4^{\ominus}$	525	—
	$BF_4^{\ominus}$	525	—
SO_2	$Cl^{\ominus}$	539	—
	$I^{\ominus}$	540	—
	$MeSO_4^{\ominus}$	537	—
	$BF_4^{\ominus}$	534	—
$CDCl_3$	$Cl^{\ominus}$	560	878
	$Br^{\ominus}$	—	882
	$I^{\ominus}$	—	887
	$MeSO_4^{\ominus}$	555	—

rather limited solubility and the easy saturation of their ^{14}N resonance signals. The $Me_4N^{\oplus}$ shift is not suitable, either, as the origin of the screening-constant scale since it would render negative most of the shifts in organic compounds. A screening-constant scale for nitrogen chemical shifts with the origin at the resonance of nitromethane, $MeNO_2$, or the nitrate ion, $NO_3^{\ominus}$, has been proposed [40] (Table 4.4) and subsequently used for a large number of organic compounds [20, 39, 41, 44, 47, 50, 57, 60, 69, 85, 106, 107, 114, 121–126, 128] with $MeNO_2$ or $NO_3^{\ominus}$ as internal standards. The relative shift between the resonances of nitromethane and the nitrate ion has been shown [40, 107] to be negligible for all practical purposes. Recently, Becker [113] has measured the resonance frequency of ^{15}N in $Me^{15}NO_2$ (Table 4.5) by double-resonance methods and referred it to the resonance of ^{1}H in internal TMS. Thus, the screening-constant scale based on the $MeNO_2/NO_3^{\ominus}$ standard may be

TABLE 4.4.   Internal Standards for Screening-Constant Scale of ^{14}N Chemical Shifts (Data from ref. 40)

Standard	Chemical shift	Applications
MeNO$_2$	0	organic solvents and acidic aqueous solutions
NO$_3^{\ominus}$(K$^{\oplus}$, Na$^{\oplus}$)	0	neutral and basic aqueous solutions
C(NO$_2$)$_4$	+ 48	auxiliary standard for nitro compounds
Dimethylformamide (DMF)	+ 276	auxiliary standard for organic and aqueous solutions

related to the TMS resonance or to the frequency scale based on the Me$_4$N$^{\oplus}$ standard. Nitromethane is a convenient internal standard because of its solubility in organic compounds. Its relatively narrow ^{14}N signal is also very suitable for the calibration of ^{14}N spectra containing resonance signals of different widths [60]. The NO$_3^{\ominus}$ ion has similar advantages as a standard. We shall use the resonances of MeNO$_2$ and NO$_3^{\ominus}$ as the origin of the screening-constant scale for nitrogen chemical shifts.

So far, the discussion has been concerned with internal standards which are dissolved in the sample examined and experience the same magnetic field. Solvent effects on the standard may be eliminated by the use of external standards, either in a system of concentric sample tubes or by sample replacement. However, this introduces bulk-susceptibility

TABLE 4.5. ^{15}N Resonance Frequencies of Me^{15}NO$_2$ (Data from ref. 113)

Solvent (MeNO$_2$ concentration about 0.1M)	^{15}N frequency (Hz) at field corresponding to proton resonance of internal TMS at exactly 220 MHz	Relative shift (frequency scale) (p.p.m.)	Shift from Me$_4$N$^{\oplus}$I$^{\ominus}$ (frequency scale) (p.p.m.)
DMSO-d_6	22 300 922	6.2	338.2
D$_2$O	915	5.9	339.2
SO$_2$	866	3.7	336.1
CDCl$_3$	783	0 (arbitrary)	331.1
CDCl$_3$ (~3.5M)	813	1.3	332.5
Benzene-d_6	793	0.4	—

effects and signal distortion in the former, or effects of spectrometer system instability in the latter case. These are probably of the order of uncertainty in the shift measurements due to solvent effects on the resonance of the standard, as shown in Table 4.10. One should note that the range of nitrogen chemical shifts is such that differences of a few p.p.m. resulting from non-uniform calibration techniques are not very significant and not worthy of attention at all if quite different molecular structures are examined in nitrogen NMR. For groups of closely related compounds, it is more important to carry out the measurements under uniform and clearly specified conditions.

Other standards, external or internal, have so far been used [131]. A very popular external standard has been NH_4NO_3 in acidified aqueous solutions [22, 24, 31, 35, 37, 43, 45, 46, 51, 52, 55, 56, 59, 71, 109, 116, 120]. Other external reference substances reported are HNO_3 [23, 32, 123], NH_3 [65, 70, 103, 104], NMe_3 [89], $MeNO_2$ [92, 118, 119], $NaNO_3$ [82], Me_2NNO [91, 92, 108], and $NaNO_2$ [108]. Some of them, like Me_2NNO and $NaNO_2$, are of dubious value for accurate studies of ^{14}N spectra because of their signal width.

It may be worthwhile to consider the possibility of relating the 'experimental' scale of nitrogen chemical shifts to the theoretical 'absolute' scale which has its origin at the resonance of the bare nitrogen nucleus. At present, it may be done only by the theoretical calculation of magnetic shielding in simple molecules. Such calculations have been made for the N_2 molecule on the basis of molecular beam experiments [19] and the results suggest that the nitrogen resonance signal of N_2 is shifted by 100 p.p.m. to higher frequencies (lower magnetic fields) from that of the bare nitrogen nucleus. So far, two values of the shift for N_2 relative to $NO_3^{\ominus}$ have been reported, 14 p.p.m. [4] and 70 p.p.m. [35] to lower frequencies (higher magnetic fields). The discrepancy is large and there is considerable doubt [131] about the accuracy of the measurements. The uncertainty of the molecular beam experiments and chemical shift measurements makes attempts [93] to set up an absolute scale of nitrogen chemical shifts rather premature. However, a rough estimate of the absolute zero on the nitrogen shift scale may be made and it is interesting that the origin seems to be located in the central region of the total range of shifts.

4.4   Structural Correlations of Nitrogen Chemical Shifts in Representative Groups of Organic Molecules

4.4.1   Amines

Aliphatic amines and ammonia are characterized by nitrogen NMR signals at the high-field (low-frequency) limit of the normal range of shifts, +380 to +320 p.p.m. on the screening-constant scale ($MeNO_2/NO_3^{\ominus}$) or -50 to $+15$ p.p.m. on the frequency scale ($Me_4N^{\oplus}$), as shown in Table 4.6. The only substances which may give resonance signals in this area are ammonium ions and alkyl or aryl isocyanates. There are evident regularities [69] in the chemical shifts of alkylamines (Table 4.6). The increasing alkyl substitution of the nitrogen atom in the series

$$RNH_2 \rightarrow R_2NH \rightarrow R_3N$$

results in a downfield (high-frequency) shift of the nitrogen resonance signal. A similar trend is observed for increasing alkyl substitution of the carbon atoms which are directly bonded to the amino nitrogen atom. Generally, the shifts may be represented by the following scheme where the arrows denote the downfield (high-frequency) direction.

$$CH_3NH_2 \;\rightarrow\; (CH_3)_2NH \;\rightarrow\; (CH_3)_3N$$
$$\downarrow \qquad\qquad\quad \downarrow$$
$$RCH_2NH_2 \;\rightarrow\; (RCH_2)_2NH \;\rightarrow\; (RCH_2)_3N$$
$$\downarrow$$
$$R_2CHNH_2$$
$$\downarrow \qquad\qquad \left\downarrow \xrightarrow{} \text{downfield shift}\right.$$
$$R_3CNH_2$$

So far, there has been no satisfactory theoretical explanation of this trend in the shifts of alkylamines [69]. It is clearly inconsistent with the order of decreasing energy of the lowest electronic excitations observed in the absorption spectra, and it may serve as a good illustration of the fact that there is no direct connection between such experimental excitation energies and the average excitation energy used in theoretical calculations (Chapter 1). The direction of the shifts is also opposite to that expected from changes in the so-called *diamagnetic* term in the expression for the screening constant and seems to cast some doubt on the validity of some calculations.

There is rather good agreement between the available ^{14}N (Table 4.6) and ^{15}N (Table 4.7) data for simple alkylamines. Interesting investigations [78, 88, 89, 103] of solvent effects and liquid-vapour shifts in the ^{15}N resonance of amines is discussed in Section 4.5. It should be noted that the shift for liquid NH_3 is 18.0 p.p.m. downfield [89] from gaseous NH_3 instead of the previously reported value of 15.9 p.p.m. which results from an error in the bulk-susceptibility correction.

Nitrogen chemical shifts are useful for the identification of primary, secondary and tertiary amines. The shift [69] for $R-CH_2-NH_2$ amines, +360 ± 5 p.p.m. (screening-constant scale) or −25 ± 5 p.p.m. (frequency scale), is very characteristic since it is out of the range of di- and tri-alkylamines and ammonium ions. Double-resonance studies [116] of ^{14}N chemical shifts in protonated amines (CF_3COOH solutions) afford a simple method of distinguishing various types of alkylamines in mixtures. The corresponding shifts are considered in the next section, which is devoted to ammonium ions.

Arylamines give nitrogen resonance signals at low fields (high frequencies), +330 to +290 p.p.m. (screening-constant scale) or +5 to +45 p.p.m. (frequency scale), compared with the shifts for alkylamines. There is some overlapping of the spectral ranges with ammonium ions, hydrazines and ureas. Increasing the number of aryl groups at the amino nitrogen atom is accompanied by a downfield shift. A fairly linear correlation between the nitrogen chemical shifts of substituted arylamines and the π-electron densities and the substituent Hammet constants [109, 127] for *meta-* and *para-*substitution has been found. In either case the resonance moves to a lower field (higher frequency) with an increasingly positive value of the constant. This corresponds to an increase in the screening with increasing π-charge density at the nitrogen atom. It is roughly in accord with the expected changes in both the *diamagnetic* and *paramagnetic* terms in the general expression for the nuclear screening constant since an electron-attracting substituent on the phenyl ring should give rise to a decrease in the electron charge density at the nitrogen atom and an increase in the π-bond order of the C−N bond. The combined

TABLE 4.6.   ¹⁴N Chemical Shifts and Linewidths of Some Aliphatic Amines (data from ref. 69)

Compound	Solvent	Concentration (volume %)	¹⁴N Chemical shift (p.p.m.)		¹⁴N Resonance half-height width (Hz)
			Screening-constant scale referred to internal $MeNO_2$ or KNO_3	Frequency scale[a] referred to $Me_4N^{\oplus}I^{\ominus}$	
NH_3	neat	—	+383 ± 0.5	−48	22
	H_2O (5% NaOH)	30	+375 ± 1	−40.5	110
$MeNH_2$	neat	—	+378 ± 1	−43	65
	dioxane	5	+378 ± 2	−43	75
	H_2O (5% NaOH)	30	+370 ± 2	−35.5	350
$EtNH_2$	neat	—	+355 ± 1	−20	70
	dioxane	30	+355 ± 1	−20	90
	H_2O (5% NaOH)	30	+351 ± 2	−16.5	350
$n\text{-}PrNH_2$	neat	—	+359 ± 1	−24	85
	dioxane	50	+360 ± 1	−25	110
$n\text{-}BuNH_2$	neat	—	+359 ± 2	−24	200
$i\text{-}PrNH_2$	neat	—	+334 ± 1	+1	100
	dioxane	50	+335 ± 1	0	100
cyclohexylamine	neat	—	+337 ± 1	−2	185
	dioxane	50	+340 ± 2	−5	185
$t\text{-}BuNH_2$	neat	—	+319 ± 1	+16	98
	dioxane	50	+320 ± 1	+15	115
	H_2O (5% NaOH)	30	+316 ± 3	+18.5	270
Me_2NH	neat	—	+371 ± 1	−36	98
	dioxane	20	+370 ± 1	−35	110
	H_2O (5% NaOH)	20	+362 ± 2	−27.5	400

Compound	Solvent		Shift		
Me_3N	neat	—	$+365 \pm 1$	−30	100
	dioxane	30	$+366 \pm 2$	−31	150
	H_2O (5% NaOH)	15	$+356 \pm 3^d$	−21	650
Et_2NH	neat	—	$+332 \pm 1$	+3	180
	dioxane	50	$+332 \pm 1$	+3	195
Et_3N	neat	—	$+327 \pm 2$	+8	320
$(n\text{-}Bu)_2NH$	neat	—	$+351 \pm 3$	−16.5	450
$CH_2{=}CHCH_2NH_2$	neat	—	$+360 \pm 1$	−25	100
benzylamine	neat	—	$+356 \pm 2$	−21	220
$H_2NCH_2CH_2NH_2$	neat	—	$+363 \pm 1$	−28	160
	dioxane	30	$+365 \pm 1$	−30	160
$H_2N(CH_2)_4NH_2$	neat	—	$+360 \pm 1$	−25	235
$MeCH(NH_2)CH_2NH_2$	neat	—	$+354 \pm 2$	−19	285
	dioxane	50	$+357 \pm 2$	−22	280
$HOCH_2CH_2NH_2$	neat	—	$+364 \pm 5$	−29	900
pyrrolidine	neat	—	$+345 \pm 3^b$	−10	290
hexamethylenetetramine	MeOH	satd.	$+328 \pm 4^c$	+7	680

a. Neat $MeNH_2$ is 43.0 p.p.m. to higher fields from $Me_4N^{\oplus}I^{\ominus}$ (refs. 102 and 88); this gives a shift of 335 p.p.m. to higher fields for $Me_4N^{\oplus}I^{\ominus}$ relative to $MeNO_2$. The shift for aqueous $Me_4N^{\oplus}Cl^{\ominus}$ from $NO_3^{\ominus}$ is +334.5 p.p.m. (ref. 69).

b Data from ref. 106.

c Data from ref. 121.

d A value of +340 p.p.m. was erroneously inserted in ref. 69.

TABLE 4.7 ^{15}N Chemical Shifts of some Simple Amines

Coumpound	Solvent or state	^{15}N Shifts (p.p.m.) ref. to $Me_4N^{\oplus}I^{\ominus}$ (frequency scale)
NH_3	vapour	-64.9^a
	liquid	-46.9^a, -45^c
$MeNH_2$	vapour	-50.4^a, -48.6^b
	liquid	-43.0^b
	benzene (0.2M)	-41.5^b
	CCl_4 (0.2M)	-39.2^b
	DMF (0.2M)	-41.3^b
	DMSO (0.2M)	-39.3^b
Me_2NH	vapour	-38.8^a
	liquid	-36.2^a
Me_3N	vapour	-36.2^a
	liquid	-32.0^a
	benzene	-30.9^b
	chloroform	-27.4^b
	methanol	-26.6^b, -27.0^d
	H_2O	$-25.1^{d'}$
NH_2D	liquid	-47.6^e
NHD_2	liquid	-48.2^e
ND_3	liquid	-48.9^e

^{a} data from ref. 103; originally referred to gaseous NH_3 (with bulk-susceptibility corrections), recalculated to $30°$ and related to $Me_4N^{\oplus}I^{\ominus}$ through neat $MeNH_2$ (43.0 p.p.m. to higher fields, refs. 88 and 102).

^{b} data from ref. 88; originally referred to TMS proton resonance, recalculated according to the frequency for $Me_4N^{\oplus}I^{\ominus}$ in $CDCl_3$ (ref. 102, Table 4.3).

^{c} data from ref. 25; related to $Me_4N^{\oplus}I^{\ominus}$ through neat $MeNH_2$.

^{d} data from ref. 89; originally referred to gaseous NMe_3, recalculated according to data in ref. 103 as in footnote (a).

^{e} data from ref. 65; originally referred to liquid NH_3.

effect should lead to a decrease in the *diamagnetic* term and to an increase in the absolute value of the *paramagnetic* term. This is a particular case where the expected changes in the two terms are acting in the same direction, and the agreement with the observed direction of the nitrogen chemical shift is not surprising.

For a guide to nitrogen chemical shifts in amines, see refs 3, 4, 9, 11, 18, 23, 24, 25, 31, 42, 48, 61, 65, 69, 72, 78, 88, 103, 109, 117, and 127.

4.4.2   Ammonium Ions

The nitrogen resonance signals of ammonium ions derived from alkylamines are always shifted to lower fields (higher frequencies) relative to the corresponding amines (Tables 4.6 and 4.9), but the shifts are not large enough to separate the spectral ranges. A simple explanation of this magnetic deshielding invokes the decreased electron density at the nitrogen atom in the ion but this is not necessarily correct. The nitrogen chemical shifts of alkylammonium ions follow the same structural correlations as in amines since the low-field (high-frequency) direction is observed in the following series:

$$(NH_4^\oplus) \rightarrow RNH_3^\oplus \rightarrow R_2NH_2^\oplus \rightarrow R_3NH^\oplus \rightarrow R_4N^\oplus$$

$$R = (H) \rightarrow Me \rightarrow Et \rightarrow i\text{-}Pr \rightarrow t\text{-}Bu \qquad \xrightarrow{\text{downfield shift}}$$

The shift of $NH_4^\oplus$ may depart from the above sequence because it is considerably pH-dependent [104]. The trend in the shifts does not agree with the simple magnetic deshielding model based on a decreased electron density relative to

TABLE 4.8.   Nitrogen Chemical Shifts of some Arylamines

		Chemical shift (p.p.m.)		
Compound	Solvent	Screening-constant scale referred to $MeNO_2$ or $NO_3^\ominus$	Frequency scale[c] referred to $Me_4N^\oplus X^\ominus$	Hammett[c] substituent constants
aniline	neat	+319 ± 3[a], 321[f]	+13.5[a], +11.5[f]	0.0
	acetone	+320.6[b], +321.7[c]	+11.9[b], +11.8[c]	
	benzene (0.1M)	+324.7[d]	+ 7.8[d]	
	CCl$_4$ (0.1M)	+322.5[d]	+10.0[d]	
	dioxane (0.1M)	+324.1[d]	+ 8.4[d]	
	pyridine (0.1M)	+319.5[d]	+13.0[d]	
	DMF (0.1M)	+319.9[d]	+12.0[d]	
	DMSO (0.1M)	+317.0[d]	+15.5[d]	
substituted anilines				
3-NO$_2$	acetone	+317.7[c]	+14.8[c]	+0.71
	DMSO	+312.1[h]	+20.4[h]	
4-NO$_2$	acetone	+306.4[b], +306.5[c]	+26.1[b], +26.2[c]	+0.78
	DMSO	+297.3[h]	+35.2[h]	
2,4-(NO$_2$)$_2$	DMSO	+284.3[h]	+48.2[h]	
3,5-Me$_2$,4-NO$_2$	DMSO	+307.6[h]	+24.9[h]	
3-COMe	acetone	+319.4[c]	+13.1[c]	+0.38
4-COMe	acetone	+312.7[c]	+19.8[c]	+0.50
2-OCOMe	acetone	+312.6[c]	+19.9[c]	+0.45
4-F	DMSO	+320.3[h]	+12.2[h]	
3-Cl	DMSO	+314.5[h]	+18.0[h]	
4-Cl	acetone	+320.5[c]	+12.0[c]	+0.23
	DMSO	+316.7[h]	+15.8[h]	
3-Br	DMSO	+314.7[h]	+17.8[h]	
4-Br	acetone	+320.8[b], +320.1[c]	+11.7[b], +12.4[c]	+0.23
	DMSO	+316.0[h]	+16.5[h]	

TABLE 4.8.—continued.

Compound	Solvent	Chemical shift (p.p.m.)		Hammett[c] substituent constants
		Screening-constant scale referred to $MeNO_2$ or $NO_3^{\ominus}$	Frequency scale[c] referred to $Me_4N^{\bullet}X^{\ominus}$	
2,4-Br$_2$	DMSO	+311.2[h]	+21.3[h]	
2,4,6-Br$_3$	DMSO	+305.5[h]	+27.0[h]	
3-I	DMSO	+315.3[h]	+17.2[h]	
4-I	acetone	+317.3[b]	+15.2[b]	
	DMSO	+315.6[h]	+16.9[i]	
3-CF$_3$	acetone	+319.8	+12.7[c]	+0.43
	DMSO	+314.3[h]	+18.2[h]	
3-COOMe	acetone	+320 ± 2[c]	+12.5[c]	+0.37
3-Me	acetone	+322.2[c]	+10.3[c]	−0.07
	DMSO	+317.8[h]	+14.7[h]	
4-Me	acetone	+323.8[b], +323.7[c]	+ 8.7[b], +8.8[c]	−0.17
	CDCl$_3$	+323.8[b]	+ 8.7[b]	
	DMSO	+319.6[h]	+12.9[h]	
3,5-Me$_2$	DMSO	+318.2[h]	+14.3[h]	
3-OMe	DMSO	+316.3[h]	+16.2[h]	
4-OMe	DMSO	+322.8[h]	+ 9.7[h]	
3,5-(OMe)$_2$	DMSO	+316.7[h]	+15.8[h]	
4-NH$_2$	acetone	+330 ± 1[c]	+ 2.5[c]	−0.66
N-methylaniline	neat	+320 ± 4[a], +325 ± 2[c]	+12.5[a], +7.5[c]	
diphenylamine	acetone	+289 ± 3[c]	+43.5[c]	
substituted diphenylamines				
3-OMe	acetone	+293 ± 7[c]	+39.5[c]	
4-OMe	acetone	+293 ± 7[c]	+39.5[c]	
4,4'-(NO$_2$)$_2$	acetone	+286 ± 15[c]	+46.5[c]	
substituted pyridines				
2-NH$_2$	acetone	+315 ± 4[a], +313 ± 10[g]	+17.5[a], +19.5[g]	
3-NH$_2$	acetone	+326 ± 4[a], +328 ± 4[g]	+ 6.5[a], + 4.5[g]	
4-NH$_2$	acetone	+317 ± 4[a], +321 ± 10[g]	+15.5[a], +11.5[g]	

[a] Ref. 69; ^{14}N direct measurements relative to internal $MeNO_2$.

[b] Ref. 72; ^{15}N double resonance data, originally referred to TMS; recalculated by assumption that $Me^{15}NO_2$ resonance frequency is 22,300,813 Hz if that of ^1H in TMS is at exactly 220 HMz (Table 4.5).

[c] Ref. 109; ^{14}N double resonance data referred to external $NO_3^{\ominus}$ in 4.5M NH_4NO_3 in 3M HCl.

[d] Ref. 88; ^{15}N double resonance data referred to TMS; recalculated as in footnote b.

[e] Recalculated by means of the shift of Me_4NCl in $CDCl_3$ (332.5 p.p.m. to higher fields from $MeNO_2$, Table 4.5).

[f] Ref. 25; ^{15}N direct measurements referred to external 70% HNO_3; recalculated by means of the shift of neat $MeNH_2$.

[g] Ref. 24; ^{14}N direct measurements with $NO_3^{\ominus}$ as external standard in 4.5M NH_4NO_3 in 3M HCl.

[h] Ref. 127; ^{15}N double resonance data referred originally to aniline in DMSO.

amines. A slight isotope effect is observed [36, 54] for deuterium-substituted ammonium ions, about 0.31 p.p.m. downfield per each deuterium atom.

For the same alkyl substituents, there is [116] a very characteristic difference in the shifts between ammonium ions derived from primary, secondary and tertiary alkylamines with the exception of the methyl group which gives much smaller

TABLE 4.9.   Nitrogen Chemical Shifts of some Ammonium Ions

Compound	^{14}N Chemical shift (p.p.m.)	
	Screening-constant scale referred to internal MeNO$_2$ or NO$_3^{\ominus}$	Frequency scale referred to Me$_4$N$^{\bullet}$X$^{\ominus}$
NH$_4^{\oplus}$	+353.5[a], +354.5[b]	−17.5[a], −18.5[b]
	+352.8[d], +354.6[c]	−16.8[d], −18.6[c]
MeNH$_3^{\oplus}$	+351 ± 1[b], +354.3[a]	−15 ± 1[b], −18.3[a]
	+351.2[d], +356.2	−15.2[d], −20.2
Me$_2$NH$_2^{\oplus}$	+348 ± 1[b], +351.6[a]	−12 ± 1[b], −15.6[a]
	+350.3[d], +354.1[c]	−14.3[d], −18.1[c]
Me$_3$NH$^{\oplus}$	+349.3[a], +346.5[d]	−13.3[a], −10.5[d]
	+346.9[c],	−10.9[c]
Me$_4$N$^{\oplus}$	+333.5 ± 0.5[b]	
	+332.5[e], +336[d]	0
EtNH$_3^{\oplus}$	+336 ± 1[b], +335.5[a]	0 ± 1[b], +0.5[a]
Et$_2$NH$_2^{\oplus}$	+320 ± 1[b], +323.0[a]	+16 ± 1[b], +13.0[a]
Et$_3$NH$^{\oplus}$	+314 ± 1[b], +316.8[a]	+22 ± 1[b], +19.2[a]
Et$_4$N$^{\oplus}$	+311 ± 0.5[b], +312 ± 2[f]	+25 ± 0.5[b], +24 ± 2[f]
n-PrNH$_3^{\oplus}$	+346.3[a]	−10.3[a]
(n-Pr)$_2$NH$_2^{\oplus}$	+327.2[a]	+ 8.8[a]
(n-Pr)$_3$NH$^{\oplus}$	+316.8[a]	+ 19.2[a]
n-BuNH$_3^{\oplus}$	+343.8[a]	− 7.8[a]
(n-Bu)$_2$NH$_2^{\oplus}$	+327.9[a]	+ 9.1[a]
(n-Bu)$_3$NH$^{\oplus}$	+319.5[a]	+16.5[a]
CH$_3$(CH$_2$)$_4$NH$_3^{\oplus}$	+342.3[a]	− 6.3[a]
[CH$_3$(CH$_2$)$_4$]$_2$NH$_2^{\oplus}$	+327.7[a]	+ 8.3[a]
[CH$_3$(CH$_2$)$_4$]$_3$NH$^{\oplus}$	+319.5[a]	+16.5[a]
[CH$_3$(CH$_2$)$_5$]$_2$NH$_2^{\oplus}$	+326.1[a]	+ 9.9[a]
CH$_3$(CH$_2$)$_{17}$NH$_3^{\oplus}$	+341.1[a]	− 5.1[a]
Me$_2$CHCH$_2$NH$_3^{\oplus}$	+338.0[a]	− 2.0[a]
(Me$_2$CHCH$_2$)$_2$NH$_2^{\oplus}$	+332.7[a]	+ 3.3[a]
Me$_2$CHCH$_2$CH$_2$NH$_3^{\oplus}$	+340.8[a]	− 4.8[a]
(Me$_2$CHCH$_2$CH$_2$)$_2$NH$_2^{\oplus}$	+328.6[a]	+ 7.4[a]
(Me$_2$CHCH$_2$CH$_2$)$_3$NH$^{\oplus}$	+315.7[a]	+20.3[a]
CH$_2$=CHCH$_2$NH$_3^{\oplus}$	+340.8[a]	− 4.8[a]
(CH$_2$=CHCH$_2$)$_2$NH$_2^{\oplus}$	+326.1[a]	+ 9.9[a]
(CH$_2$=CHCH$_2$)$_3$NH$^{\oplus}$	+315.6[a]	+20.4[a]
PhCH$_2$NH$_3^{\oplus}$	+336.3[a]	− 0.3[a]
(PhCH$_2$)$_3$NH$^{\oplus}$	+307.7[a]	+28.3[a]
i-PrNH$_3^{\oplus}$	+325.1[a]	+10.9[a]
(i-Pr)$_2$NH$_2^{\oplus}$	+301.3[a]	+34.7[a]
(cyclohexyl)NH$_3^{\oplus}$	+328.0[a]	+ 8.0[a]
(cyclohexyl)$_2$NH$_2^{\oplus}$	+301.9[a]	+34.1[a]
t-BuNH$_3^{\oplus}$	+314 ± 1[b], +316.1[a]	+22 ± 1[b], +19.9[a]
$\begin{matrix} CH_2-CH_2 \\ \quad\quad\quad NH_2^{\oplus} \\ CH_2-CH_2 \end{matrix}$	+329.3[a], +329 ± 2[f]	+ 6.7[a], + 7 ± 2[f]
$\begin{matrix} CH_2-CH_2 \\ CH_2 \quad\quad NH_2^{\oplus} \\ CH_2-CH_2 \end{matrix}$	+334.1[a]	+ 1.9[a]
$\begin{matrix} CH_2-CH_2 \\ O \quad\quad NH_2^{\oplus} \\ CH_2-CH_2 \end{matrix}$	+338.6[a]	− 2.6[a]
(NH$_3$CH$_2$CH$_2$NH$_3$)$^{2\oplus}$	+342 ± 2[b], +344.1[a]	− 6 ± 2[b], − 8.1[a]
(NH$_3$CH$_2$CH$_2$CH$_2$NH$_3$)$^{2\oplus}$	+344.1[a]	− 8.1[a]
[NH$_3$(CH$_2$)$_6$NH$_3$]$^{2\oplus}$	+342.6[a]	− 6.6[a]
[NH$_3$(CH$_2$)$_{10}$NH$_3$]$^{2\oplus}$	+341.7[a]	− 5.7[a]

TABLE 4.9.—continued.

	^{14}N Chemical shift (p.p.m.)	
Compound	Screening-constant scale referred to internal $MeNO_2$ or $NO_3^\ominus$	Frequency scale referred to $Me_4N^\bullet X^\ominus$
$NH_3^\oplus SO_2^\ominus$	$+285 \pm 2^b$, $+285 \pm 2^f$	$+51 \pm 2^b$, $+51 \pm 2^f$
$HONH_3^\oplus$	$+293 \pm 5^b$	$+43 \pm 5^b$
$HONMe_3^\oplus$	$+269 \pm 3^f$	$+67 \pm 3^f$
$PhNH_3^\oplus$	$+327.7^a$, $+332.4^g$	$+8.3^a$, $+3.6^g$
(anilinium ion)	$+330^h$, $+328^h$	$+6^h$ $+8^h$
	$+326^h$	$+10^h$
$MeCOCH=CHNMe_2^\oplus$ (H_2O)	$+326^i$	$+10^i$
$MeCOCH=CHNEt_2^\oplus$ (MeCN)	$+305^i$	$+31^i$
$EtCOCH=CHNEt_2^\oplus$ (H_2O)	$+307^i$	$+29^i$
$n\text{-PrCOCH}=CHNMe_2^\oplus$ (H_2O)	$+327^i$	$+ 9^i$
$n\text{-PrCOCH}=CHNEt_2^\oplus$ (H_2O)	$+303^i$	$+33^i$
$n\text{-}C_5H_{11}COCH=CHNMe_2^\oplus$ (H_2O–MeCN)	$+327^i$	$+ 9^i$

[a] Ref. 116, double resonance ^{14}N data, 0.05M amine in CF_3COOH; referred to external 4.5M $NH_4^\oplus NO_3^\ominus$ in 3M HCl.

[b] Ref. 69, direct resonance ^{14}N data, referred to internal $MeNO_2$; saturated solution in 10M HCl.

[c] Ref. 104, ^{15}N data at $30°$; 5M in amine hydrochloride, 1M in HCl; referred to external NH_3 liquid, recalculated to $NMe_4^\oplus$.

[d] Ref. 121, ^{14}N direct measurements referred to internal $NMe_4^\oplus Cl^\ominus$; 3M in amine hydrochloride, 0.8M in $NMe_4^\oplus Cl^\ominus$, 8M in HCl; shift between $MeNO_2$ and $NMe_4^\oplus$ is 336.0 p.p.m. under these conditions.

[e] Ref. 113, $CDCl_3$ solutions, Tables 4.3 and 4.5.

[g] Ref. 72, ^{15}N double resonance data in CF_3COOH; originally referred to TMS; recalculated relative to $NMe_4^\oplus Cl^\ominus$ frequency in D_2O (Table 4.3) and then to $MeNO_2$ (336 p.p.m. downfields in acidic solution, Table 4.10).

[h] Ref. 22, direct ^{15}N measurements recalculated by means of neat $MeNH_2$ shift (43.0 p.p.m. to higher fields from $Me_4N^\oplus$) and $Me_4N^\oplus$ shift (336 p.p.m. to higher fields from $MeNO_2$).

[i] Ref. 115, direct ^{14}N measurements referred to external $MeNO_2$; experimental errors not reported.

contributions (3–2 p.p.m.). This may be useful in the analysis of mixtures, as has been shown for mono-, di-, and tri-n-propylamine [116]. It is more convenient to measure the shifts

primary $\xrightarrow[\text{12–26 p.p.m. downfield}]{}$ secondary $\xrightarrow[\text{8–13 p.p.m. downfield}]{}$ tertiary

for the ammonium ions than those for the parent amines because of the slow proton exchange in the former which makes possible some double-resonance measurements of nitrogen shifts from the corresponding proton spectra.

There are some discrepancies in the recently published nitrogen chemical shifts for ammonium ions [69, 104, 116]. The shift previously reported [69] for $Me_3NH^\oplus$ has turned out to be erroneous according to later measurements (Table 4.9), but even the most recent (Table 4.10) results for ^{15}N and ^{14}N show some differences which may reflect the uncertainty in the shifts introduced by sample replacement methods of calibration. However, the differences are not significant enough to affect the trend generally observed in the nitrogen chemical shifts of ammonium ions. The precise data from Table 4.10 show that at approximately the same $Cl^\ominus$ concentrations, the sequence of shifts in the direction of decreasing screening (increasing frequency) is in agreement

$$MeNH_3^\oplus \rightarrow Me_2NH_2^\oplus \rightarrow Me_3NH^\oplus \rightarrow Me_4N^\oplus$$

with that given at the beginning of this section. The shift of $NH_4^\oplus$ may lie either between those of $MeNH_3^\oplus$ and $Me_2NH_2^\oplus$ at low $Cl^\ominus$ concentrations, or between $Me_2NH_2^\oplus$ and $Me_3NH^\oplus$ at higher $Cl^\ominus$ concentrations (Table 4.35). It is also sensitive to the nature of the accompanying anion, since $NH_4^\oplus NCS^\ominus$ (5M in 2M HCl) [105] it has been found at 6.7 ± 0.6 p.p.m. downfield from its position in 15M $NH_4^\oplus NO_3^\ominus$ in 2M HCl.

Whilst the nitrogen resonance signals of alkylammonium ions are shifted downfield from those of alkylamines, the

TABLE 4.10.   Comparison of Nitrogen Chemical Shift Calibration for Simple Alkylammonium Ions

Compound	Chemical shifts referred to $Me_4N^\oplus Cl^\ominus$ (p.p.m.) (frequency scale)		Total $Cl^\ominus$ molarity
	^{15}N data[a] external standard NH_3 (liquid)	^{14}N data[b] internal standard $Me_4N^\oplus Cl^\ominus$	
$NH_4^\oplus Cl^\ominus$	-17.8 ± 0.3	-16.8 ± 0.3	7.0
$MeNH_3^\oplus Cl^\ominus$	-17.3 ± 0.3	-15.2 ± 0.3	12.4
$Me_2NH_2^\oplus Cl^\ominus$	-15.5 ± 0.3	-14.3 ± 0.3	12.0
$Me_3NH^\oplus Cl^\ominus$	-10.4 ± 0.3	-10.5 ± 0.3	11.0
$Me_4N^\oplus Cl^\ominus$	0	0	

[a] Ref. 104; originally referred to NH_3 (liquid) at 30° by sample replacement method. The shift was found to be independent of the concentration of the alkylammonium chloride, but dependent of the total $Cl^\ominus$ concentration. Concentration 5M in amine hydrochloride. The values of chemical shift given, corresponding to the molarity in the last column of the table, differ somewhat from those in Table 4.9.

[b] Ref. 121; direct ^{14}N measurements. Clear multiplets due to NH coupling were observed. Concentrations 3M in amine hydrochloride (2M for NH_4Cl), 0.5M in $Me_4N^\oplus Cl^\ominus$. Total $Cl^\ominus$ mlarity is given in the last column. The shift of $Me_4N^\oplus Cl^\ominus$ (0.5M) in HCl (total 11M in $Cl^\ominus$) was found to be 336.0 p.p.m. to higher fields from internal $MeNO_2$.

opposite effect is observed for aniline and the corresponding phenylammonium ion.

$i\text{-PrNH}_2 \xrightarrow{\text{downfield}} i\text{-PrNH}_3^{\oplus}$

+335 p.p.m.   +325 p.p.m. (screening-constant scale)

$\text{PhNH}_2 \xrightarrow{\text{upfield}} \text{PhNH}_3^{\oplus}$

+320 p.p.m.   +328 p.p.m.

A rather simple explanation of this shift may be offered. The nitrogen resonance signal of arylamines is shifted to low field compared with that from alkylamines, and there is certainly a contribution of the lone electron pair delocalization in the aromatic ring system. The protonation of aniline should remove this effect and increase magnetic screening but there should also be an opposite, deshielding contribution such as the one observed for alkylammonium ions. If we compare the shifts for aniline and isopropylamine and the corresponding ions, it turns out that the shifts for the latter are almost the same. This is compatible with the explanation which assumes the structure of phenylammonium ions to be similar to that of protonated isopropyl amine. A similar upfield shift has been observed [115] upon quaternization of enaminoketones to $\text{RCOCH=CHNR}_3^{\oplus}$ and theoretically explained on simplified molecular models.

The ammonium ions derived from amino acid methyl esters [120], $\text{RCH(NH}_3^{\oplus})\text{COOMe}$, give nitrogen resonance signals in the normal range for ammonium ions (Table 4.11). A downfield shift upon replacing R = H with R = Me or $\text{CH}_2\text{R}'$ in these ions is analogous to that found in alkylammonium ions, for example $\text{MeNH}_3^{\oplus}$ and $\text{EtNH}_3^{\oplus}$ (Table 4.9).

For a guide to nitrogen chemical shifts in ammonium ions, see refs. 1, 2, 3, 4, 6, 14, 18, 22, 24, 25, 31, 36, 54, 69, 74, 97, 104, 105, 115, 116, 117, 120, and 121.

4.4.3   Amine Oxides

The small amount of data which are available from ^{14}N spectra (Table 4.12) indicate that the substitution of the lone electron pair in tertiary amines with an oxygen atom results in a deshielding of the nitrogen nuclei by about 70–100 p.p.m. Hydroxylammonium ions which may be derived from

TABLE 4.11. ^{15}N Natural Abundance Measurements of Chemical Shifts of $RCH(NH_3^{\oplus})COOMe\ Cl^{\ominus}$ in Aqueous (5–9M) Solutions with pH 0.5–2.0 (ref. 120)

R	^{15}N Chemical shifts of $-NH_3^{\oplus}$	
	Screening-constant scale[a] referred to $NO_3^{\ominus}$ (p.p.m.)	Frequency scale[b] referred to $Me_4N^{\oplus}X^{\ominus}$ (p.p.m.)
H	+343.5	−9.0
Me	+330.2	+4.3
Me_2CH	+335.2	−0.7
Me_2CHCH_2	+331.5	+3.0
$Et(Me)CH$	+334.6	−0.1
$PhCH_2$	+333.1	+1.4
$4\text{-}HO\text{--}C_6H_4CH_2$	+334.1	+0.4
![imidazole]$\text{HN}\diagdown\overset{\text{--CH}_2}{\underset{C^{\diagup}\overset{}{\oplus}}{}}\text{NH}$ (imidazole)	+331.8	+2.7
	+209.3, +207.2	+125.2, +127.3
$^{\oplus}NH_3(CH_2)_4$	+332.0	+2.5
(add. $NH_3^{\oplus}$)	+347.7	−13.2
$H_2NCNH(CH_2)_3$	+332.5	+2.0
$\overset{\|}{\underset{NH_2^{\oplus}}{}}$ (guanidine NH_2)	+310.2	+24.3
(guanidine NH)	+298.7	+35.8

[a] Originally referred to $NH_4^{\oplus}$ signal in 5M NH_4NO_3 in 2M HCl, recalculated to $NO_3^{\ominus}$ reference [353.5 p.p.m. downfield from $NH_4^{\oplus}$, Table 4.9, footnote (a)].
[b] Referred to $Me_4N^{\oplus}Cl^{\ominus}$ shift (334.5 p.p.m. upfield from $NO_3^{\ominus}$, ref. 69, Table 4.6).

either hydroxylamines or amine oxides do not differ significantly in their nitrogen chemical shifts from amine oxides, but there is a downfield shift relative to ammonium ions. The low-field (high-frequency) shift for amine oxides relative to amines may be interesting from a theoretical point of view. It is opposite to the low-field (high-frequency) shift of 123 p.p.m. for NF_3 as compared with ONF_3 [94]. The latter has been explained [94] (Chapter 1) as being due to a reduction

$$Me_3N \xrightarrow{\text{downfield}} Me_3N^{\oplus} - O^{\ominus}$$
+365 p.p.m.   +273 p.p.m. (screening-constant scale)

$$F_3N \xrightarrow{\text{upfield}} F_3NO$$
+8 p.p.m.   +131 p.p.m.

of the orbital angular momentum (reduction of p-orbital coefficients at the nitrogen atom) which prevails over the increased number of excitations in F_3NO. In the case of amines and amine oxides, the small electronegativity of carbon atoms as compared with oxygen and fluorine may reverse the relative importance of the two opposing effects and lead to an increase in the absolute value of the paramagnetic term and, consequently, to an increased magnetic deshielding.

It should be noted that the downfield shift upon replacing the lone electron pair at a nitrogen atom with a bond to an oxygen atom is characteristic of amine oxides. If the nitrogen atom is involved in a multiple-bond system, an upfield shift of the nitrogen resonance signal is observed, as in fulminates (Section 4.4.9), azine N-oxides (Section 4.4.12), nitrones, azoxy compounds (Section 4.4.13), and nitro compounds (Section 4.4.14) relative to nitriles, azines, imines, azo compounds, and nitroso compounds, respectively.

TABLE 4.12. ^{14}N Chemical Shifts of some Amine Oxides

Compound	Solvents	^{14}N Chemical shift (p.p.m.)		^{14}N Resonance half-height width (Hz)
		Screening-constant scale referred to internal $MeNO_2$ or $NO_3^{\ominus}$	Frequency scale[a] referred to $Me_4N^{\oplus}X^{\ominus}$	
$Me_3N^{\oplus}-O^{\ominus}$	acetone (satd.)	$+273 \pm 4$[b]	$+60 \pm 4$[b]	170 ± 20
	$H_2O/NaOH$	$+264 \pm 10$[c]	$+69 \pm 10$[c]	?
$Ph-\overset{\displaystyle Me}{\underset{\displaystyle Me}{N^{\oplus}}}-O^{\ominus}$	acetone (1 : 8 w/w)	$+265 \pm 1$[b]	$+68 \pm 1$[b]	145 ± 5
$Me_3N^{\oplus}-OH\ Cl^{\ominus}$	HCl aq.	$+269 \pm 3$[c]	$+67 \pm 3$[c]	?
$H_3N^{\oplus}-OH\ Cl^{\ominus}$	HCl aq.	$+276 \pm 20$[c]	$+60 \pm 20$[c]	?
	HCl aq.	$+293 \pm 5$[d]	$+43 \pm 5$[d]	500 ± 30

[a] Recalculated as in Tables 4.6 and 4.9
[b] Ref. 122, ^{14}N direct measurements relative to internal $MeNO_2$.
[c] Ref. 24, from external NH_4NO_3.
[d] Ref. 69, from internal $MeNO_2$.

4.4.4   Hydrazines

Very little is known about the nitrogen chemical shifts in hydrazine derivatives (Table 4.13). The resonance signals seem to be shifted to lower fields (higher frequencies) compared with those of amines but the displacement is not sufficiently large to separate the spectral range from that for alkyl- and

TABLE 4.13. Nitrogen Chemical Shifts of some Hydrazines

Compound	Solvent	^{14}N Chemical shift (p.p.m.)	
		Screening-constant scale referred to internal MeNO$_2$ or NO$_3^{\ominus}$	Frequency scale referred to Me$_4$N$^{\oplus}$X$^{\ominus}$
H$_2$NNH$_2$	neat	+312 ± 20[a]	+22[a]
H$_2$NNH$_2$	H$_2$O	+326.8 ± 0.3[b]	+7.7[b]
PhNHNH$_2$	neat	+291.7 ± 0.3[b]	+42.8[b]
		+316.6 ± 0.3[b]	+17.9[b]
PhNHNHPh (hydrazobenzene)	DMF	+284 ± 1[c]	+51[c]

[a] Ref. 4, ^{14}N data originally referred to NO$_2^{\ominus}$ (237 p.p.m. downfield from internal NO$_3^{\ominus}$, ref. 60.

[b] Ref. 105, natural abundance ^{15}N measurements, referred to NH$_4$ in 8m NH$_4$NO$_3$ in 2M HCl, recalculated with NH$_4^{\oplus}$ shift (353.5 p.p.m. upfield from NO$_3^{\ominus}$) and NMe$_4^{\oplus}$ shift (334.5 p.p.m. upfield from NO$_3^{\ominus}$), Table 4.9.

[c] Ref. 25, ^{15}N direct measurements, recalculated with neat MeNH$_2$ shift (378 p.p.m. upfield from internal MeNO$_2$).

arylamines. The values of the shifts in Table 4.13 suggest that for phenylhydrazine, PhNHNH$_2$, the signal at the lower field (high frequency) is due to the NH group rather than NH$_2$, similar to amines where phenyl substituents on the nitrogen atoms result in deshielding. This assignment is opposite to that reported in the literature [105].

4.4.5. Amides, Thioamides, Ureas, Thioureas and Related Compounds

The nitrogen chemical shifts in compounds of general formula

$$R^1 - \underset{\underset{X}{\|}}{C} - N \big<^{R^2}_{R^3} \qquad (R = H, \text{alkyl, aryl, } NR_2 \text{ group; } X = O, S, Se)$$

are to lower fields (higher frequencies) than those of amines or ammonium ions. The corresponding spectral range starts at about +300 p.p.m. on the screening constant scale (+35 p.p.m., frequency scale) and extends to about +210 p.p.m. (+125, frequency scale). A further subdivision of the range may be made according to molecular structure:

	Screening-constant scale $(MeNO_2/NO_3^{\ominus})$ (p.p.m.)	Frequency scale $(Me_4N^{\oplus})$ (p.p.m.)
ureas, $(R_2N)_2CO$	+305 to +280	+30 to +55
amides, $RCONR_2$ and thioureas, $(R_2N)_2CS$	+280 to +235	+55 to +100
thioamides, $RCSNR_2$	+240 to +210	+95 to +125

There are considerable solvent effects on the nitrogen shifts of the amido-group (tables 4.14, 4.32 and 4.34) so that caution is necessary in the interpretation of spectral changes in terms of molecular structure. A replacement of a hydrogen atom in NH with a methyl group does not seem to give any straightforward change in the nitrogen chemical shift, but there is a definite downfield (high-frequency) shift upon replacing Me with Et at the nitrogen atom. The phenyl group also gives a shift to lower fields. Changes in group R^1, attached to the carbonyl group, do not seriously affect the shift provided that R^1 is an aryl or an alkyl group.

Substitution	Nitrogen resonance displacement
$R^{2,3} = H \rightarrow R^{2,3} = Me$	little effect
$R^{2,3} = Me \rightarrow R^{2,3} = Et, Pr, Bu$	ca. 15 p.p.m. downfield
$R^{2,3} = Me \rightarrow R^{2,3} = Ph$	ca. 15 p.p.m. downfield
analogous changes in R^1	no appreciable effect

The downfield shift mentioned here may be reasonably explained [60] by structural changes involving an increase in π-bonding of the nitrogen atom through its lone electron pair when either a stronger electron donor is introduced (Me $\rightarrow$ Et) or the π-orbital system is extended over the R^2 and R^3 groups. This should result in an increase in the absolute value of the *paramagnetic* term. Thus, primary and N-methylamides may be spectrally distinguished from secondary and tertiary amides with larger alkyl groups or phenyl groups at the amido nitrogen atom.

The vinylogues of amides, $R^1COCH=CHNR^2R^3$, give ^{14}N resonance signals [81] in the normal range for amides, about

+250 to +300 p.p.m. upfield from $MeNO_2$, but the reported experimental data display such a large random spread of values that their accuracy is dubious and no detailed structural correlations are possible. The quarternization at the nitrogen atom to $R^1 COCH=CHN^{\oplus}R^2 R^3 R^4$ seems to result in an upfield shift [115] (Table 4.9), in agreement with the shift observed for the phenylammonium ion relative to aniline (Section 4.4.2).

For a guide to nitrogen chemical shifts in amides and related compounds, see refs. 11, 17, 22, 23, 24, 25, 31, 37, 40, 46, 55, 59, 60, 75, 88, 100, 105, 107, 115, 117 and 124.

4.4.6   Cyanamides and Carbodiimides

These isomeric structures

$$\begin{matrix} R^1 \\ R^2 \end{matrix} \Big\rangle N - C \equiv N\colon \qquad\qquad R^1 \underset{\cdot\cdot}{N} = C = \underset{\cdot\cdot}{N} R^2$$

cyanamide   carbodiimide

were investigated early by means of ^{14}N NMR spectra [8] which clearly distinguished between the structures for $R^1 = R^2$. For diethylcyanamide shifts of 146 (NR_2) and 258 p.p.m. (CN) upfield from $NO_3^{\ominus}$ were reported, and 240 p.p.m. for N,N'-dicyclohexylcarbodiimide. However, the accuracy of the measurements was very low and the shifts measured recently [124] (Table 4.15) deviate appreciably from the old data. The nitrogen resonance of the amido group in cyanamides occurs at very high fields (low frequencies), slightly higher than in the corresponding amines. The CN resonance is also at much higher fields than in nitriles (120–140 p.p.m. upfield from $MeNO_2$, Section 4.4.9). The CN resonance in the cyanamides lies very close to that of their oxygen analogues, the cyanates (R—O—CN, Table 4.18). The isomeric carbodiimide structure is characterized by a shift which is about half-way between the shifts for the amido and the cyano group in cyanamides.

4.4.7   Miscellaneous NR_3 Groups

The nitrogen chemical shift for the NR_2 moiety in guanidine derivatives, $(R_2 N)_2 C=NR$, is similar to those in amines, but it

TABLE 4.14.   Nitrogen Chemical Shifts of some $R_1C(=X)NR_2R_3$ Compounds

Substance	Solvent	Chemical shift (p.p.m.)					
		Screening-constant scale referred to $NO_3^\ominus$ or $MeNO_2$			Frequency scale[j] referred to $NMe_4^\oplus X^\ominus$		
		(NH_2)	(NH)	(N)	(NH_2)	(NH)	(N)
H_2NCONH_2	H_2O	+299a	—	—	+35.5a	—	—
		+302b			+32.5b		
$MeNHCONH_2$	H_2O	+303a	+303a	—	+31.5a	+31.5a	—
$MeNHCONHMe$	H_2O		+305a	—		+29.5a	—
$EtNHCONH_2$	H_2O	+303a	+283a	—	+31.5a	+51.5a	—
$EtNHCONHEt$	H_2O		+284a	—		+50.5a	—
$i\text{-PrNHCONHPr-}i$	acetone		+284a	—		+50.5a	—
$n\text{-BuNHCONH}_2$	H_2O	+306a	+284a	—	+28.5a	+50.5a	—
$PhNHCONH_2$	acetone	+305a	+275a	—	+29.5a	+59.5a	—
$PhNHCONHPh$	acetone		+270a	—		+64.5a	—
$Me_2NCONMe_2$	neat			$+320\pm2$b			+14.5b
NH_2CSNH_2	H_2O	+267a		—	+67.5a		—
$MeNHCSNH_2$	H_2O	+272a	+270a	—	+62.5a	+64.5a	—
$MeNHCSNHMe$	H_2O		+275a	—		+59.5a	—
$EtNHCSNH_2$	H_2O	+275a	+252a	—	+59.5a	+82.5a	—
$n\text{-PrNHCSNH}_2$	H_2O	+276a	+250a	—	+58.5a	+84.5a	—
$PhNHCSNH_2$	acetone	+272a	+250a	—	+62.5a	+84.5a	—
$PhNHCSNHPh$	acetone		+250a	—		+84.5a	—
$Me_2NCSNMe_2$	acetone			$+291\pm3$b			+43.5b
$HCONH_2$	acetone	+268c		—	+66.5c		—
	neat	+262e		—	+72.5e		—
	neat	+265.5h		—	+69.0h		—

Compound	Solvent		
MeCONH2	inf. dil. acetone	+274 ± 2[h]	+60.5[h]
MeCONH2	inf. dil. MeOH	+268 ± 1.5[h]	+66.5[h]
EtCONH2	CDCl3	+270[c]	+64.5[c]
n-PrCONH2	CDCl3	+273[c]	+61.5[c]
i-PrCONH2	CDCl3	+272[c]	+62.5[c]
ClCH2CONH2	CDCl3	+273[c]	+61.5[c]
Cl2CHCONH2	CDCl3	+279[c]	+55.5[c]
Cl3CCONH2	CDCl3	+281[c]	+53.5[c]
PhCH2CONH2	CDCl3	+283[c]	+51.5[c]
Ph2CHCONH2	CDCl3	+274[c]	+60.5[c]
CH2=CHCONH2	CDCl3	+273[c]	+61.5[c]
PhCH=CHCONH2	CDCl3	+276[c]	+58.5[c]
PhCONH2	CDCl3	+277[c]	+57.5[c]
HCONHMe	CDCl3	+282[c]	+52.5[c]
HCONHMe	neat	+264[c]	+70.5[c]
HCONHMe		+266.3[e]	+68.2[e]
HCONHMe		+266.4[h]	+68.1[h]
HCONHEt	inf. dil. acetone	+272 ± 2[h]	+62.5[h]
HCONHEt	inf. dil. MeOH	+267 ± 1.5[h]	+67.5[h]
HCONHEt	neat	+249[c]	+85.5[c]
HCONHPh	inf. dil. acetone	+247.1[h]	+87.4[h]
HCONHPh	inf. dil. MeOH	+253 ± 2[h]	+81.5[h]
HCONHPh	CDCl3	+247 ± 1.5[h]	+87.5[h]
HCONHPh	CDCl3	+235[c]	+99.5[c]
MeCONHMe	dioxane (5M)	+269[c]	+65.5[c]
MeCONHMe	inf. dil. dioxane	+271.5[f]	+63.0[f]
MeCONHMe	neat	+279.5[f]	+55.0[f]
MeCONHMe	CCl4 (0.1M)	+266.7[g]	+67.8[g]
MeCONHMe	CCl4 (0.25M)	+275.5[g]	+59.0[g]
MeCONHMe	DMSO (0.25M)	+274.2[g]	+60.3[g]
MeCONHMe		+274[g]	+60.5[g]

TABLE 4.14.—continued

Substance	Solvent	Chemical shift (p.p.m.)					
		Screening-constant scale referred to $NO_3^{\ominus}$ or $MeNO_2$			Frequency scale[j] referred to $NMe_4^{\oplus}X^{\ominus}$		
		(NH₂)	(NH)	(N)	(NH₂)	(NH)	(N)
MeCONHEt	CDCl₃	—	+253c	—	—	+81.5c	—
MeCONHPr-n	CDCl₃	—	+247c	—	—	+87.5c	—
MeCONHBu-n	CDCl₃	—	+249c	—	—	+85.5c	—
MeCONHPh	CDCl₃	—	+243c	—	—	+91.5c	—
o-Nitroacetanilide	CDCl₃	—	+253c	—	—	+81.5c	—
m-Nitroacetanilide	CDCl₃	—	+246c	—	—	+88.5c	—
2,4-dinitroacetanilide	CDCl₃	—	+254c	—	—	+80.5c	—
EtCONHMe	CDCl₃	—	+270c	—	—	+64.5c	—
EtCONHPh	CDCl₃	—	+244c	—	—	+90.5c	—
PhCONHPh	CDCl₃	—	+252c	—	—	+82.5c	—
MeCSNH₂	CDCl₃	+224i	—	—	+110.5i	—	—
PhCH₂CSNH₂	CDCl₃	+229i	—	—	+105.5i	—	—
PhCSNH₂	CDCl₃	+239i	—	—	+95.5i	—	—
MeCSNHPh	CDCl₃	—	+211i	—	—	+123.5i	—
PhCSNHPh	CDCl₃	—	+217i	—	—	+117.5i	—
PhCSeNH₂	CDCl₃	+235i	—	—	+99.5i	—	—
H₂NCSeNH₂	CDCl₃	+259a	—	—	+75.5a	—	—
HCONMe₂		—	—	+276d	—	—	+58.5d
	neat	—	—	+272.3h	—	—	+62.2h
	H₂O	—	—	+271d	—	—	+63.5d
	inf. dil. acetone	—	—	+273h	—	—	+61.5h

HCONEt$_2$	inf. dil. MeOH	—	+264[h]	—	+70.5[h]
	neat	—	+246.5[h]	—	+88.0[h]
	inf. dil. acetone	—	+247[h]	—	+87.5[h]
	inf. dil. MeOH	—	+237[h]	—	+97.5[h]

[a] Ref. 55; ^{14}N double resonance data, standard (external) 4.5M NH$_4$NO$_3$ in 3M HCl, aqueous or acetone solutions ($\pm$2 p.p.m.).

[b] Ref. 124; unpublished results, direct ^{14}N measurements, ref. to KNO$_3$ or MeNO$_2$.

[c] Ref. 37; experimental conditions as in footnote (a).

[d] Ref. 40 and 107; direct ^{14}N measurements, ref. to internal MeNO$_2$ and KNO$_3$.

[e] Ref. 105; natural abundance ^{15}N measurements ref. to external NH$_4^{\oplus}$ in 15M NH$_4$NO$_3$ (+353.5 p.p.m. upfield from NO$_3^{\ominus}$).

[f] Ref. 100; double resonance ^{15}N data, dioxane solutions, dilution study 0–5M.

[g] Ref. 88; double resonance ^{15}N data, referred originally to TMS, recalculated with neat MeNH$_2$ shift [43.0 p.p.m. upfield from NMe$_4^{\oplus}$ shift (334.5 p.p.m. upfield from NO$_3^{\ominus}$, Table 4.6.

[h] Ref. 59; double resonance ^{14}N data ref. to 4.5M NH$_4$NO$_3$ in 3M HCl.

[i] Ref. 46; double resonance ^{14}N data, referred to external 4.5M NH$_4$NO$_3$ in 3M HCl.

[j] Recalculated for Me$_4$N$^{\oplus}$X$^{\ominus}$ shift, 334.5 p.p.m. upfield from NO$_3^{\ominus}$ (Table 4.6) and 332.5 p.p.m. upfield from MeNO$_2$ (Table 4.5).

TABLE 4.15. ^{14}N Chemical Shifts of some Cyanamide and Carbodiimide Structures

Compound	Solvent	^{14}N Chemical shift (p.p.m.)	
		Screening-constant[a] scale referred to internal MeNO$_2$ or NO$_3^{\ominus}$	Frequency scale[b] referred to Me$_4$N$^{\oplus}$X$^{\ominus}$
H$_2$NCN	DMSO	+366 ± 2 (HN$_2$)	−28
		+196 ± 2 (CN)	+142
Me$_2$NCN	neat	+386 ± 2 (NMe$_2$)	−54
		+184 ± 2 (CN)	+148
Ph$_2$NCN	acetone (satd.)	+335 ± 10 (NPh$_2$)	−3
		+193 ± 10 (CN)	+139
$\begin{array}{c}\text{CH}_2\text{--CH}_2\\ \text{H}_2\text{C}\quad\quad\text{CH--N=C=N--HC}\\ \text{CH}_2\text{--CH}_2\end{array}\begin{array}{c}\text{CH}_2\text{--CH}_2\\ \quad\quad\text{CH}_2\\ \text{CH}_2\text{--CH}_2\end{array}$	CCl$_4$ (satd.)	+275 ± 4	+57

[a] Ref. 124, direct ^{14}N measurements, referred to internal MeNO$_2$.
[b] Recalculated according to Table 4.5 by setting NMe$_4^{\oplus}$Cl$^{\ominus}$ − MeNO$_2$ shift at 338 p.p.m. for DMSO solutions and 322 p.p.m. otherwise.

is to far higher fields (lower frequencies) than the shift for the imino group, C=NR (Tables 4.16 and 4.25). Guanidine is a nitrogen analogue of urea and thiourea and its nitrogen resonance signal occupies the highest-field position in the sequence.

	$(R_2N)_2C{=}NR$	$(R_2N)_2C{=}O$	$(R_2N)_2C{=}S$
nitrogen shift (screening-constant scale) (p.p.m.)	$ca. +330$	$ca. +300$	$ca. +270$

The NR_2 group in nitramines, $R^1R^2NNO_2$, gives a nitrogen resonance signal at considerably lower fields (higher frequencies) than amines, and the effect is even larger in nitrosamines, R^1R^2NNO (Tables 4.16 and 4.6). It may be quite simply explained as being due to the strong delocalization of the lone electron pair on the NR_2 group caused by strong electron-acceptors like the nitro or nitroso group. The combined effects of an increase in π-bond order and a decrease in the total electron density should give downfield-shift contributions from both the *diamagnetic* and *paramagnetic* terms in the screening constant. The explanation is analogous to that for the shifts in aniline derivatives (Section 4.4.1). The same argument may be used for an explanation of the low-field (high-frequency) displacement of the NH nitrogen resonance in succinimide (Table 4.16) as compared with amides.

The nitrogen resonance signals of pyridones which are tautomers of the corresponding hydroxypyridines and structurally resemble amides, may be found in the vicinity of the

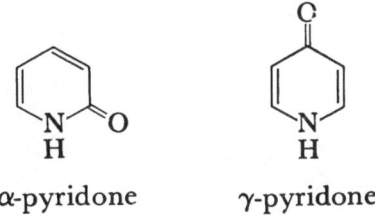

α-pyridone   γ-pyridone

low-field (high-frequency) limit of the spectral range for amides (Tables 4.16 and 4.14).

TABLE 4.16.   Nitrogen Resonance of some Miscellaneous NR_3 Groups

Compound	Solvent	^{14}N Chemical shift (p.p.m.)		^{14}N Resonance half-height width (Hz)
		Screening-constant scale referred to internal $MeNO_2$ or $NO_3^{\ominus}$	Frequency scale[e] referred to $Me_4N^{\oplus}X^{\ominus}$	
$MeNHNO_2$	$MeOCH_2CH_2OMe$	$+208 \pm 8$[a] (NHMe)	$+124$[a]	540
$(MeNNO)^{\ominus}Na^{\oplus}$	H_2O	$+145 \pm 15$[a] (NMe)	$+187$[a]	1000
Me_2NNO	neat	$+141 + 5$[b] (NMe$_2$)	$+193$[b]	
		$+140 + 10$[c] (NMe$_2$)	$+194$[c]	
$(Me_2N)_2C{=}NH$	neat	$+332 \pm 2$[b] (NMe$_2$)	$+2$	700 ± 30
		$+327 \pm 3$[d] (NMe$_2$)	$+5$	360 ± 30
		$+209 \pm 3$[d] (NH)	$+123$	
CH$_2$—CO NH CH$_2$—CO	H_2O	$+199 \pm 4$[b]	$+135$	
α-pyridone (2-hydroxypyridine)	acetone	$+209 \pm 2$[d]	$+123$	90
γ-pyridone (4-hydroxypyridine)	acetone	$+201 \pm 2$[d]	$+131$	390

a Ref. 86, referred to external $MeNO_2$, ^{14}N direct measurements.
b Ref. 24, direct ^{14}N measurements, referred to external $NO_3^{\ominus}$ in NH_4NO_3.
c Ref. 91, direct ^{14}N data referred to external $NO_3^{\ominus}$.
d Ref. 124, direct ^{14}N data, referred to internal $MeNO_2$.
e Recalculated to $Me_4N^{\oplus}$ shift, 332 p.p.m. upfield from $MeNO_2$ and 334 p.p.m. upfield from $NO_3^{\ominus}$, Tables 4.5 and 4.9.

4.4.8   Isocyanates, Isothiocyanates and Azides

This group of compounds of general formula

$$R\diagdown \underset{\cdot\cdot}{N}=X=Y$$

where $X = C$ or $N^{\oplus}$ and $Y = O$, S or $N^{\ominus}$, is characterized by nitrogen shifts of +360 to +270 p.p.m. on the screening-constant scale (−25 to +65 on the frequency scale), and the spectral range may be divided further according to the molecular structure:

	Screening-constant scale $(MeNO_2/NO_3^{\ominus})$ (p.p.m.)	Frequency scale $(Me_4N^{\oplus})$ (p.p.m.)
isocyanates, R—NCO	+360 to +330	−25 to +5
isothiocyanates, R—NCS	+290 to +260	+45 to +75
azides, R—NNN (RN group)	+320 to +300	+15 to +35

The corresponding sub-ranges are rather narrow so that it is easy to distinguish spectrally between these groups of compounds. In all of them, there is a considerable downfield shift of the nitrogen resonance when R is changed from Me to Et, like in amines, ammonium ions, amides, isonitriles and nitro compounds (see the corresponding sections). An attempt has been made [60] to explain this shift in terms of changes in the π-bond orders and π-charge densities at the nitrogen atom which may result from changes in the electronegativity of group R. The low-field shift for aryl derivatives relative to alkyl derivatives in the group of R—N=X=Y compounds (Table 4.17) may be simply explained, as for amides, by an increasing π-bond order at the nitrogen atom due to an extension of the π-electron system over the phenyl group in Ph—N=X=Y. In all cases, the nitrogen resonance of the corresponding $(N=X=Y)^{\ominus}$ ion is shifted downfield from that of R—N=X=Y. Thus, we may consider such ions derived from R—N=X=Y compounds, where R is a lone electron pair, the strongest 'electron-donor', and then the downfield direction or the nitrogen shifts is parallel to that of decreasing electronegativity:

R = Me → Et → lone electron pair

→ downfield (high-frequency) shift

Alkyl azides give three nitrogen resonance signals corresponding to the three non-equivalent nitrogen atoms [15, 60]. An assignment of the shifts [60] may be made from a

$$R–N=N^{\oplus}=N^{\ominus}$$

screening-constant scale (p.p.m.)	*ca.* +300, *ca.* +130, *ca.* +170
frequency scale (p.p.m.)	*ca.* +35, *ca.* +105, *ca.* +65

comparison with the shifts observed for the azide ion, $N_3^{\ominus}$ and the changes in the position of the highest-field (lowest-frequency) resonance upon changing group R.

Isocyanates and isothiocyanates may be easily distinguished in nitrogen NMR spectra from the isomeric structures of cyanates and thiocyanates (Section 4.4.9):

	Screening-constant scale (p.p.m.)	Frequency scale (p.p.m.)
isothiocyanates (R–N=C=S)	*ca.* +270	*ca.* +65
thiocyanates (R–S–C≡N)	*ca.* +100	*ca.* +235
isocyanates (R–N=C=O)	*ca.* +350	*ca.* −15
cyanates (R–O–C≡N)	*ca.* +200	*ca.* +135
fulminates (R–C≡N$^{\oplus}$–O$^{\ominus}$)	*ca.* +170	*ca.* +165

The fulminate structure (R–CNO), isomeric with those of cyanates and isocyanates, is also spectroscopically distinct from the other two.

For a guide to nitrogen chemical shifts in isocyanates, isothiocyanates and azides, see refs. 3, 4, 9, 11, 15, 23, 24, 25, 35, 60, 82, 90, 96, 98, 105, and 124.

4.4.9   Nitriles, Isonitriles, Cyanates, Thiocyanates and Fulminates

In this group of compounds, the nitrogen atom constitutes a part of a linear group of atoms which may be represented by classical formulae

$$X–C≡N: \text{or}   X'–N^{\oplus}≡C–Y$$

where X = alkyl or aryl in nitriles; X = –O– in cyanates; X = –S– in thiocyanates; X′ = –O$^{\ominus}$ in fulminates (nitrile oxides) while Y = alkyl or aryl; X′ = alkyl or aryl, Y = lone

TABLE 4.17. Nitrogen Chemical Shifts of some Isocyanates, Isothiocyanates and Azides

Compound	Solvent		Nitrogen chemical shifts (p.p.m.)		^{14}N Resonance half-height width (Hz)
			Screening-constant scale referred to internal MeNO$_2$ or NO$_3^{\ominus}$	Frequency scaled[d] referred to Me$_4$N$^{\oplus}$X$^{\ominus}$	
MeNCO	neat		+363 ± 0.5[a]	−30.5[a]	50[a]
			+361 ± 1[b]	−26.5[b]	35[b]
	benzene		+361 ± 1[b]	−26.5[b]	35[b]
EtNCO	neat		+346 ± 1[a]	−13.5[a]	72[a]
			+343 ± 1[b]	−8.5[b]	37[b]
n-PrNCO	neat		+346 ± 1[b]	−11.5[b]	70[b]
PhNCO	neat		+330 ± 3[b]	+4.5[b]	50[b]
(NCO)$^{\ominus}$K$^{\oplus}$	H$_2$O		+300 ± 0.5[a]	+34.5[a]	13[a]
HNCO	ether		+351 ± 2[b]	−16.5[b]	20[b]
	cyclohexane		+353 ± 2[b]	−18.5[b]	35[b]
MeNCS	neat		+291 ± 1[a]	+41.5[a]	40[a]
			+285[f]	+49.5[f]	40[f]
			+287[h]	+47.5[h]	
EtNCS	neat		+273 ± 1[a]	+59.5[a]	55[a]
PhNCS	neat		+266 ± 2[c]	+66.5[c]	220[c]
			+267[g]	+67.5[g]	240[g]
			+265[f]	+69.5[f]	110[f]
HNCS	benzene		+265[f]	+69.5[f]	25[f]
(NCS)$^{\ominus}$K$^{\oplus}$	H$_2$O		+168 ± 1[a]	+166.5[a]	147[a]
(NCS)$^{\ominus}$NH$_4^{\oplus}$	5M in 2M HCl		+168 ± 0.5[e]	+166.5[e]	
MeNNN	neat	(MeN)	+320 ± 1[a]	+12.5[a]	101[a]
		(central)	+128 ± 0.5[a]	+204.5[a]	17[a]
		(terminal)	+170.5 ± 0.5[a]	+162.0[a]	19[a]
EtNNN	neat	(EtN)	+305 ± 1[a]	+27.5[a]	122[a]
		(central)	+129.5 ± 0.5[a]	+203.0[a]	22[a]
		(terminal)	+167.5 ± 0.5[l]	+165.0[a]	28[a]
PhNNN	benzene	(PhN)	+286[i]	+48.5[i]	1200[i]
		(central)	+135[i]	+199.5[i]	180[i]
		(terminal)	+194[i]	+140.5[i]	55[i]
HNNN	ether	(HN)	+300[i]	+34.5[l]	100[i]
		(central)	+129[i]	+205.5[i]	24[i]
		(terminal)	+165[i]	+169.5[i]	100[i]
(NNN)$^{\ominus}$Na$^{\oplus}$	H$_2$O	(central)	+128 ± 0.5[a]	+206.5[a]	20[a]
		(terminal)	+277 ± 1[a]	+57.5[a]	60[a]
PhCONNN	benzene	(R−N)	+322[i]	+12.5[i]	430[i]
		(central)	+140[i]	+194.5[i]	300[i]
		(terminal)	+237[i]	+97.5[i]	73[i]
p-MeC$_6$H$_4$SO$_2$NNN	neat	(R−N)	+304[i]	+30.5[i]	410[i]
		(central)	+152[i]	+182.5[i]	250[i]
		(terminal)	+209[i]	+125.5[i]	120[i]
p-NO$_2$C$_6$H$_4$NNN	benzene	(R−N)	+310[i]	+24.5[i]	270[i]
		(central)	+139[i]	+195.5[i]	75[i]
		(terminal)	+211[i]	+123.5[i]	45[i]

a Ref. 60, direct ^{14}N measurements, referred to MeNO$_2$ (KNO$_3$ for aqueous solutions).
b Ref. 90, direct ^{14}N measurements, referred to external (NCO)$^{\ominus}$, recalculated to NO$_3^{\ominus}$.
c Ref. 124, direct ^{14}N measurements, referred to internal MeNO$_2$.
d Recalculated by assuming Me$_4$N$^{\oplus}$X$^{\ominus}$, 334.5 p.p.m. to higher field from NO$_3^{\ominus}$ and 332.5 p.p.m. from internal MeNO$_2$.
e Ref. 105, natural abudance ^{15}N measurements, referred to external MeNO$_2$.
f Ref. 98, direct ^{14}N measurements, referred to external NO$_3^{\ominus}$.
g Ref. 96, direct ^{14}N measurements, referred to external KNO$_3$.
h Ref. 56, direct ^{15}N measurements, recalculated from the MeNH$_2$ shift (378 p.p.m. upfield from MeNO$_2$, 43.0 p.p.m. from MeN$^{\oplus}$).
i Ref. 133, direct ^{14}N measurements, referred to external NO$_3^{\ominus}$ by the sample replacement method.

electron pair in isonitriles. The corresponding nitrogen resonance signals lie at lower magnetic fields (higher frequencies) than in isocyanates, isothiocyanates and the R—N moiety in azides, and the shifts are within +220 to +110 p.p.m on the screening-constant scale or +115 to +225 p.p.m. on the frequency scale, as shown in Table 4.18.

The spectral range for nitriles, R—CN, is quite narrow

$$R-C\equiv N \qquad \text{+140 to +120 p.p.m.} \qquad \text{+195 to +215 p.p.m.}$$
$$\text{(screening-constant scale)} \qquad \text{(frequency scale)}$$

TABLE 4.18. Nitrogen Chemical Shifts of some Nitriles, Isonitriles, Cyanates, Thiocyanates and Fulminates

Compound	Solvent	Screening-constant scale referred to $MeNO_2$ or $NO_3^{\ominus}$	Frequency scale[o] referred to $Me_4N^{\oplus}X^{\ominus}$	^{14}N Resonance half-height width (Hz)
Me_3CCN	neat	$+139 \pm 1$[a]	$+193.5$[a]	190[a]
EtCN	neat	$+137 \pm 1$[a]	$+195.5$[a]	122[a]
MeCN	neat	$+137 \pm 1$[a]	$+195.5$[a]	88[a]
		$+137.2$[c]	$+195.3$[c]	—
		$+136.4$[n]	$+196.1$[n]	—
$ClCH_2CN$	neat	$+126 \pm 1$[a]	$+206.5$[a]	175[a]
Cl_3CCN	CCl_4	$+122 \pm 1$[a]	$+210.5$[a]	220[a]
BrCN	acetone	$+124 \pm 1$[a]	$+208.5$[a]	165[a]
$CH_2(CN)_2$	acetone	$+128 \pm 2$[a]	$+204.5$[a]	280[a]
$NC(CH_2)_2CN$	acetone	$+134 \pm 2$[a]	$+198.5$[a]	370[a]
$NC(CH_2)_3CN$	acetone	$+136 \pm 2$[a]	$+196.5$[a]	400[a]
$EtCH(CN)_2$	acetone	$+132 \pm 2$[a]	$+200.5$[a]	420[a]
$HOCH_2CH_2CN$	acetone	$+136 \pm 2$[a]	$+196.5$[a]	370[a]
$MeOOCCH_2CN$	acetone	$+130 \pm 2$[a]	$+202.5$[a]	460[a]
$CH_2=CHCH_2CN$	neat	$+133 \pm 1$[a]	$+199.5$[a]	166[a]
$CH_2=CHCN$	neat	$+124 \pm 1$[a]	$+208.5$[a]	110[a]
$EtOCH_2CN$	neat	$+126 \pm 2$[l]	$+206.5$[l]	255[l]
$HOCH_2CN$	neat	$+135 \pm 1$[l]	$+197.5$[l]	275[l]
$Me_2C(OH)CN$	neat	$+130 \pm 4$[l]	$+202.5$[l]	840[l]
$PhCH_2CN$	acetone	$+133 \pm 3$[a]	$+199.5$[a]	300[a]
PhCN	acetone	$+124 \pm 2$[a]	$+208.5$[a]	290[a]
$(CN)^{\ominus}K^{\oplus}$	H_2O	$+99 \pm 1$[a]	$+235.5$[a]	52[a]
		$+101$[b]	$+233.5$[b]	—
		$+97$[m]	$+239.5$[m]	—
Me_3CNC	neat	$+181.2$[f]	$+153.3$[f]	ca. 1[f]
i-PrNC	neat	$+189.7$[f]	$+144.8$[f]	ca. 1[f]
$C_6H_{11}NC$	neat	$+194$[e]	$+138.5$[e]	—
n-PrNC	neat	$+202.3$[f]	$+132.2$[f]	0.6[f]

TABLE 4.18—*continued*.

Compound	Solvent	Nitrogen chemical shifts (p.p.m.)		
		Screening-constant scale referred to $MeNO_2$ or NO_3	Frequency scale[o] referred to $Me_4N^{\oplus}X^{\ominus}$	^{14}N Resonance half-height width (Hz)
Me_3CCH_2NC	neat	+207.6[f]	+126.9[f]	0.6[f]
$PhCH_2NC$	neat	+211[e]	+121.5[e]	—
EtNC	neat	+203 ± 0.5[a]	+129.5[a]	—
		+205.1[d]	+127.4[d]	0.31[d]
MeNC	neat	+218 ± 0.5[a]	+114.5[a]	—
		+219.6[d]	+112.9[d]	0.26[d]
PhNC	neat	+200[e]	+132.5[e]	—
p-ClC_6H_4NC	EtOH/MeOH	+201 ± 5[g]	+131.5[g]	—
p-$C_6H_4(NC)_2$	EtOH/MeOH	+200 ± 5[g]	+132.5[g]	—
p-$MeOC_6H_4NC$	benzene	+201[e]	+131.5[e]	—
EtOCN	Et_2O	+222 ± 1[h]	+110.5[h]	45[h]
PhOCN	neat	+208[i]	+125.5[i]	—
EtSCN	neat	+99 ± 3[g]	+233.5[g]	—
		+103 ± 2[j]	+229.5[j]	140[j]
Me—⟨⟩(Me, Me)—CNO	CH_2Cl_2	+165[i]	+169.5[i]	—
O_2N—⟨⟩—CNO	acetone	+179 ± 1[k]	+153.5[k]	30[k]
	benzene	+170 ± 3[k]	+162.5[k]	50[k]
$(CNO)^{\ominus}Na^{\oplus}$	H_2O	+176[i]	+158.5[i]	—

a Ref. 44; direct ^{14}N measurements, referred to $MeNO_2$ or KNO_3 (aqueous solutions).
b Ref. 23; direct ^{15}N measurements, recalculated from $MeNH_2$ shift.
c Ref. 105; direct natural abundance ^{15}N measurements, ref. to external $MeNO_2$.
d Ref. 64; direct ^{14}N measurements, high resolution spectra, ref. to external $MeNO_2$.
e Ref. 83; direct ^{14}N measurements, ref. to external aqueous $NaNO_3$.
f Ref. 77; direct ^{14}N high resolution spectra, ref. to external $NO_3^{\ominus}$.
g Ref. 24; direct ^{14}N measurements, ref. to external 4.5M NH_4NO_3 in 3M HCl.
h Ref. 90; direct ^{14}N measurements, originally referred to external $(NCO)^{\ominus}$, recalculated to $NO_3^{\ominus}$.
i Ref. 82; direct ^{14}N measurements, ref. to external satd. aqueous $NaNO_3$.
j Ref. 60; direct ^{14}N measurements, ref. to internal $MeNO_2$.
k Ref. 122; direct ^{14}N measurements, ref. to internal $MeNO_2$.
l Ref. 125; direct ^{14}N measurements, ref. to internal $MeNO_2$.
m Ref. 35; direct ^{14}N measurements, ref. to external satd. aqueous NH_4NO_3.
n Ref. 103; measurements of $MeC^{15}N$ referred to gaseous $^{15}NH_3$, recalculated to $NO_3^{\ominus}$ according to Table 4.6 and 4.7.
o Recalculated relative to $NMe_4^{\oplus}$ shift, 334.5 p.p.m. upfield from $NO_3^{\ominus}$ and 332.5 p.p.m. upfield from $MeNO_2$.

and there seems to be [44] a downfield shift of the resonance with increasing electronegativity of the group R, which is opposite to those observed in the NR_3, $NR_4^\oplus$, $RCONR_2$, and $R-N=X=Y$ structures. Unsaturated centres (phenyl or vinyl groups) also seem to shift the resonance to lower fields (higher frequencies) and this may be due to conjugative effects alone or to the increased electronegativity of the sp^2-hybridized carbon atoms as compared with sp^3. The cyano group bonded to elements other than carbon, e.g. N, O, and S, gives resonance signals in the vicinity of the nitrile range but far enough away for an easy spectral distinction:

	Screening-constant scale (p.p.m.)	Frequency scale (p.p.m.)
cyanamides, CN group (R_2N-CN)	+200 to +180	+135 to +155
cyanates ($R-O-CN$)	+220 to +200	+115 to +135
thiocyanates ($R-S-CN$)	*ca.* +100	*ca.* +235

The nature of these shifts may be complicated by the large differences in electronegativity between the atoms and possible conjugative effects of the lone electron pairs at N, O, and S.

Isonitriles, R—NC, give nitrogen resonance signals within this spectral range which clearly distinguishes them from isomeric nitriles. The ^{14}N signals of isonitriles are very narrow

$$R-N^\oplus \equiv C^\ominus \qquad +220 \text{ to } +180 \text{ p.p.m.} \qquad +115 \text{ to } +155 \text{ p.p.m.}$$
(screening-constant scale)   (frequency scale)

and display a true high-resolution structure due to spin–spin coupling [77]. Early double-resonance measurements [38] of the ^{14}N shifts in MeNC, EtNC and *t*-BuNC gave values which agree very well with the relative shifts observed later in direct measurements from three independent sources [44, 64, 77] but differ by about 20 p.p.m. upfield from the latter, if recalculated relative to the frequency of such standards as $MeNO_2$ or $Me_4N^\oplus$, according to Tables 4.3 and 4.5. It seems that the early data [38] may contain an error in the sign of the modulation frequency used the recalculation of the observed frequencies to those corresponding to the ^1H frequency of TMS at exactly 100 MHz. The nitrogen resonance signal of isonitriles moves [44] to lower fields (higher frequencies) with increasing electronegativity of the group R in R—NC and upon replacing an alkyl with an aryl group. The shift is very regular, about 15 p.p.m. downfield per each H atom in MeNC replaced with an alkyl group. This resembles

R—NC   R = Me → Et → i-Pr → t-Bu

R = alkyl → aryl $\xrightarrow{\text{downfield shift}}$

the shifts observed for amines, ammonium ions, amides, isocyanates and related compounds, and nitro groups, but is opposite to that in nitriles. The opposing trends in the nitrogen chemical shifts in nitriles and isonitriles may be simple [44] explained in terms of changes in π-bond orders and charge densities within the average excitation energy approximation (Chapter 1).

Fulminates, R—CN$^{\oplus}$—O$^{\ominus}$, which may be considered as oxides of nitriles, show a considerable high-field (low-frequency) nitrogen chemical shift relative to nitriles which is

R—C≡N	R—C≡N$^{\oplus}$—O$^{\ominus}$	
+120 to +140 p.p.m.	+160 to +180 p.p.m.	(screening-constant scale)
+215 to +195 p.p.m.	+175 to +155 p.p.m.	(frequency scale)

similar to that in the N-oxides of pyridine-type hetero-aromatic compounds (Section 4.4.11). Such upfield shifts are also observed for nitrones, azoxy compounds and nitro compounds which may be formally considered as N-oxides of imines, azo- and nitro compounds, respectively.

A similar upfield shift was found upon protonation or alkylation of a nitrile to the corresponding nitrilium ion (Table 4.19).

(CN)$^{\ominus}$ $\xrightarrow{R^{\oplus}}$ R—C≡N: $\xrightarrow{R'^{\oplus}}$ R—C≡N$^{\oplus}$—R'

cyanide ion   nitrile   nitrilium ion

+99 p.p.m. $ca.$ +130 p.p.m. $ca.$ +240 p.p.m. (screening-constant scale)

+236 $ca.$ +205 $ca.$ + 95 (frequency scale)

The shift resembles, in both direction and magnitude, that observed upon protonation or alkylation of pyridine-type heterocycles (Section 4.4.11). The nitrogen chemical shifts of nitrilium ions differ only slightly from the shifts in isonitriles where the electronic structure near the nitrogen atom is

TABLE 4.19. Nitrogen Chemical Shifts of some Nitrilium Ions
(Data from ref. 62)

Ion	Nitrogen chemical shift[a] (in p.p.m.)	
	Screening-constant scale referred to internal MeNO$_2$ or NO$_3^{\ominus}$	Frequency scale referred to Me$_4$N$^{\oplus}$X$^{\ominus}$
(HC ^{15}NH)$^{\oplus}$	+235	+100
(MeC ^{15}NH)$^{\oplus}$	+239	+96
(MeC ^{14}NMe)$^{\oplus}$	+248	+87

[a] Double resonance INDOR measurements from proton spectra, originally referred to aqueous NH$_4^{\oplus}$ (external), recalculated by setting NH$_4^{\oplus}$ shift at 353 p.p.m. upfield from NO$_3^{\ominus}$ and 18 p.p.m. from NMe$_4^{\oplus}$; measurements in FSO$_3$H–SbF$_5$–SO$_2$ solutions at −60° or in SO$_2$ solutions.

similar. The increased magnetic screening of nitrogen nuclei in the CN$^{\ominus}$, R—CN, R—CN$^{\oplus}$—R′ series seems to indicate a dominant role for changes in the *paramagnetic* term due to multiple-bond effects rather than to the *diamagnetic* shielding due to the total electron density.

For a guide to nitrogen chemical shifts in nitriles, isonitriles, cyanates, thiocyanates, and fulminates, see refs. 3, 4, 11, 22, 23, 24, 25, 31, 32, 35, 38, 44, 48, 60, 62, 63, 64, 77, 82, 83, 90, 103, 105, 117, 122 and 125.

4.4.10 Azoles

We shall consider here various hetero-aromatic five-membered ring structures containing from one to four nitrogen atoms (azoles, diazoles, triazoles, tetrazoles) and eventually oxygen and sulphur atoms (oxazoles and thiazoles). Two types of nitrogen atoms in such compounds may be found, and there are two corresponding chemical shift ranges in nitrogen NMR. The first type, found always in azoles but not in oxazoles and

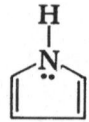

Diagram 4.1

thiazoles, involves direct bonding to three other atoms in a plane or very closely to a plane. This will be called here the pyrrole-type nitrogen atom, and its conventional representation is that in the preceding formula of pyrrole. In the approximation that divides the electrons in a molecule into the π- and σ-bonding systems, the pyrrole-type nitrogen atom supplies two electrons to the former. The other type of structure involves a nitrogen atom bonded to only two neighbouring atoms and having a lone electron pair which does not constitute a part of the π-bond system. This will be called the pyridine-type nitrogen atom and may be conventionally represented as in the following formula of oxazole. It supplies only one electron to the π-bonding system of

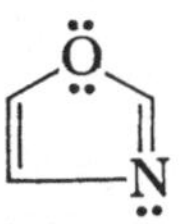

Diagram 4.2

the aromatic ring. There is a large difference between the nitrogen chemical shifts [126] of the two types of atoms (Table 4.20). If both structures are present in a molecule then the pyrrole-type nitrogen atom gives a resonance signal shifted by about 100 p.p.m. to higher fields (lower frequencies) from the resonance of the other type.

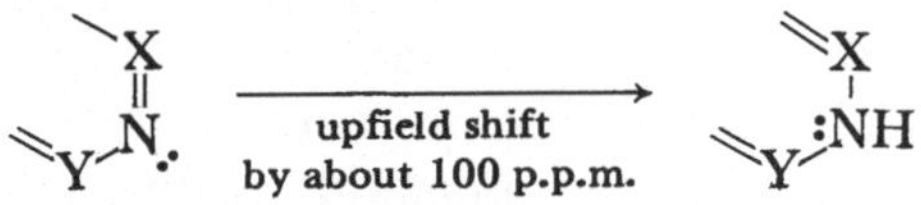

Diagram 4.3

The entire spectral range of the pyrrole-type nitrogen atoms is from +100 to +260 p.p.m. on the screening-constant scale (+235 to +175 p.p.m. on the frequency scale) and the nitrogen shifts are very useful as a means of identification of azole ring systems. Moreover, the shifts are very different

	Screening-constant scale (p.p.m.)	Frequency scale (p.p.m.)
azoles	+230 to +260	+105 to +75
diazoles	+160 to +230	+175 to +105
triazoles	+120 to +220	+215 to +115
tetrazoles	+100 to +150	+235 to +185

within isomeric pairs of azoles, e.g. (Table 4.20) and this may make nitrogen chemical shifts a very convenient molecular property for an examination of isomer mixtures or tautomeric equilibria.

	Screening-constant scale (p.p.m.)	Frequency scale (p.p.m.)
1-methyltetrazole	+150	+182
2-methyltetrazole	+103	+230
1-methyl-1,2,4-triazole	+170	+162
2-methyl-1,2,4-triazole	+220	+112
1-methyl-1,2,3-triazole	+143	+190
2-methyl-1,2,3-triazole	+130	+202
1-methylbenzotriazole	+148	+184
2-methylbenzotriazole	+118	+214

If the conjugated system of an azole is expanded over additional benzene rings, as in benzo- and dibenzo-derivatives, the resonance signals of the pyrrole-type nitrogen atoms are moved to higher fields (lower frequencies), for example, in

pyrrole	indole	carbazole	
ca. +230 p.p.m.	ca. +250 p.p.m.	ca. +260 p.p.m.	(screening-constant scale)
+105	+85	+75	(frequency scale)

and in the benzo-derivatives of diazoles and triazoles (Table 4.20). The same effect may be observed in azines (Section 4.4.11). An exception here is the case where the structure of the resulting benzo-derivative is quinoid, as in Diagram 4.4. There is rather little effect on the nitrogen

2-methylbenzopyrazole   2-methylbenzotriazole

Diagram 4.4

chemical shift due to changing NH to NMe at the pyrrole-type nitrogen atom, certainly not larger than solvent effects. This may be important for an examination of tautomeric equilibria by employing the corresponding N-methyl derivatives as model compounds for reference shifts.

The pyridine-type nitrogen shifts in azoles fall into a range of +140 to 0 p.p.m. on the screening-constant scale (+195 to +325 p.p.m., frequency scale), and this includes also oxazoles and thiazoles. The existing data from ^{14}N spectroscopy [126] are less accurate here because the corresponding ^{14}N resonance signals are broader than for the pyrrole type. There is an increased magnetic shielding for structures where another nitrogen atom or an oxygen atom occupies an adjacent position.

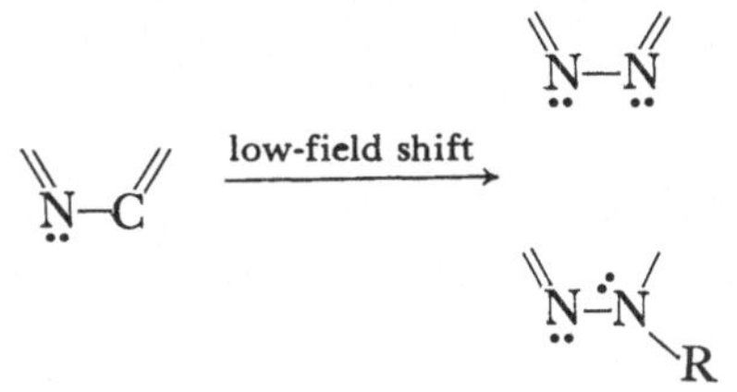

Diagram 4.5

The pyridine-type nitrogen shifts are also very different and characteristic of individual isomeric azoles (Table 4.20) they make possible an unambiguous spectroscopic differentiation between pyrazoles and imidazoles, oxazoles and isoxazoles, thiazoles and isothiazoles, etc., respectively.

Azoles which contain both the pyrrole- and pyridine-type nitrogen atoms and the NH group may exhibit tautomerism. Such tautomeric equilibria may be investigated by means of nitrogen chemical shifts. The question is simple if the spectrum is a resolvable superposition of the spectra of individual tautomers, but a fast proton migration may lead to an

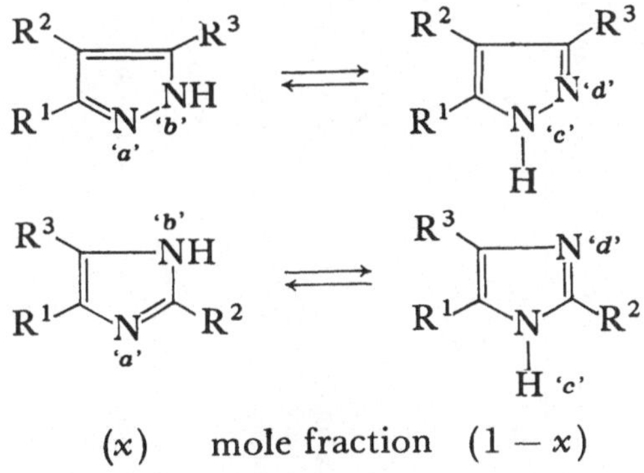

(x) mole fraction $(1 - x)$

Diagram 4.6

TABLE 4.20.   Nitrogen Chemical Shifts of some Azoles (Ref. 126 if not stated otherwise)

Compound	Solvent (concentration w/w if not stated otherwise)	Nitrogen chemical shift (p.p.m.)		^{14}N signal half-height width (Hz)	SCF–PPP densities	π-charge at N
		Screening-constant scale referred to internal MeNO$_2$ or NO$_3^{\ominus}$	Frequency scale^{e} referred to Me$_4$N$^{\oplus}$X$^{\ominus}$			
pyrrole	neat	+235 ± 2	+ 98	172	1.5258	
	CCl$_4$ (0.01M)	+228a	+104a	—		
	dioxane (1:1)	+230b	+102b	—		
	MeOH (1:1)	+233a	+100a	—		
2-acetylpyrrole	CDCl$_3$ (0.01M)	+232 ± 2	+100	50		
2,5-diacetylpyrrole	CDCl$_3$ (0.01M)	+238 ± 3	+100	40		
N-methylpyrrole	neat	+228a	+104a	—		
		+232a	+100a	—		
	neat	+231 ± 2	+102	104	1.5702	
	MeOH (1:1)	+227b	+104b	—		
N-2,5-trimethylpyrrole	neat	+233 ± 1	+100	90		
		+230b	+102b	—		
pyrazole	dioxane (1:1)	+135 ± 3	+198	730	1.4824 (NH)	
	MeOH (1:1)	+133 ± 3	+200	670	1.3270 (N)	
N-methylpyrazole	CCl$_4$ (1:2)	+68 ± 2 (N)	+264	325	1.3306 (N)	
		+178 ± 2 (NMe)	+154	142	1.5126 (NMe)	
	MeOH+CCl$_4$ (1:1:2)	+78 ± 3 (N)	+254	385		
		+178 ± 2 (NMe)	+154	185		
imidazole	dioxane (satd.)	+171 ± 5	+162	1200	1.3010 (N)	
	MeOH (1:1)	+171 ± 3	+162	600	1.5166 (NH)	

Compound	Solvent					
N-methylimidazole	neat	+123 ± 2 (N) +221 ± 1 (NMe)	325 150	+210 +112	1.2974 (N) 1.5546 (NMe)	
	CCl₄ (1:5)	+116 ± 3 (N) +218 ± 1 (NMe)	300 125	+216 +114		
1,2,4-triazole	dioxane (satd.)	+134 ± 2	540	+198	1.4764 (1-NH) 1.3266 (2-N) 1.3068 (4-N)	1.2290 (1-N) 1.2290 (2-N) 1.5152 (4-NH)
	MeOH (1:1)	+136 ± 3	940	+196		
1-methyl-1,2,4-triazole	neat	+126 ± 3 (N, N) +170 ± 2 (NMe)	450 160	+206 +162	1.5084 (1-Me) 1.3386 (2-N) 1.3030 (4-N)	
	MeOH (1:1)	+130 ± 5 (N, N) +170 ± 1 (NMe)	550 170	+202 +162		
4-methyl-1,2,4-triazole	MeOH (satd.)	+80 ± 4 (N, N) +220 ± (NMe)	600 190	+252 +112	1.2186 (1-N) 1.2186 (2-N) 1.5490 (4-NMe)	
1,2,3-triazole	neat	+60 ± 8 (N, N) +132 ± 4 (NH)	300 150	+272 +200	1.4606 (1-NH) 1.2304 (2-N) 1.2138 (3-N)	1.2624 (1-N) 1.4570 (2-NH) 1.2624 (3-N)
	MeOH (1:1)	+60 ± 8 (N, N) +128 ± 6 (NH)	250 200	+272 +204		
1-methyl-1,2,3-triazole	neat	+22 ± 1 (N, N) +143 ± 1 (NMe)	440 120	+310 +190	1.4874 (1-NMe) 1.2306 (2-N) 1.2060 (3-N)	
	MeOH (1:1)	+28 ± 1 (N, N) +144 ± 1 (NMe)	480 125	+304 +188		

TABLE 4.20—continued

Compound	Solvent (concentration w/w if not stated otherwise)	Nitrogen chemical shift (p.p.m.)		^{14}N signal half-height width (Hz)	SCF–PPP densities	π charge at N
		Screening-constant scale referred to internal MeNO$_2$ or NO$_3^{\ominus}$	Frequency scale[c] referred to Me$_4$N$^{\oplus}$ X$^{\ominus}$			
2-methyl-1,2,3-triazole	neat	+51 ± 1 (N, N) +130 ± 1 (NMe)	+282 +202	235 102	1.2916 (1-N) 1.4488 (2-NMe) 1.2916 (3-N)	
	MeOH	+53 ± 2 (N, N) +132 ± 2 (NMe)	+280 +200	250 105		
tetrazole	acetone (satd.)	+15 ± 3 +106 ± 2	+318 +226	445 300	1.4586 (1-NH) 1.2204 (2-N) 1.1228 (3-N) 1.2172 (4-N)	1.2872 (1-N) 1.4034 (2-NH) 1.1892 (3-N) 1.2188 (4-N)
	MeOH (satd.)	+25 ± 5 +106 ± 4	+308 +226	480 350		
1-methyltetrazole	CHCl$_3$ (1:1)	+17 ± 5 (N, N, N) +150 ± 2 (NMe)	+316 +182	600 160	1.4860 (1-NMe) 1.2230 (2-N) 1.1138 (3-N) 1.2118 (4-N)	
2-methyltetrazole	CCl$_4$ (1:1)	+5 ± 6 (N–N) +44 ± 3 (N–NMe) +101 (NMe) +10 ± 6 (N–N)	+328 +288 +232 +322	300 200 115 300	1.3002 (1-N) 1.4246 (2-NMe) 1.1920 (3-N) 1.2116 (4-N)	
	MeOH (2:1)	+55 ± 4 (N–NMe) +103 ± 2 (NMe)	+278 +230	200 105		

Compound	Condition				
oxazole	CCl$_4$ (1:1)	+124 ± 1	+208	106	1.2848
	CCl$_4$ + MeOH (1:1:1)	+125 ± 2	+208	170	
isoxazole	neat	-2 ± 1	+334	220	1.3058
	DMF (1:1)	-4 ± 2	+336	290	
	MeOH (1:1)	+6 ± 2	+326	335	
thiazole	neat	+56 ± 2	+276	150	1.2906
	MeOH	+68 ± 2	+264	240	
isothiazole	neat	+80 ± 1	+252	108	1.3342
	MeOH (1:1)	+85 ± 2	+248	135	
indole	dioxane (satd.)	+251 ± 5	+82	730	1.5568
	MeOH (satd.)	+246 ± 5	+86	380	
	CCl$_4$ (0.01M)	+254.0[a]	+78[a]	–	
N-methylindole	neat	+248 ± 5	+84	900	1.6024
2-methylindole	benzene (0.01M)	+246 ± 1[a]	+86[a]	–	
3-methylindole	benzene (0.01M)	+256 ± 1[a]	+76[a]	–	
benzopyrazole	acetone (satd.)	+75 ± 8	+258	500	1.5138 (1-NH) 1.3660 (1-N)
		+197 ± 5	+136	360	1.2932 (2-N) 1.4276 (2-NH)

TABLE 4.20—continued

Compound	Solvent (concentration w/w if not stated otherwise)	Nitrogen chemical shift (p.p.m.)		^{14}N signal half-height width (Hz)	SCF–PPP densities	π-charge at N
		Screening-constant scale referred to internal MeNO$_2$ or NO$_3^{\ominus}$	Frequency scale[e] referred to Me$_4$N$^{\oplus}$X$^{\ominus}$			
1-methylbenzopyrazole	acetone (satd.)	+71 ± 8 (N) +197 ± 5 (NMe)	+262 +136	500 350	1.5464 (NMe) 1.2996 (N)	
2-methylbenzopyrazole	acetone (satd.)	+85 ± 5 (N) +162 ± 2 (NMe)	+248 +170	700 350	1.3714 (N) 1.4544 (NMe)	
benzimidazole	acetone (satd.) MeOH (satd.)	+185 ± 5 +192 ± 5	+148 +140	400 750	1.5434 (NH) 1.3238 (N)	
N-methylbenzimidazole	acetone (satd.)	+130 ± 8 (N) +228 ± 5 (NMe)	+202 +104	350 300	1.5848 (NMe) 1.3178 (N)	

compound		solvent					
benzotriazole		dioxane (satd.)	+81 ± 7	+252	1500	1.4836 (1-NH) 1.1870 (2-N) 1.2206 (3-N)	1.3072 (1-N) 1.3650 (2-NH) 1.3072 (3-N)
		MeOH	+89 ± 7	+244	1600		
1-methylbenzotriazole		acetone (satd.)	+40 ± 8 (N, N) +148 ± 5 (NMe)	+292 +184	600 300	1.5182 (NMe) 1.1830 (2-N) 1.2120 (3-N)	
2-methylbenzotriazole		neat	+50 ± 8 (N, N) +118 ± 2 (NMe)	+282 +214	500 250	1.3194 (1-N) 1.3788 (2-NMe) 1.3194 (3-N)	
benzoxazole		neat	+140 ± 3	+192	400	1.3002	
benzothiazole		neat	+60 ± 2	+272	700	1.3130	

TABLE 4.20—continued

Compound	Solvent (concentration w/w if not stated otherwise)	Nitrogen chemical shift (p.p.m.)		^{14}N signal half-height width (Hz)	SCF−PPP densities	π-charge at N
		Screening-constant scale referred to internal MeNO$_2$ or NO$_3^{\ominus}$	Frequency scale[e] referred to Me$_4$N$^{\oplus}$ X$^{\ominus}$			
2-methylbenzothiazole	neat	+63 ± 5	+270	750		
benzisothiazole	neat	+119 ± 2	+214	640	1.8708	
carbazole	acetone (satd.)	+260 ± 5	+72	1500	1.5998	
	acetone (0.01M)	+264 ± 1[a]	+68[a]	—		
N-methylcarbazole borazole	acetone (satd.)	+260 ± 4	+72	1000	1.6740	
(BF−NH)$_3$	CCl$_4$ (1:3)	+262 ± 1[c]	+72[c]	80[c]		
(BCl−NH)$_3$	toluene, benzene	+297 ± 1[d]	+36[d]	—		
(BBr−NH)$_3$	cyclohexane, benzene	+266 ± 1[d]	+66[d]	—		
	benzene	+258 ± 1[d]	+74[d]	—		

[a] Ref. 99, double resonance ^{14}N data referred to external NO$_3^{\ominus}$ in NH$_4$NO$_3$; ± 0.5 p.p.m if not stated otherwise.
[b] Ref. 24, direct ^{14}N measurements, referred to external 4.5M NH$_4$NO$_3$ in 3M HCl.
[c] Ref. 121, direct ^{14}N measurements relative to internal MeNO$_2$.
[d] Ref. 42, direct ^{14}N measurements referred to shift of unsubstituted borazole, recalculated to MeNO$_2$.
[e] Recalculated to Me$_4$N$^{\oplus}$ shift, 332.5 p.p.m. upfield from MeNO$_2$, Table 4.5.

averaging of the corresponding resonance signals. For diazoles the situation is rather straightforward since the averaged shift A is a weighted mean of 'a' and 'c', and the other shift B is a weighted mean of 'b' and 'd'.

$$A = xa + (1 - x)c \tag{4.2}$$

$$B = xb + (1 - x)d \tag{4.3}$$

$$x = \frac{(A + B) - (c + d)}{(a + b) - (c + d)} \tag{4.4}$$

If the substitution is symmetrical ($R^1 = R^3$) then $a = d$, $b = c$ and $A = B$ (only one averaged resonance signal is observed).

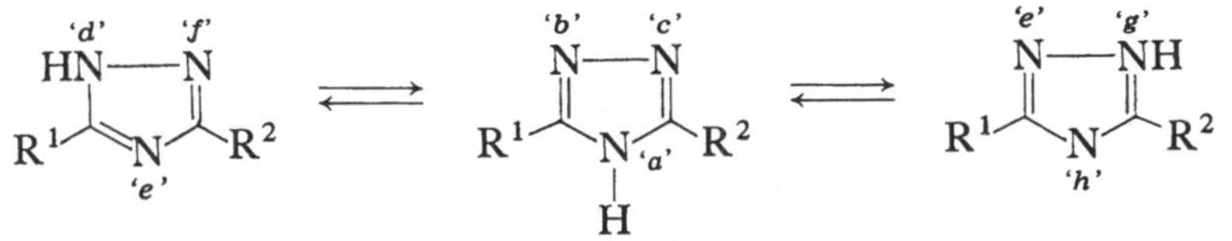

(x) (1 − x − y) (y) mole fraction

Diagram 4.7

For triazoles, in the fast proton exchange limit, the averaged shifts are:

$$A = xe + (1 - x - y)a + yh \tag{4.5}$$

$$B = xd + (1 - x - y)b + yi \tag{4.6}$$

$$C = xf + (1 - x - y)c + yg \tag{4.7}$$

If the substitution is symmetrical ($R^1 = R^2$), only two averaged signals are observed, since $x = y$, $b = c$, $d = g$, $f = i$, $e = h$, and $B = C$, so that

$$x = \frac{1}{2} \cdot \frac{(A + 2B) - (a + 2b)}{(d + f + e) - (a + 2b)} \tag{4.8}$$

For tetrazoles, if there is a fast proton migration, only two resonance signals should be observed in nitrogen NMR. The averaged shifts are:

$$A = x(a + d) + 0.5(1 - 2x)(f + g) \tag{4.9}$$

$$B = x(b + c) + 0.5(1 - 2x)(e + h) \tag{4.10}$$

$$x = \frac{(A + B) - 0.5(e + f + g + h)}{(a + b + c + d) - (e + f + g + h)} \tag{4.11}$$

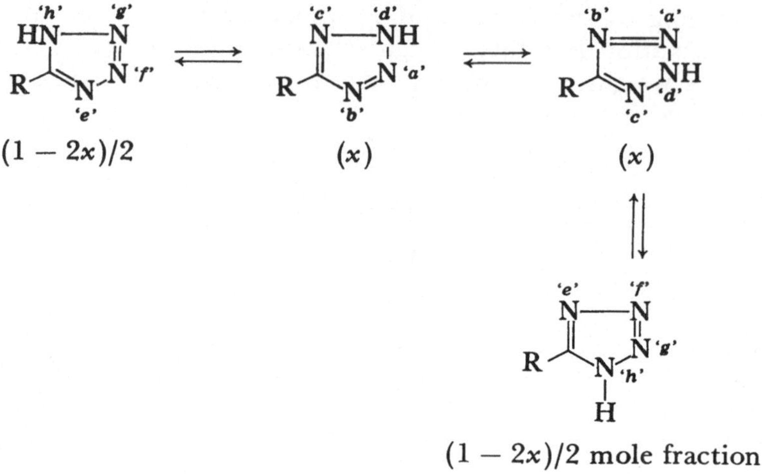

Diagram 4.8

An approximate consideration of the ^{14}N spectra [126] of unsubstituted 1,2,3-triazole, 1,2,4-triazole and tetrazole in polar solvents (MeOH, dioxane, acetone) indicate that the

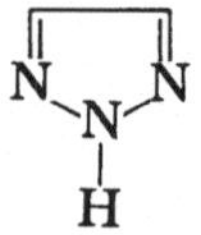

Diagram 4.9

symmetrical isomer of 1,2,3-triazole should prevail (70–100%) in the equilibrium, and that tetrazole exists mainly as indicated in Diagram 4.10, but no conclusive answer was obtained for 1,2,4-triazole.

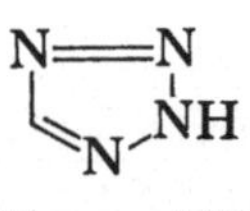

Diagram 4.10

The tautomerism in 2-substituted benzothiazoles may also be observed by means of nitrogen chemical shifts. Double-resonance ^{14}N data [45] indicate (Table 4.21) that if X = NH$_2$ or NHMe then the thiazole form prevails in solution (the observed shift is in the range characteristic of amines) but if X = OH or SH then the thiazoline form should dominate in the equilibrium, since the shift is just within the spectral range characteristic of amides. Solvent effects on the nitrogen chemical shifts of pyrrole and related compounds are discussed in Section 4.5.

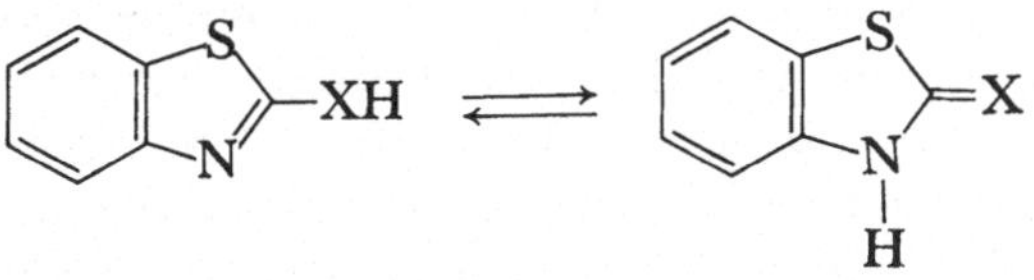

thiazole form   thiazoline form

Diagram 4.11

It has been shown [126] that the nitrogen chemical shifts of both types of nitrogen atoms in azoles, oxazoles, thiazoles, and their benzo-derivatives display a roughly linear correlation with SCF—MO calculated π-charge densities. The resonance is shifted to higher magnetic fields (lower frequencies) with increasing π-charge density rather than with increasing total

TABLE 4.21.   Nitrogen Chemical Shifts of some Benzothiazole Derivatives (Data from ref. 45)

| | | ^{14}N chemical shift (p.p.m.) | |
| | | Screening-constant[a] scale referred to internal MeNO$_2$ or NO$_3^{\ominus}$ | Frequency scale[b] referred to Me$_4$N$^{\oplus}$X$^{\ominus}$ |
Compound	Solvent		
benzothiazole (B.T.)	acetone	+60 ± 2	+274
4-hydroxy-B.T.	acetone	+66 ± 8	+268
6-nitro-B.T.	acetone	+67 ± 4	+268
2-amino-B.T.	acetone	+315 ± 2 (NH$_2$)	+20
2-methylamino-B.T.	DMSO	+315 ± 15 (NHMe)	+20
2-hydroxy-B.T.	CDCl$_3$	+243 ± 2	+92
2-mercapto-B.T.	acetone	+211 ± 3	+124

[a] Double resonance ^{14}N data, referred to external NO$_3^{\ominus}$ or NH$_4$NO$_3$.
[b] Recalculated to Me$_4$N$^{\oplus}$ shift, 334.5 p.p.m. upfield from NO$_3^{\ominus}$ [Table 4.6, footnote (a)].

electron charge. Thus, pyrrole-type nitrogen atoms are characterized by the highest π-charge densities, the lowest total charge densities, and the highest magnetic screening of their nitrogen nuclei. A consideration of the shifts in azoles [126] in terms of the average excitation energy approximation (Chapter 1) and the calculated π-charge densities and π-bond orders gives an excellent correlation between the calculated changes in the *paramagnetic* term and the observed shifts for the pyrrole-type nitrogen atoms. A reasonable relationship has been found for the pyridine-type nitrogen atoms in azoles.

A similar correlation of the shifts in borazoles [42] with Hückel–MO calculated π-charge densities has been reported.

For a guide to nitrogen chemical shifts in azoles and related compounds, see refs. 24, 45, 99, 106, 117, 121 and 126.

4.4.11   Azines

Azines are heterocyclic aromatic structures containing a number of nitrogen atoms in six-membered rings. The bonding system at the nitrogen atom in azines corresponds to that described as the pyridine-type structure in the preceding Section., and may be conventionally represented as in the classical formula of pyridine. The nitrogen atom supplies one

Diagram 4.12

electron to the π-orbital system of the aromatic structure and has a lone electron pair which is not involved in the π system. The nitrogen chemical shifts of azines fall approximately within the same spectral range as the shifts for pyridine-type nitrogen atoms in azoles, +120 to −30 p.p.m on the screening-constant scale (+215 to +365 p.p.m., frequency scale). The values for most typical azine structures are given [107] in Table 4.22. The sequence of shifts from low to high fields (high to low frequencies) in each series containing one, two or three six-membered rings, respectively, is shown in Diagram 4.13. For the same number and position of nitrogen atoms, the nitrogen resonance is shifted to higher fields (lower frequencies) with an increasing number of fused six-membered rings.

The increasing magnetic screening of the nitrogen nuclei in azines has been shown [107] to follow almost linearly the π-charge densities calculated by SCF–MO methods. For unsubstituted or methyl-substituted ring systems, the relationship is

$$q_N^\pi = (1.153 + 0.001014\delta_N) \pm 0.006 \qquad (4.12)$$

TABLE 4.22.   Nitrogen Chemical Shifts of some Typical Azine Structures (Ref. 107 if not stated otherwise)

Compound	Solvent	Chemical shift (p.p.m.)		^{14}N resonance half-height width (Hz)	π-charge density at N
		Screening-constant scale referred to MeNO$_2$	Frequency scaled referred to Me$_4$N$^\oplus$		
pyridine	neat CCl$_4$ (2M) MeOH	+63 ± 2b +57 ± 2b +83c	+270b +276b +250	175 170	1.2142
quinoline	neat benzene (0.1M) CDCl$_3$ (0.1M) MeOH (0.1M)	+72 ± 3 +60.6a +71.1a +83.4a	+260 +271.9 +261.4 +249.1	650	1.2338
isoquinoline	neat	+68 ± 4	+264	680	1.2294
acridine	CH$_2$Br$_2$	+94 ± 8	+238	800	1.2516
pyridazine	DMF CHCl$_3$ benzene	−20 ± 2	+352	380 410 320	1.1362

TABLE 4.22—continued

| Compound | Solvent | Chemical shift (p.p.m.) | | ^{14}N resonance half-height width (Hz) | π-charge density at N |
		Screening-constant scale referred to $MeNO_2$	Frequency scale[d] referred to $Me_4N^{\oplus}$		
pyrimidine	DMF	+82 ± 2	+250	205	1.2356
	MeOH	+86 ± 2	+246	210	
pyrazine	DMF	+42 ± 2	+290	220	1.1880
	MeOH	+48 ± 3	+284	470	
phthalazine	dioxane	+11 ± 4	+321	800	1.1598
cinnoline	dioxane	−36 ± 5	+368	950	1.1240 1.1152
quinazoline	CH_2Br_2	+90 ± 3	+240	950	1.2352 1.2472

quinoxaline		dioxane MeOH	$+46 \pm 3$	+286	950	1.1920
phenazine		dioxane CH_2Br_2	$+68 \pm 3$ $+67 \pm 3$	+264 +265	660 750	1.1952
1,3,5-triazine		dioxane	$+98 \pm 1$	+234	255	1.2628
1,2,4,5-tetrazine		acetone	-5 ± 1[e]	+337	250[e]	1.0922[e]

[a] Ref. 88, double resonance ^{15}N data, referred originally to TMS, recalculated from $MeNH_2$ (neat) shift, Tables 4.6 and 4.7.

[b] Ref. 106, direct ^{14}N measurements, referred to internal $MeNO_2$.

[c] Ref. 123, double resonance ^{15}N data, referred to neat pyridine.

[d] Recalculated to $NMe_4^{\oplus}$ shift, 332.5 p.p.m. upfield from $MeNO_2$, Table 4.5.

[e] Ref. 121, direct ^{14}N measurements, referred to internal $MeNO_2$.

where q_N^π is the calculated π-charge density at the nitrogen atom and δ_N is the corresponding nitrogen chemical shift on the screening-constant scale relative to $MeNO_2$. Effects of substituents at the pyridine ring [128] (Table 4.23) are consistent with this trend since electron-attracting substituents

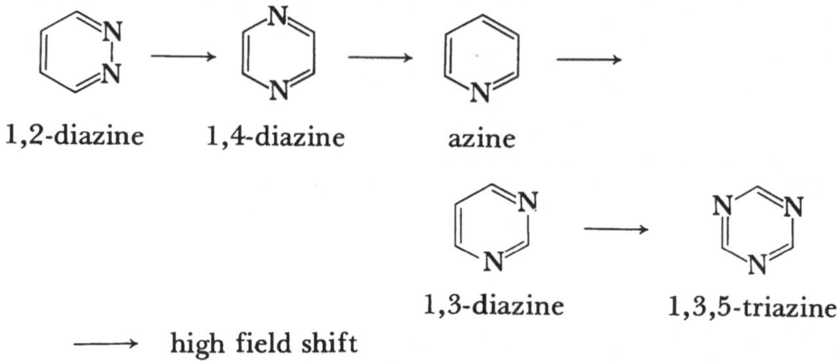

1,2-diazine   1,4-diazine   azine

1,3-diazine   1,3,5-triazine

⟶   high field shift

Diagram 4.13

like $^{\ominus}NO_2$, CN, COR give downfield shifts whilst electron-releasing groups like NH_2, OH, OR, give upfield shifts relative to that of pyridine itself. It should be noted that the older

TABLE 4.23.   Nitrogen Chemical Shifts of some Substituted Azines

| Compound | Solvent | Nitrogen chemical shift (p.p.m.) | | ^{14}N Resonance half-height width (Hz) |
		Screening-constant scale referred to internal $MeNO_2$ or $NO_3^{\ominus}$	Frequency scale[f] referred to $Me_4N^{\oplus}X^{\ominus}$	
pyridine	neat	+63 ± 2[a]	+270[a]	175[a]
	CCl$_4$	+57 ± 2[a]	+276[a]	170[a]
2-NH$_2$	acetone	+128 ± 4[b]	+205[b]	650[b]
3-NH$_2$	acetone	+88 ± 4[b]	+245[b]	680[b]
4-NH$_2$	acetone	+105 ± 4[b]	+228[b]	680[b]
2-OH	acetone	+209 ± 2[b]	+124[b]	90[b]
3-OH	acetone	+85 ± 4[b]	+248[b]	600[b]
4-OH	acetone	+201 ± 2[b]	+132[b]	390[b]
2-OMe	acetone	+110 ± 3[c]	+223[c]	360[c]
3-OMe	acetone	+59 ± 4[c]	+274[c]	680[c]
4-OMe	acetone	+90 ± 6[c]	+242[c]	1100[c]
2-Me	acetone	+72 ± 3[b]	+261[b]	325[b]
3-Me	acetone	+68 ± 3[b]	+265[b]	330[b]
4-Me	acetone	+74 ± 3[b]	+259[b]	480[b]
2-F	neat	+106 ± 5[d]	+223[d]	—

TABLE 4.23—*continued*

Compound	Solvent	Nitrogen chemical shift (p.p.m.)		^{14}N Resonance half-height width (Hz)
		Screening-constant scale referred to internal $MeNO_2$ or $NO_3^{\ominus}$	Frequency scale[f] referred to $Me_4N^{\oplus}X^{\ominus}$	
3-F	neat	+59 ± 5[d]	+274[d]	—
2-Cl	neat	+67 ± 3[c]	+266[c]	390[c]
3-Cl	neat	+59 ± 3[c]	+274[c]	400[c]
4-Cl	neat	+69 ± 4[c]	+264[c]	670[c]
2-Br	neat	+61 ± 3[c]	+271[c]	560[c]
3-Br	neat	+55 ± 3[c]	+278[c]	560[c]
4-Br	neat	+56 ± 5[c]	+277[c]	1100[c]
2-CN	acetone	+62 ± 4[c]	+271[c]	—
3-CN	acetone	+64 ± 4[c]	+269[c]	—
4-CN	acetone	+54 ± 4[c]	+279[c]	—
2-COMe	neat	+72 ± 5[c]	+261[c]	1000[c]
3-COMe	neat	+48 ± 12[c]	+285[c]	> 1300[c]
4-COMe	neat	+38 ± 12[c]	+295[c]	> 1300[c]
2-CHO	neat	+53 ± 4[c]	+280[c]	670[c]
3-CHO	neat	+52 ± 6[c]	+281[c]	> 1300[c]
4-CHO	neat	+34 ± 6[c]	+299[c]	> 1300[c]
2-NO$_2$	acetone	+80 ± 5[c]	+253[c]	450[c]
3-NO$_2$	acetone	+56 ± 5[c]	+277[c]	400[c]
4-NO$_2$	acetone	+35 ± 6[c]	+298[c]	490[c]
2,6-Cl, Cl	benzene	+81 ± 5[d]	+252[d]	—
3,5-Cl, Cl	benzene	+52 ± 5[d]	+281[d]	—
2,5-Cl, Cl	benzene	+75 ± 5[d]	+258[d]	—
2,3-Cl, Cl	benzene	+73 ± 5[d]	+260[d]	—
2,6-Br, Br	benzene	+71 ± 5[d]	+262[d]	—
2,5-Br, Br	benzene	+68 ± 5[d]	+225[d]	—
quinoline	neat	+72 ± 3[a]	+260[a]	—
8-OH	acetone	+95 ± 4[e]	+238[e]	—
5-Cl, 8-OH	acetone	+93 ± 7[e]	+240[e]	—
8-OH, 7-Me	acetone	+91 ± 4[e]	+242[e]	—
8-OH, 5-Me	acetone	+90 ± 5[e]	+243[e]	—
8-OH, 6-Cl	acetone	~+90[e]	~+243[e]	—
8-OH, 6-NO$_2$	acetone	~+90[e]	~+243[e]	—
2-OH	CDCl$_3$	+238 ± 3[e]	+95[e]	—
2-OH, 4-Me	CDCl$_3$	+238 ± 4[e]	+95[e]	—
2-OH, 3-OMe	acetone	+ +229 ± 4[e]	+104[e]	—
4-Cl, 2-OH	CDCl$_3$	~+230[e]	~+103[e]	—

[a] See Table 4.22.

[b] Ref. 107, direct ^{14}N measurements, referred to internal $MeNO_2$.

[c] Ref. 128, direct ^{14}N measurements, referred to internal $MeNO_2$.

[d] Ref. 76, direct ^{14}N measurements, originally referred to external pyridine; accuracy quoted of ±0.5 to ±0.8 p.p.m. seems to be unrealistic, as may be inferred from linewidths and comparison with recent data, ref. 128.

[e] Ref. 43, double resonance ^{14}N data, referred to external NH_4NO_3 (4.5M) in 3M HCl.

[f] Recalculated to $NMe_4^{\oplus}$ shift, 332.5 p.p.m. upfield from $MeNO_2$, Table 4.5.

literature [24] does not show this trend for mono-substituted pyridines. A correlation with the π-charge densities calculated by the simple Hückel–MO method has been found [76] for a number of halogeno-derivatives of pyridine.

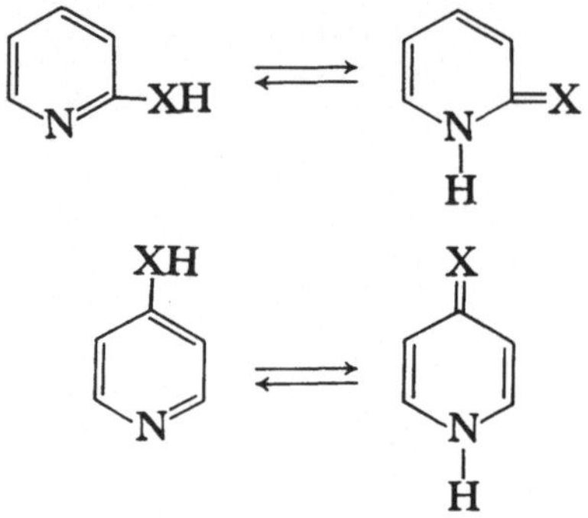

Diagram 4.14

Some hydroxy- and amino-derivatives of azines may exhibit tautomerism, as in the examples shown in Diagram 4.14. The nitrogen chemical shifts of 3-hydroxypyridine and all three

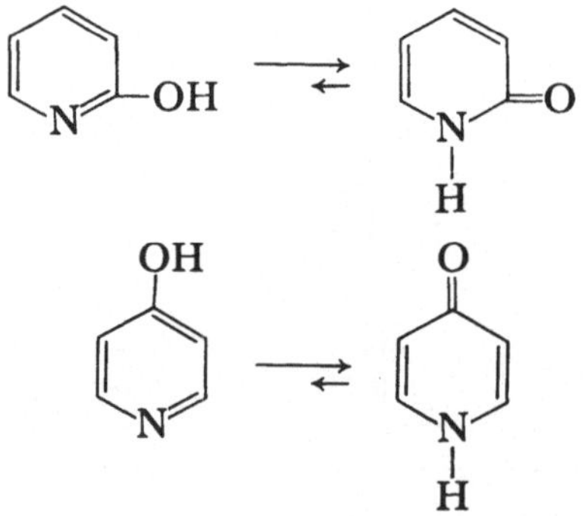

| | Chemical shifts (screening-constant scale in p.p.m.) | |
	Observed	Calculated for OH form
	+209	+120
	+201	+104

Diagram 4.15

isomeric aminopyridines are consistent with the general relationship between the shifts and the π-charge densities in azines [107]. A local linear relationship

$$q_N^\pi(\text{PyNH}_2, \text{PyOH}) = (0.9678 + 0.0028755\delta_N) \pm 0.006 \quad (4.13)$$

analogous to equation 4.12 has been derived [107]. The shifts for 2-hydroxy- and · 4-hydroxypyridine deviate significantly from the correlation and from the shifts of the corresponding methoxy-derivatives (Table 4.23) and indicate that in solution

Chemical shifts (screening-constant scale in p.p.m.)

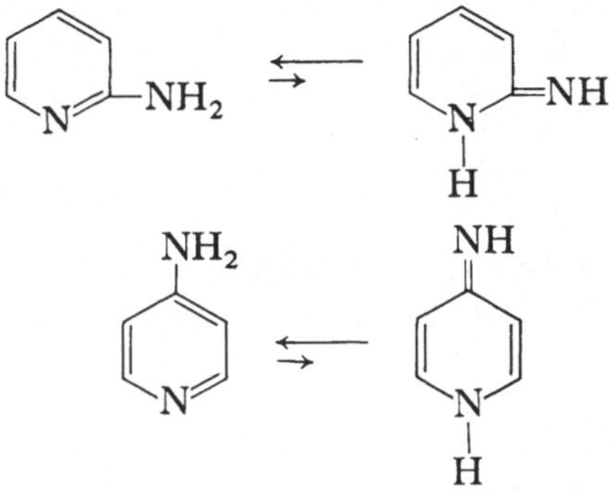

+128

+105

Diagram 4.16

they are largely isomerized to the corresponding pyridone forms. On the other hand, the aminopyridines exist mainly as such rather than in the corresponding imino-forms. A similar situation exists in 2-hydroxyquinolines [43] (Tables 4.23), where the ^{14}N shifts are clearly different from those in other quinoline derivatives.

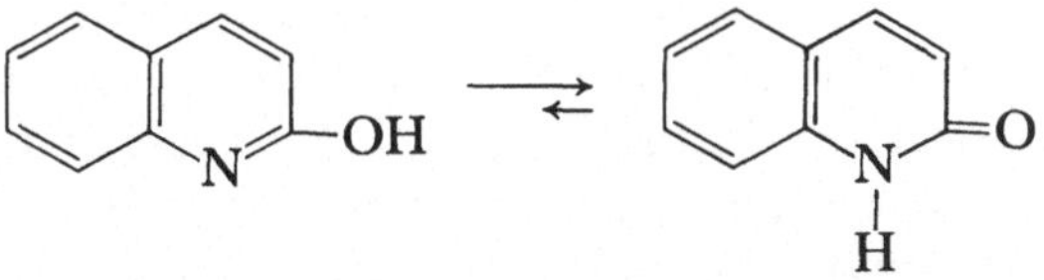

about +230 p.p.m.,
screening-constant scale

Diagram 4.17

The high-resolution ^{15}N NMR spectrum of the adenine ring portion of ^{15}N-enriched adenosine triphosphate (ATP) displays [34] five nitrogen resonance signals. They may be assigned as shown below, according to the observed ^{15}N–^{1}H spin–spin couplings. The high-field position of the resonances of N-3 and N-1 relative to pyridine (Table 4.22) is in agreement with the observed increased magnetic screening of

	Screening-constant scale (ref. to external $NO_3^{\ominus}$, 353.5 p.p.m. downfield from $NH_4^{\oplus}$) (p.p.m.)	Frequency scale (ref. to external $Me_4 N^{\oplus}$, 334.5 p.p.m. upfield from $NO_3^{\ominus}$) (p.p.m.)
NH_2	+295.0	+39.5
N-9	+204.8	+133.7
N-7	+142.7	+191.8
N-3	+148.8	+185.7
N-1	+157.9	+176.6

nitrogen   nuclei   in   pyrimidine   (1,3-diazine)   systems (Table 4.22) and with the effect of the NH_2 group on the pyridine nitrogen resonance (Table 4.23).

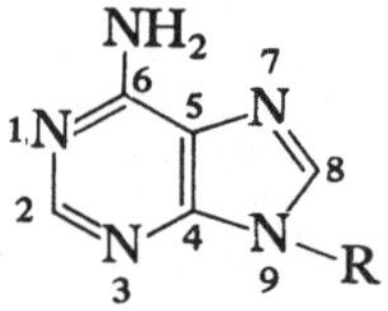

Diagram 4.18

It   has   been   shown   [107]   for   unsubstituted   azine   ring systems that the average excitation energy approximation in the theory of chemical shifts can account satisfactorily for the observed nitrogen shifts. An earlier attempt [68] to apply the different excitation energy approximation, partially successful for the simplest azines, turned out to be unsatisfactory [107] in view of the upfield shift observed for the polycyclic azine structures. An earlier consideration of the average excitation energy approximation for pyridine derivatives [49, 66] con- tained an erroneously derived equation (Chapter 1) for the term including ground-state orbitals in an expression for the *paramagnetic* term contribution to the screening constant, as has been shown in later work [107, 131].

For a guide to nitrogen chemical shifts in azines, see refs. 4, 13, 18, 21, 24, 28, 29, 30, 31, 34, 43, 49, 53, 60, 66, 68, 76, 84, 88, 95, 106, 107, 114, 117, 123 and 128.

4.4.12   Azinium Ions and Azine Oxides

If the lone electron pair at the nitrogen atom in an azine is replaced with a bond to another atom, as in Diagram 4.19, where X = H, alkyl or aryl (to give the corresponding azinium ion), or $X = O^\ominus$ (to yield the corresponding azine N-oxide) or $X = OH$ (to give the corresponding N-hydroxyazinium ion which is also the conjugate acid of the azine N-oxide), the nitrogen resonance signal is shifted to higher fields (lower frequencies). The ^{14}N signals of azinium ions and azine

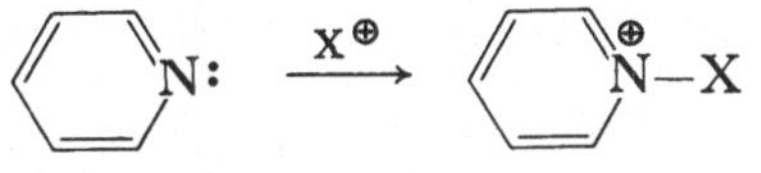

Diagram 4.19

N-oxides are usually much sharper than those of the parent azines. The upfield shift is analogous to the magnetic shielding effects observed in arylammonium ions (Section 4.4.2), nitrilium ions and nitrile N-oxides (fulminates) (Section 4.4.9), nitrones and azoxy-compounds (Section 4.4.13). The largest screening effect relative to parent azine structures is exhibited by azinium ions. Within the azinium ions a low-field (high-frequency) shift is observed in the R = H, Me, Et sequence (Table 4.24) like in amines, ammonium ions, isonitriles, amides, isocyanates, isothiocyanates, azides, and nitro compounds.

TABLE 4.24. Nitrogen Chemical Shifts of some Cations and Oxides Derived from Azine Structures

Molecule or ion	Solvent	^{14}N Chemical shift (p.p.m.)		^{14}N Resonance half-height width (Hz)
		Screening-constant scale referred to internal $MeNO_2$ or $NO_3^\ominus$	Frequency scale[f] referred to $Me_4N^\oplus X^\ominus$	
pyridinium ion (PyH$^\oplus$)	conc. HCl	+181 ± 1[a]	+155[a]	20[a]
quinolinium ion	conc. HCl	+185 ± 2[a]	+151[a]	50[a]
isoquinolinium ion	conc. HCl	+188 ± 2[a]	+148[a]	55[a]
N-methylpyridinium ion	H_2O	+174.7[b]	+159.8[b]	−
3-Me,-	H_2O	+175.1[b]	+159.4[b]	−
4-Me,-	H_2O	+181.3[b]	+153.2[b]	−
3,5-Me, Me,-	H_2O	+175.3[b]	+159.3[b]	−
3,4-Me, Me,-	H_2O	+181.3[b]	+153.2[b]	−
4-Et,-	H_2O	+180.9[b]	+153.6[b]	−
3-NH$_2$,-	H_2O	+175.0[b]	+159.5[b]	−
4-COMe,-	H_2O	+168.4[b]	+166.1[b]	−
3-COMe,-	H_2O	+173.6[b]	+160.9[b]	−
3-OH,-	H_2O	+173.1[b]	+161.4[b]	−
3-COOMe,-	H_2O	+173.2[b]	+161.3[b]	−
4-COOMe,-	H_2O	+167.4[b]	+167.1[b]	−
3-COOET,-	H_2O	+173.9[b]	+161.2[b]	−
4-COOEt,-	H_2O	+167.6[b]	+166.9[b]	−
3-CHO,-	H_2O	+173.0[b]	+161.5[b]	−
3-Cl,-	H_2O	+170.9[b]	+163.6[b]	−
3-Br,-	H_2O	+170.7[b]	+163.8[b]	−
4-CN,-	H_2O	+166.5[b]	+168.0[b]	−

TABLE 4.24.—*continued.*

		^{14}N Chemical shift (p.p.m.)		
Molecule or ion	Solvent	Screening-constant scale referred to internal MeNO$_2$ or NO$_3^{\ominus}$	Frequency scale[f] referred to Me$_4$N$^{\oplus}$X$^{\ominus}$	^{14}N Resonance half-height width (Hz)
N-ethylpyridinium ion	H$_2$O	+161.0[c]	+173.5[c]	10[c]
2-F,-	H$_2$O	+178.6[c]	+155.9[c]	80[c]
2-Cl,-	H$_2$O	+161.3[c]	+173.2[c]	18[c]
2-Br,-	H$_2$O	+158.7[c]	+175.8[c]	24[c]
2-I,-	H$_2$O	+156.1[c]	+178.4[c]	30[c]
2-CN,-	H$_2$O	+160.4[c]	+174.1[c]	16[c]
2-OMe,-	H$_2$O	+183.5[c]	+151.0[c]	70[c]
2-Me,-	H$_2$O	+161.3[c]	+173.2[c]	15[c]
2-Et,-	H$_2$O	+162.0[c]	+172.5[c]	40[c]
2-CH$_2$=CH,-	H$_2$O	+166.0[c]	+168.5[c]	70[c]
2-NNN,-	H$_2$O	+180[c]	+154.5[c]	80[c]
2-CHO,-	H$_2$O	+164[c]	+170.5[c]	80[c]
2-COMe,-	H$_2$O	+164.5[c]	+170[c]	15[c]
2-COOEt,-	H$_2$O	+162.8[c]	+171.7[c]	22[c]
2,6-Me, Me,-	H$_2$O	+161.2[c]	+173.3[c]	50[c]
2,6-Cl, Cl,-	H$_2$O	+162.3[c]	+172.2[c]	70[c]
2,6-Br, Br,-	H$_2$O	+158[c]	+176.5[c]	65[c]
pyridine N-oxide	acetone (2 : 1 w/w)	+85 ± 1[d]	+248[d]	66[d]
	CHCl$_3$ (2 : 3 w/w)	+85 ± 1[d]	+228[d]	90[d]
	MeOH (satd.)	+94 ± 1[d]	+238[d]	80[d]
	H$_2$O	+101 ± 1[d]	+234[d]	190[d]
N-hydroxypyridinium ion	HCl (conc.)	+133 ± 5[d]	+203[d]	570[d]
isoquinoline N-oxide	acetone (satd.)	+89 ± 1[d]	+244[d]	74[d]
	CHCl$_3$ (satd.)	+93 ± 1[d]	+240[d]	325[d]
	MeOH (satd.)	+110 ± 2[d]	+222[d]	360[d]
	dioxane (satd.)	+98 ± 3[d]	+234[d]	330[d]
quinoline N-oxide	acetone (satd.)	+94 ± 1[d]	+238[d]	47[d]
	CHCl$_3$ (satd.)	+95 ± 1[d]	+238[d]	180[d]
	MeOH (satd.)	+105 ± 1[d]	+228[d]	180[d]
2-methylpyridine N-oxide	acetone (1 : 8 w/w)	+85 ± 1[d]	+248[d]	29[d]
	MeOH (1 : 4 w/w)	+97 ± 2[d]	+236[d]	97[d]
	MeOH	+102 ± 2[e]	+230[e]	–
3-methylpyridine N-oxide	acetone (1 : 8 w/w)	+82 ± 1[d]	+250[d]	38[d]
	MeOH (1 : 4 w/w)	+95 ± 2[d]	+238[d]	96[d]
	MeOH	+94 ± 3[e]	+238[e]	–
4-methylpyridine N-oxide	acetone (satd.)	+87 ± 1[d]	+246[d]	41[d]
	MeOH (1 : 4 w/w)	+105 ± 1[d]	+228[d]	155[d]
	MeOH	+103 ± 3[e]	+230[e]	–
2-methoxypyridine N-oxide	acetone (satd.)	+145 ± 10[d]	+188[d]	300[d]
3-methoxypyridine N-oxide	acetone (satd.)	+95 ± 1[d]	+238[d]	360[d]
4-methoxypyridine N-oxide	acetone (satd.)	+102 ± 1[d]	+230[d]	68[d]
4-nitropyridine N-oxide	CHCl$_3$	+65 ± 3[e]	+268[e]	–
4-cyanopyridine N-oxide	CHCl$_3$	+83 ± 3[e]	+250[e]	–
N-hydroxy-2-methylpyridinium ion	HCl	+146 ± 5[e]	+188[e]	–
N-hydroxy-4-methylpyridinium ion	HCl	+149 ± 10[e]	+186[e]	–

[a] Ref. 60, Direct ^{14}N measurements, referred to internal MeNO$_2$, doublets observed due to ^{14}N^{1}H coupling.

[b] Ref. 97, iodides, double resonance (INDOR) ^{14}N data, referred to internal NH$_4$NO$_3$, error ±0.5 p.p.m.

[c] Ref. 130, tetrafluoroborates and iodides, experimental technique as in ref. 97, footnote (b), error ±0.5 p.p.m.

[d] Ref. 122, direct ^{14}N measurements, referred to MeNO$_2$ or KNO$_3$ (aqueous solution).

[e] Ref. 24, direct ^{14}N measurements, referred to external 4.5 M NH$_4$NO$_3$ in 3M HCl.

[f] Recalculated to Me$_4$N$^{\oplus}$ shift, 332.5 p.p.m. upfield from MeNO$_2$ (Table 4.5), 334.5 p.p.m. upfield from NO$_3^{\ominus}$ [Table 4.6, footnote (a)], 336 p.p.m. upfield from MeNO$_2$ in acidic aqueous solution [Table 4.10, footnote (b)].

The upfield shift of the nitrogen resonance signal in the pyridinium ion relative to pyridine has been the subject of several theoretical considerations [21, 53, 60, 95] which have

	Screening-constant scale (p.p.m.)	Frequency scale (p.p.m.)
$\overset{\diagup}{\underset{\diagdown}{N}}{}^{\oplus}-R$	+160 to +190	+175 to +145
$\overset{\diagup}{\underset{\diagdown}{N}}{}^{\oplus}-OH$	+130 to +150	+205 to +185
$\overset{\diagup}{\underset{\diagdown}{N}}{}^{\oplus}-O^{\ominus}$	+80 to +140	+255 to +195

ascribed it to changes in the *paramagnetic* contribution to the nuclear screening constant. The most recent [95] of them involves all-valence-electron calculations and gives excellent agreement with the observed shift.

Azine N-oxides give resonance signals at slightly higher fields (lower frequencies) than azines. It is interesting that changes in the nitrogen chemical shifts of azines and their N-oxides seem to be parallel (Tables 4.22, 4.23, and 4.24). Protonation at the oxygen atom to give the corresponding N-hydroxyazinium ion results in a further magnetic shielding by about 50 p.p.m.

For a guide to nitrogen chemical shifts in azinium ions and azine N-oxides, see refs. 13, 18, 21, 24, 53, 60, 95, 97, 122, 123 and 130.

4.4.13   Oximes, Imines, Nitrones, Azo and Azoxy Compounds

In this group of molecules there is a double bond at the nitrogen atom in a structure which may be conventionally written as

$$\overset{Y}{\diagdown}\underset{\cdot\cdot}{N} = X$$

where X = C, Y = OH or OR for oximes and their ethers, respectively; X = N, Y = aryl or alkyl for azo compounds; X = C, Y = alkyl, aryl for imines. The corresponding N-oxide structure

$$\overset{Y}{\underset{\ominus O \diagup}{\diagdown}} N^{\oplus} = X$$

is characteristic of nitrones (X, Y = alkyl, aryl) and azoxy compounds (X = N, Y = alkyl, aryl). These structures resemble those in azines and their N-oxides, respectively, and the corresponding nitrogen chemical shifts are also alike, as can be seen from Table 4.25. There is also a downfield shift for compounds with the N=N structure as compared with the C=N structure. The nitrogen resonance signals of the N-oxide structures such as azoxy-benzene, nitrone, furoxan, are shifted upfield relative to the parent structures of azobenzene, imines and furazan, respectively, similar to azine N-oxides relative to azines.

$$-\overset{..}{N}=\overset{..}{N}- \xrightarrow[\text{shift}]{\text{upfield}} -\overset{\oplus}{\underset{\underset{O^{\ominus}}{|}}{N}}=\overset{..}{N}-$$

azo   azoxy

$$-\overset{..}{\underset{..}{O}}-\overset{..}{N}=\underset{|}{C}- \xrightarrow[\text{shift}]{\text{upfield}} -\overset{..}{\underset{..}{O}}-\overset{\oplus}{\underset{\underset{O^{\ominus}}{|}}{N}}=\underset{|}{C}-$$

oxime (furazan)   furoxan

$$R-\overset{..}{N}=\underset{|}{C}- \xrightarrow[\text{shift}]{\text{upfield}} R-\overset{\oplus}{\underset{\underset{O^{\ominus}}{|}}{N}}=\underset{|}{C}-$$

imine   nitrone

Protonation at the nitrogen atom results in a still larger upfield shift (Table 4.25), as is found for cations derived from azobenzene and diphenylketimine ($Ph_2 C=NH$). This is also in

TABLE 4.25.   Nitrogen Chemical Shifts of some Oximes, Imines, Nitrones, Azo- and Azoxy Compounds and Related Heterocyclic Structures

Compound	Solvent	Nitrogen chemical shift (p.p.m.)		
		Screening-constant scale referred to internal MeNO$_2$ or NO$_3^{\ominus}$	Frequency scale[j] referred to Me$_4$N$^{\oplus}$X$^{\ominus}$	^{14}N Resonance half-height width (Hz)
MeCH=NOH	CCl$_4$	+27 ± 4[a]	+306[a]	900[a]
Me$_2$C=NOH	CCl$_4$	+44 ± 5[a]	+288[a]	1100[a]
PhCH=N$\overset{\oplus}{\underset{O^{\ominus}}{}}$Me	acetone (satd.)	+103 ± 1[b]	+230[b]	70[b]
PhCH=N$\overset{\oplus}{\underset{O^{\ominus}}{}}$Ph	acetone (satd.)	+94 ± 1[b]	+238[b]	100[b]

TABLE 4.25.—*continued*

Compound	Solvent	Nitrogen chemical shift (p.p.m.)		
		Screening-constant scale referred to internal $MeNO_2$ or $NO_3^{\ominus}$	Frequency scale[j] referred to $Me_4N^{\oplus}X^{\ominus}$	^{14}N Resonance half-height width (Hz)
dimethylfurazan	neat	-26 ± 2[a]	$+358$[a]	460[a]
		-26 ± 5[f]	$+358$[f]	—
Me—N–O–N—Me (furazan ring)				
benzofurazan	CH_2Cl_2	-28 ± 5[f]	$+360$[f]	—
dimethylfuroxan	neat	$+10 \pm 1$[a]	$+322$[a]	140[a]
Me—N–O–N$^{\oplus}$(O$^{\ominus}$)—Me (furoxan ring)				
benzofuroxan	acetone (satd.)	$+26 \pm 3$[a]	$+306$[a]	370[a]
PhCH=NMe	neat	$+55 \pm 1$[d]	$+278$[d]	—
PhCH=NPh	ether	$+55 \pm 5$[i]	$+278$[i]	—
p-MeC$_6$H$_4$CH=NPh	$CDCl_3$	$+51.3$[c]	$+281.2$[c]	—
PhCH=NC$_6$H$_4$-p-Me	$CDCl_3$	$+54.0$[c]	$+278.5$[c]	—
p-ClC$_6$H$_4$CH=NPh	$CDCl_3$	$+56.7$[c]	$+275.8$[c]	—
p-NO$_2$C$_6$H$_4$CH=NPh	$CDCl_3$	$+68.6$[c]	$+263.9$[c]	—
Ph$_2$C=NH	neat	$+72 \pm 1$[g]	$+260$[g]	—
Ph$_2$C=NH$_2^{\oplus}$Cl$^{\ominus}$	SO_2	$+212 \pm 1$[g]	$+120$[g]	—
trans PhN=NPh	ether	-130 ± 1[d]	$+462$[d]	—
	dioxane	-144 ± 5[a]	$+476$[a]	1100[a]
	MeOH	-135 ± 5[a]	$+467$[a]	750[a]
conjugate acid of *trans* azobenzene	H$_2$SO$_4$, EtOH, H$_2$O	$+20 \pm 1$[d]	$+312$[d]	—
EtN=NEt	neat	-154 ± 10[e]	$+486$[e]	260[e]
F$_3$CN=NCF$_3$	neat	-140 ± 10[e]	$+472$[e]	—
$^{\ominus}$ON=NO$^{\ominus}$	NaOH, H$_2$O	-82 ± 10[e]	$+414$[e]	3000[e]
HON=NOH	HClO$_4$, H$_2$O	-86 ± 10[e]	$+418$[e]	3750[e]
trans FN=NF	CCl$_3$F (0.2M)	-68[h]	$+400$[h]	—
cis FN=NF	CCl$_3$F (0.2M)	-6[h]	$+338$[h]	—
trans PhN=NPh ($\overset{\oplus}{\underset{O^{\ominus}}{N}}$)	ether	$\{+56$[d] / $+52\}$	$\{+276$[d] / $+280\}$	—

[a] Ref. 121, direct ^{14}N measurements, referred to internal MeNO$_2$.

[b] Ref. 122, direct ^{14}N measurements, referred to internal MeNO$_2$.

[c] Ref. 129, unpublished results, direct ^{15}N measurements, relative to PhCH=NPh.

[d] Ref. 23, direct ^{15}N measurements, recalculated to MeNO$_2$ with neat MeNH$_2$ shift.

[e] Ref. 108, direct ^{14}N measurements, referred to NH$_4^{\oplus}$, originally measured from Me$_2$NNO; results are rather inacurate due to method of calibration.

[f] Ref. 12, direct ^{14}N measurements, referred to external NH$_4$NO$_3$.

[g] Ref. 25, direct ^{15}N measurements, recalculated as in footnote (d).

[h] Recalculated from earlier work [J. H. Noggle, J. D. Baldeschwieler, and C. B. Colburn, *J. Chem. Phys.*, 37, 128 (1962)].

[i] Ref. 24, direct ^{14}N measurements, referred to external 4.5M NH$_4$NO$_3$ in 3M HCl.

[j] Recalculated to Me$_4$N$^{\oplus}$ shift, 332.5 p.p.m. upfield from MeNO$_2$ (Table 4.5).

accord with the corresponding spectral changes for azines and azinium ions (Section 4.4.12).

Nitrogen chemical shifts are clearly different for the isomeric structures of nitrones, oximes and nitroso compounds (Section 4.4.16).

	Screening-constant scale (p.p.m.)	Frequency scale (p.p.m.)
oxime, $R_2C{=}N{-}OR$	*ca.* +30	*ca.* +300
nitrone, $R_2C{=}N^{\oplus}{-}R$ $\quad\quad\quad\mid$ $\quad\quad\quad O^{\ominus}$	*ca.* +100	*ca.* +230
nitroso compound, $R_3C{-}N{=}O$	−400 to −500	+700 to +800

An investigation of nitrogen chemical shifts in systems involving tautomeric equilibria between azo- and hydrazone forms [56] with various substituents at the benzene ring has indicated that the hydrazone form prevails in the equilibrium

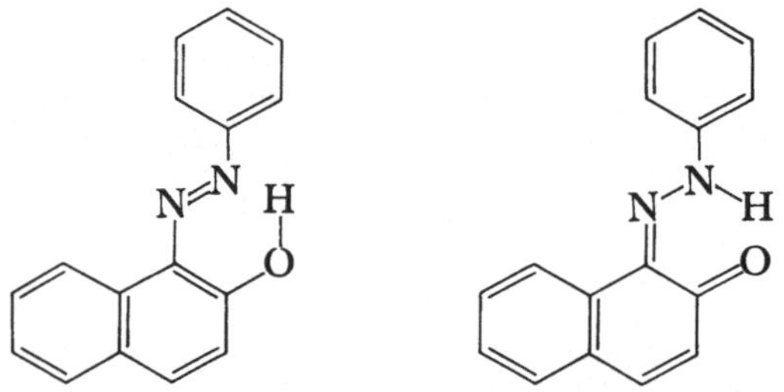

Diagram 4.20

mixture since the observed ^{14}N shifts (double-resonance measurements) are in the range of +166 to +81 p.p.m. (screening-constant scale, referred to external $NO_3^{\ominus}$) which is quite different from that for azo compounds (Table 4.25).

For azo compounds, $R{-}N{=}N{-}R$, it has been suggested [108] that the nitrogen resonance signal moves to higher fields (lower frequencies) with increasing electronegativity of the groups R, but the low accuracy of the ^{14}N measurements in this case and the rather arbitrary classification of substituents R according to electronegativity makes the conclusion rather premature.

No satisfactory theoretical explanation of the shifts in this group of compounds has so far been found. An attempt [108] to relate the shifts, within the average excitation energy approximation, to the experimentally observed low-energy transitions in the electronic absorption spectra did not give useful results (Chapter 1, Table 1.2), and the method itself seems to be theoretically unsound.

For a guide to nitrogen chemical shifts in oximes, imines, nitrones, azo and azoxy compounds, see refs, 12, 23, 24, 25, 27, 108, 117, 121, 122 and 129.

4.4.14   Nitro Compounds

The nitrogen resonance signals of the nitro group occur in a rather narrow range of +60 to −30 p.p.m. on the screening-constant scale (+275 to +365 p.p.m., frequency scale). In ^{14}N NMR spectra, their resonance signals are usually sharp. The nitrogen chemical shifts of nitroalkanes occupy almost the entire range, from about +50 to −30 p.p.m. on the screening-constant scale (+285 to +365 p.p.m., frequency scale). A simple spectral distinction [20] may be made between

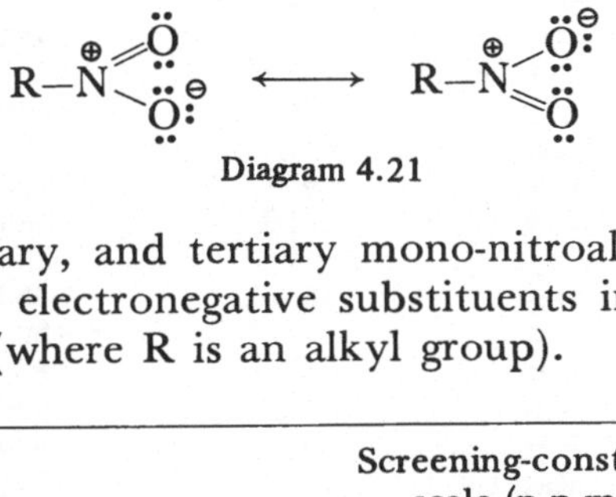

Diagram 4.21

primary, secondary, and tertiary mono-nitroalkanes which do not contain any electronegative substituents in the vicinity of the NO_2 group (where R is an alkyl group).

	Screening-constant scale (p.p.m.)	Frequency scale (p.p.m.)
$MeNO_2$	0	+332.5
RCH_2NO_2	ca. −10	ca. +342
R_2CHNO_2	ca. −20	ca. +353
R_3CNO_2	ca. −30	ca. +362

There is a very regular upfield shift of the nitrogen resonance in nitroalkanes, $R-NO_2$, with increasing electronegativity of the group R. Simple additivity rules [39] for the shifts in mono-nitroalkanes, $R^1R^2R^3C-NO_2$ (R = H, Me, Et or longer alkyl, Cl, Br), relative to $MeNO_2$ give quite accurate results with the following increments: 0 p.p.m. for each H atom, −10.4 p.p.m. per each methyl group, −9.5 p.p.m. per

TABLE 4.26.   Nitrogen Chemical Shifts of some Nitroalkanes

Compound	Solvent or state	Nitrogen chemical shift (p.p.m.)		^{14}N Resonance half-height width (Hz)
		Screening-constant scale referred to MeNO$_2$ (±0.5 p.p.m.)	Frequency scale[a] referred to Me$_4$N$^\oplus$X$^\ominus$	
MeNO$_2$	neat, CCl$_4$	0[b]	+332.5[b]	24[b], 14[d]
EtNO$_2$	neat, CCl$_4$	−12[b]	+344.5[b]	30[b]
	neat	−10.8[c]	+343.3[c]	
n-PrNO$_2$	neat, CCl$_4$	−10[b]	+342.5[b]	35[b]
	neat	−9.5[c]	+342[c]	
n-BuNO$_2$	neat, CCl$_4$	−10[b]	+342.5[b]	49[b]
Me(CH$_2$)$_4$NO$_2$	neat, CCl$_4$	−10[b]	+342.5[b]	56[b]
Me(CH$_2$)$_5$NO$_2$	neat	−10[b]	+342.5[b]	57[b]
PhCH$_2$NO$_2$	CCl$_4$	−7.5[b]	+340[b]	163[b]
i-PrNO$_2$	neat	−24[b]	+356.5[b]	38[b]
nitrocyclohexane	neat	−20[b]	+352.5[b]	92[b]
t-BuNO$_2$	neat	−30[b]	+362.5[b]	43[b]
CH$_2$(NO$_2$)$_2$	ether (20%)	+20[d]	+312.5[f]	?
MeCH(NO$_2$)$_2$	neat	+11[e]	+321.5[e]	28[e], 22[d]
	neat	+10.5[f]	+322[f]	24[f]
EtCH(NO$_2$)$_2$	neat	+11[e]	+321.5[e]	35[e]
Me$_2$C(NO$_2$)$_2$	CH$_2$Cl$_2$	0[e]	+332.5[e]	34[e]
CH(NO$_2$)$_3$	neat, CCl$_4$	+35.5[b]	+297[b]	11[b]
MeC(NO$_2$)$_3$	CH$_2$Cl$_2$	+26[c]	+306.5[c]	11[c]
EtC(NO$_2$)$_3$	neat	+27[d]	+305.5[d]	?
C(NO$_2$)$_4$	neat	+48[b]	+284.5[b]	10[b]
	neat	+47[d]	+285.5[d]	1.5[d]
C$_2$(NO$_2$)$_6$	CH$_2$Cl$_2$	+46.5[f]	+286[f]	6[f]
HOCH$_2$CH$_2$NO	neat	−5.5[c]	+338[c]	30[c]
CH$_2$BrNO$_2$	neat	+7[d]	+325.5[d]	23[d]
MeCHBrNO$_2$	neat	−3.5[d]	+336[d]	31[d]
MeCHClNO$_2$	neat	−5.5[e]	+338[e]	30[e]
	neat	−4.5[d]	+337[d]	24[d]
Me$_2$CClNO$_2$	neat	−13[b]	+345.5[b]	35[b]
EtCHBrNO$_2$	neat	−4[e]	+336.5[e]	75[e]
EtCHClNO$_2$	neat	−4.5[e]	+337[e]	50[e]
	CH$_2$Cl$_2$	−2.6[c]	+335.1[c]	?
Me(CH$_2$)$_3$CHClNO$_2$	neat	−4[e]	+336.5[e]	75[e]
MeCCl$_2$NO$_2$	neat	+3.5[d]	+329[d]	?
EtCCl$_2$NO$_2$	neat	+1[e]	+331.5[e]	31[e]
CCl$_3$NO$_2$	neat	+12.5[d]	+320[d]	?
CBr$_3$NO$_2$	neat	+13 ± 10[g]	+320[g]	(110)[g]
CF$_3$NO$_2$	neat	+23 ± 10[g]	+310[g]	(100)[g]
C$_3$F$_7$NO$_2$	neat	+23 ± 10[g]	+310[g]	(110)[g]
CF$_2$ClCFClNO$_2$	neat	+3 ± 10[g]	+330[g]	(110)[g]
CFCl(NO$_2$)COOH	neat	+16[f]	+316.5[f]	78[f]
MeCBr(NO$_2$)$_2$	neat	+13[f]	+319.5[f]	25[f]
MeCCl(NO$_2$)$_2$	neat	+13.5[f]	+319[f]	21[f]
CBr$_2$(NO$_2$)$_2$	neat	+21.5[d]	+311[d]	?
CCl$_2$(NO$_2$)$_2$	neat	+22[d]	+310.5[d]	3[d]
CBr(NO$_2$)$_3$	neat	+33[f]	+299.5[f]	6[f]

TABLE 4.26—*continued*

| | | Nitrogen chemical shift (p.p.m.) | | |
| | | Screening-consatant scale referred to MeNO$_2$ ($\pm$0.5 p.p.m.) | Frequency scale[a] referred to Me$_4$N$^{\oplus}$X$^{\ominus}$ | ^{14}N Resonance half-height width (Hz) |
Compound	Solvent or state			
CCl(NO$_2$)$_3$	neat	+36[f]	+296.5[f]	8[f]
	neat	+34.5[d]	+298.5[f]	?
CF(NO$_2$)$_3$	neat	+39[f]	+293.5[f]	5[f]
HOCH$_2$C(NO$_2$)$_3$	neat	+30[c]	+302.5[c]	10[c]
NO$_3^{\ominus}$	H$_2$O	0[h]	+334.5[h]	12[h]
NO$_2^{\ominus}$	H$_2$O	-237 ± 4[h]	+572[h]	520[h]

a Recalculated to Me$_4$N$^{\oplus}$ shifts, 332.5 p.p.m. upfield from MeNO$_2$ (Table 4.5), 334.5 p.p.m. upfield from NO$_3^{\ominus}$ [Table 4.6, footnote (a)].

b Ref. 20, direct ^{14}N measurements, referred to internal MeNO$_2$.

c Ref. 47, direct ^{14}N measurements, referred to internal MeNO$_2$.

d Ref. 118, direct ^{14}N measurements, referred to external MeNO$_2$.

e Ref. 39, direct ^{14}N measurements, referred to internal MeNO$_2$.

f Ref. 57 and 58 direct ^{14}N measurements, referred to external MeNO$_2$.

g Ref. 92, data obtained with a wide-line spectrometer and originally referred to Me$_2$NNO as external standard. The accuracy was very low and the reported linewidths are probably results of instrumental broadening.

h Ref. 60, direct ^{14}N measurements, referred to internal NO$_3^{\ominus}$.

each ethyl or longer alkyl group, +5.3 p.p.m. per each Cl or Br atom, on the screening constant scale. However, an aggregation of strongly electron-attracting substituents, for example two nitro groups on a carbon atom, cannot be accommodated within such a simple additive scheme of inductive effects and the Taft inductive constants for the entire R^{1}R^{2}R^{3}C— moiety should be compared with the chemical shifts of the nitro group. An early attempt [47] to do this gave only a rough correlation but subsequent comparisons [57, 118] of the existing experimental data with the Taft constants have shown that linear relationships exist for certain types of nitroalkanes. Thus, for mono-nitroalkanes the following correlation has been found [57]:

$$\delta_N \text{(mono-nitroalkane)} = -0.6 + 103 \; \sigma^* \text{ p.p.m.} \quad (4.14)$$

whilst for poly-nitroalkanes

$$\delta_N \text{(poly-nitroalkane)} = -25.0 + 13.8 \; \sigma^* \text{ p.p.m.} \quad (4.15)$$

where δ_N is the shift from MeNO$_2$ on the screening-constant scale and σ^* is the Taft constant for group R in R—NO$_2$. A consideration of a larger set of shifts for the nitro group suggests [118] that the Taft constants should be compared with the shifts separated into four groups:

CH_3NO_2, RCH_2NO_2, R_2CHNO_2, R_3CNO_2 (Group I)

XCH_2NO_2, $XCHRNO_2$, XCR_2NO_2 (Group II)

X_2CHNO_2, X_2CRNO_2 (Group III)

X_3CNO_2 (Group IV and IVa)

where R is a hydrocarbon group and X is an electron-attracting substituent or a hydrocarbon group (Fig. 4.1). The difference between Group IV and IVa is that in the latter all

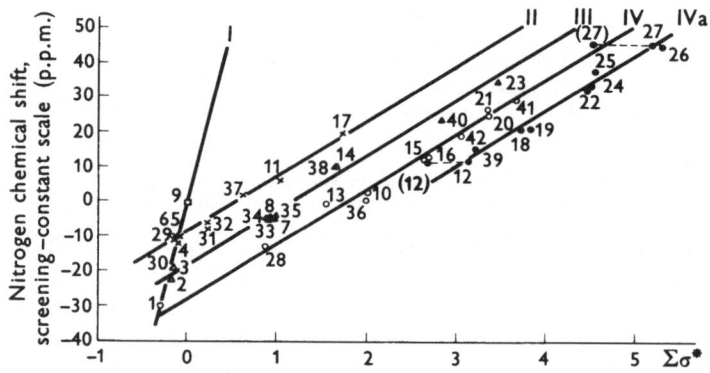

Fig. 4.1.   Correlation of ^{14}N chemical shifts in nitroalkanes and inductive Taft constants. Quoted from ref. 118.

$\Sigma_\sigma*$ is the Taft constant for group R in $R{-}NO_2$ calculated as a sum of X_i contributions for $R = X_1X_2X_3C{-}$. The numbers on the plot correspond to: 1, Me_3CNO_2; 2, Me_2CHNO_2; 3, nitrocyclohexane; 4, $EtNO_2$; 5, n-$PrNO_2$; 6, n-$BuNO_2$; 7, $MeCHClNO_2$; 8, $MeCHBrNO_2$; 9, $MeNO_2$; 10, $MeCCl_2NO_2$; 11, $BrCH_2NO_2$; 12, Cl_3CNO_2; 13, $Me_2C(NO_2)_2$; 14, $MeCH(NO_2)_2$; 15, $MeCBr(NO_2)_2$; 16, $MeCCl(NO_2)_2$; 17, $CH_2(NO_2)_2$; 18, $Br_2C(NO_2)_2$; 19, $Cl_2C(NO_2)_2$; 20, $MeC(NO_2)_3$; 21, $EtC(NO_2)_3$; 22, $BrC(NO_2)_3$; 23, $CH(NO_2)_3$; 24, $ClC(NO_2)_3$; 25, $FC(NO_2)_3$; 26, $C_2(NO_2)_6$; 27, $C(NO_2)_4$; 28, Me_2CClNO_2; 29, $Me(CH_2)_4NO_2$; 30, $Me(CH_2)_5NO_2$; 31, $PhCH_2NO_2$; 32, $HOCH_2CH_2NO_2$; 33 n-$BuCHClNO_2$; 34, $EtCHBrNO_2$; 35, $EtCHClNO_2$; 36, $EtCCl_2NO_2$; 37, $O_2NCH_2CH_2NO_2$; 38, $EtCH(NO_2)_2$; 39, $HOOCCFClNO_2$; 40, $HOCH_2CF(NO_2)_2$; 41, $HOCH_2C(NO_2)_2$: 42, $CHF(NO_2)_2$.

substituents X are electron-attracting. The steepest slope is observed upon replacing hydrogen atoms in CH_3NO_3 with alkyl groups. The shift is about 10 p.p.m. downfield [20] per each hydrogen atom substituted with an alkyl group, and the same holds for other· structural series:

$$CH_3NO_2 \rightarrow RCH_2NO_2 \rightarrow R_2CHNO_2 \rightarrow R_3CNO_2$$
$$XCH_2NO_2 \rightarrow XCHRNO_2 \rightarrow XCR_2NO_2$$
$$X_2CHNO_2 \rightarrow X_2CRNO_2$$

for the same substituent X. Each step in the above sequences is equivalent to shifting the nitrogen resonance about 10 p.p.m. upfield. Similar shifts have been found in many groups of molecules such as amines, ammonium ions, amides, iso-cyanates and related compounds, isonitriles and azinium ions.

The nitrogen chemical shifts of the nitro group in nitro-alkanes have been shown [58] to be almost linearly related to the frequency of the asymmetric stretching vibration of the NO_2 group,

$$\nu_{NO_2} \text{ (asymmetric stretch)} = 1573 + 1.147\delta_N(NO_2) \text{ cm}^{-1} \quad (4.16)$$

where $\delta_N(NO_2)$ is the nitrogen chemical shift in p.p.m. relative to $MeNO_2$ on the screening-constant scale.

The increased screening of nitrogen nuclei in the nitro group of a nitroalkane, $R-NO_2$, with the increasing electro-negativity of group R may be simply explained in terms of π-bond order and charge density changes through the average excitation energy approximation [60, 131] of the theory of chemical shifts. Here again, attempts [92] to find relation-ships between the shifts and the low-energy electronic transi-tions seem to be theoretically unsound [94] since there is no dominant contribution of any single electronic excitation to the screening constant in this type of structure. Moreover the mean excitation energy or the 'virtual' excitations considered by the theory may bear no direct relationship to the real excited states of a molecule. The most important factor determining the nitrogen chemical shifts within the nitroalkane group seems to be that involving changes in π-bond orders. The observed trend in the shifts is a reversal of that expected from electron density changes at the nitrogen atom. The total electron density should decrease with increasing electron-with-drawing ability of substituent R in $R-NO_2$ and thus contri-bute to a downfield shift whilst an upfield shift is actually found. A consideration of ^{14}N resonance linewidths [118] with corrections for viscosity and molecular volume effects suggests that in geminal polynitroalkanes the electric field gradient at the nitrogen nuclei approaches that in symmetrical ammonium ions, $R_4N^\oplus$. This is hardly to be expected if the 'inductive' loss of electron charge through the R—N bond is replaced with an excess of charge by polarization of the multiple N—O bonds, as has been suggested in the literature [92, 112].

The nitrogen chemical shifts of the nitro group in aromatic nitro compounds fall into the narrow range of +6 to +40 p.p.m. on the screening-constant scale (+327 to +293 p.p.m., frequency scale) as shown in Table 4.27. The nitrogen nuclei in the NO_2 group attached to an unsaturated system of carbon atoms or to an aromatic ring are more shielded than in nitroalkanes except those containing a strongly electron-attracting substituent and the nitro group at the same carbon atom. There is also an evident high-field (low-frequency) shift for the aromatic NO_2 group if the benzene ring is substituted with electron-attracting groups like additional NO_2 groups, CHO, COR, etc. However, there is no significant dependence on the position of the substituents. It seems that inductive rather than conjugative effects play the most important role in the determination of the magnetic screening of the NO_2 group nitrogen nuclei in aromatic systems. Attention is drawn to the fact that the shifts of *ortho*-substituted nitrobenzenes, where the nitro group may be forced out of the plane of the benzene ring, do not differ experimentally from the shifts of the corresponding *meta-* and *para*-substituted isomers. There seems to be no effect which could be due to the presence of intramolecular hydrogen bonding in *ortho*-nitrophenol [41]. Thus, from the point of view of the nitrogen chemical shifts of the NO_2 group, aromatic rings behave as electron-attracting groups R in R—NO_2 compounds with no apparent conjugative effects. Recent measurements (Table 4.27) do not show any 'hyperconjugation' effects of the methyl group on the benzene ring such as was claimed in earlier work [31] where a shift difference of 11 p.p.m. was reported for the nitro groups in *ortho-* and *para*-nitrotoluene. A discrepancy between the reported ^{14}N shifts for the aromatic nitro group [41] and the ^{15}N shifts in para-substituted nitrobenzenes [33] has been removed by a reinvestigation of the ^{15}N data [67].

The nitrogen resonance signals of the NO_2 group attached to a nitrogen or oxygen atom, as in nitramines R_2NNO_2 and

TABLE 4.27. Nitrogen Chemical Shifts of some Aromatic Nitro Compounds

| Compound | Solvent | Nitrogen chemical shift (p.p.m.) | | ^{14}N Resonance half-height width (Hz) |
		Screening-constant scale referred to $MeNO_2$	Frequency scale[a] referred to $Me_4N^{\oplus}X^{\ominus}$	
Nitrobenzene	acetone	+8[b]	+324.5[b]	60[b]
	acetone	+9.3[c]	+323.2[c]	53[c]
2-OMe,-	acetone	+6[b]	+326.5[b]	70[b]
3-OMe,-	acetone	+8[b]	+324.5[b]	90[b]
4-OMe,-	acetone	+7[b]	+325.5[b]	100[b]
	acetone	+7.8[b]	+324.7[b]	
2-OH,-	acetone	+7[b]	+325.5[b]	70[b], 60[c]
3-OH,-	acetone	+9[b]	+323.5[b]	140[b], 65[c]
4-OH,-	acetone	+7[b]	+325.5[b]	110[b], 220[c]
2-NH_2,-	acetone	+7[c]	+325.5[c]	65[c]
3-NH_2,-	acetone	+8[c]	+324.5[c]	97[c]
4-NH_2,-	acetone	+10 ± 1[c]	+322.5[c]	115[c]
	acetone	+7.7[d]	+324.8[d]	
2-Me_3Si,-	DMF	+7[e]	+325.5[e]	100[e]
3-Me_3Si,-	DMF	+9[e]	+323.5[e]	139[e]
4-Me_3Si,-	DMF	+12[e]	+320.5[e]	182[e]
2-Me,-	acetone	+7[b]	+325.5[b]	50[b]
3-Me,-	acetone	+8[b]	+324.5[b]	90[b]
4-Me,-	acetone	+7.5[b]	+325[b]	85[b]
4-F,-	acetone	+9.1[d]	+323.4[d]	
4-Cl,-	acetone	+10.3[d]	+322.2[d]	
4-Br,-	acetone	+10.7[d]	+321.8[d]	
4-CN,-	acetone	+11.3[d]	+321.2[d]	
2-CHO,-	acetone	+11[b]	+321.5[b]	75[b]
3-CHO,-	acetone	+12[b]	+320.5[b]	100[b]
4-CHO,-	acetone	+12.5[b]	+320[b]	95[b]
1,2-dinitrobenzene	acetone	+12[b]	+320.5[b]	35[b]
	acetone	+13.7[c]	+318.8[c]	35[c]
1,3-dinitrobenzene	acetone	+13[b]	+319.5[b]	63[b]
	acetone	+14.7[c]	+317.8[c]	50[c]
1,4-dinitrobenzene	acetone	+13[b]	+319.5[b]	70[b]
	acetone	+13.7[c]	+318.8[c]	50[c]
1,3,5-trinitrobenzene	acetone	+18.5[b]	+314[b]	63[b]
	acetone	+19.5[c]	+313[c]	75[c]
2,4,6-trinitrotoluene	acetone	+14[b]	+318.5[b]	100[b]
2,4,6-trinitrophenol	acetone	+18[b]	+314.5[b]	128[b]
2,4,6-trinitrobenzoic acid	acetone	+19[b]	+313.5[b]	125[b]
hexanitrobenzene	CH_2Cl_2	+38.5[c]	+294[c]	8[c]
$CH_3CH{=}CHNO_2$	neat	+3 ± 2[f]	+329.5[f]	48[f]

[a] Recalculated to $Me_4N^{\oplus}$ shift, 332.5 p.p.m. upfield from $MeNO_2$ (Table 4.5).
[b] Ref. 41, direct ^{14}N measurements, referred to internal $MeNO_2$, error ±1 p.p.m.
[c] Ref. 119, direct ^{14}N measurements, referred to external $MeNO_2$, error ±0.5 p.p.m.
[d] Ref. 67, direct ^{15}N measurements, referred originally to nitrobenzene, error ±0.2 p.p.m., recalculated to $MeNO_2$.
[e] Ref. 50, direct ^{14}N measurements, referred to internal $MeNO_2$, error ±1 p.p.m.
[f] Ref. 132, direct ^{14}N measurements, referred to internal $MeNO_2$.

nitrates $R-ONO_2$ (Table 4.28), occur at higher magnetic
fields (lower frequencies) when compared with C-nitro compounds. This is in accord with the general trend in shifts for
the NO_2 group.

There is a large difference in nitrogen chemical shifts
between nitro compounds and the isomeric nitrites (Section 4.4.16) and also between nitro and nitroso compounds.

	Screening-constant scale (p.p.m.)	Frequency scale (p.p.m.)
nitro compound, $R-NO_2$	+60 to −30	+275 to +365
nitrite, $R-O-N=O$	ca. −200	ca. +530
nitroso compound, $R-N=O$	ca. −400	ca. +700

For a guide to nitrogen chemical shifts in nitro groups, see
refs. 3, 4, 5, 7, 10, 20, 22, 23, 24, 25, 26, 31, 33, 39, 40, 41,
47, 50, 51, 52, 57, 58, 60, 67, 71, 73, 79, 80, 85, 86, 87, 92,
94, 105, 110, 111, 112, 113, 114, 117, 118, 119, 121 and 132.

4.4.15   Anions Derived from Nitro Compounds

Anions formed by removal of a proton from nitroalkanes give
nitrogen resonance signals usually at higher fields (lower

$$R^1R^2CHNO_2 \xrightarrow{-H^{\oplus}} [R^1R^2CNO_2]^{\ominus}$$

frequencies) than the parent nitro compounds (Table 4.29). In
spite of differences in the reported [85, 110] values of the
^{14}N shifts for some broad resonance signals, the sequence of
shifts from high to low fields (low to high frequencies) in
molecules of the type

$$(X-NO_2)^{\ominus}M^{\oplus}$$

follows that of substituent groups $X = Me_2C$, MeCH, CH_2,
$C(NO_2)_2$, $BrCNO_2$, $ClCNO_2$, $C(CN)NO_2$, $CHNO_2$, O. Thus,
the increasing electronegativity of the X moiety seems to be
accompanied by a downfield shift of the nitrogen resonance,
which is opposite to the trend in shifts of nitro compounds.
There is only one marked exception to this scheme, the
resonance of $M^{\oplus}C(NO_2)_3^{\ominus}$ which occurs at a higher field than

TABLE 4.28. Nitrogen Chemical Shifts of Nitro Groups in some Nitramines and Nitrates

Compound	Solvent	Nitrogen chemical shift (p.p.m.)		
		Screening-constant scale referred to internal $MeNO_2$ or $NO_3^{\ominus}$	Frequency scale[h] referred to $Me_4N^{\oplus}X^{\ominus}$	^{14}N Resonance half-height width (Hz)
$H_2NCONHNO_2$	acetone	$+40 \pm 1$[b]	$+292$[b]	30[b]
$O_2NN\langle CH_2N(NO_2)CH_2 \rangle NNO_2$ (with two $N\text{-}NO_2$ bridges)	DMSO	$+38 \pm 1$[a]	$+294$[a]	26[a]
$O_2NN\langle CH_2N(NO_2) \rangle_2 CH_2$ (cyclic nitramine)	acetone	$+36 \pm 1$[c]	$+296$[c]	35[c]
	DMSO	$+34 \pm 1$[a]	$+298$[a]	25[a]
$O_2NN\langle CH_2NCH_2 \rangle CH_2 \langle CH_2NCH_2 \rangle NNO_2$	acetone	$+27 \pm 1$[c]	$+306$[c]	30[c]
$O_2N\text{-}C_6H_2(NO_2)_2\text{-}NHNO_2$	acetone	$+34.5 \pm 1$[b] (N-nitro)	$+298$[b] (N-nitro)	32[b] (N-nitro)
		$+22 \pm 1$[b] (C-nitro)	$+310$[b] (C-nitro)	85[b] (C-nitro)
$MeNHNO_2$	$MeOCH_2CH_2OMe$	$+23.2 \pm 0.5$[d]	$+309.3$[d]	10[d]
$O_2NNHCH_2CH_2NHNO_2$	$MeOCH_2CH_2OMe$	$+25.7 \pm 0.5$[d]	$+306.8$[d]	20[d]
$EtOOCNHNO_2$	ether	$+42 \pm 10$[e]	$+292$[e]	—
Me_2NNO_2	$CDCl_3$	$+23 \pm 10$[e]	$+312$[e]	—
		$+21.8 \pm 1.5$[f]	$+312.7$[f]	—
$NCCH_2CH_2CH_2NHNO_2$	$CDCl_3$	$+28 \pm 10$[f]	$+306$[f]	—
$HOOCCH_2NHNO_2$ (di-K salt)	D_2O	$+21 \pm 4$[f]	$+314$[f]	—
$EtOOCCH_2NHNO_2$ (K salt)	D_2O	$+21 \pm 7$[f]	$+314$[f]	—
$HOOCH_2NMeNO_2$	D_2O	$+26 \pm 4$[f]	$+308$[f]	—
$NCCH_2NMeNO_2$	D_2O	$+24 \pm 4$[f]	$+310$[f]	—
$MeCHNHNO_2 \mid MeCHNHNO_2$	acetone	$+26 \pm 7$[f]	$+308$[f]	—
O_2NONO_2	CCl_4	$+60 \pm 1$[b]	$+272$[b]	80[b]
	$CHCl_3$	$+48 \pm 2$[g]	$+286$[g]	—
$HONO_2$ (97%)	—	$+47.5 \pm 0.5$[b]	$+285$[b]	17[b]
$EtONO_2$	neat, CCl_4	$+37 \pm 0.5$[b]	$+296$[b]	17[b]
$ClONO_2$	neat	$+43.6$[g]	$+291$[g]	—
myoinositol hexanitrate	acetone	$+60 \pm 4$[b]	$+272$[b]	370[b]

a Ref. 121, direct ^{14}N measurements, referred to internal $MeNO_2$.
b Ref. 60, direct ^{14}N measurements, referred to internal $MeNO_2$.
c Ref. 20, direct ^{14}N measurements, referred to internal $MeNO_2$.
d Ref. 86, direct ^{14}N measurements, referred to external $MeNO_2$.
e Ref. 92, direct ^{14}N measurements, originally referred to external Me_2NNO.
f Ref. 51, double resonance ^{14}N data, referred to external $NO_3^{\ominus}$ (4.5M NH_4NO_3 in 3M HCl).
g Ref. 71, direct ^{14}N measurements, referred to 4.5M NH_4NO_3 in 3M HCl by sample replacement method.
h Recalculated by assuming that $Me_4N^{\oplus}$ resonance is at exactly 332.5 p.p.m. to higher field from $MeNO_2$ (Table 4.5) and 334.5 p.p.m. to higher field from $NO_3^{\ominus}$ [Table 4.6, footnote (a)].

TABLE 4.29 ^{14}N Chemical Shifts of some Anions Derived from Nitro Compounds

Compound	Solvent with about 20% KOH or NaOH	Concentration (mole %)	^{14}N Chemical shift in p.p.m.			^{14}N Resonance half-height width (Hz)	Net[b] π-charge at N	Total[b] mobile bond order at N
			Screening-constant scale referred to $MeNO_2$ or $NO_3^{\ominus}$		Frequency scale[a] referred to $Me_4N^{\oplus}X^{\ominus}$			
$NO_3^{\ominus}K^{\oplus}$	H_2O	5	0		+334.5[a]		+1.0145	1.7316
$CH_2NO_2^{\ominus}Na^{\oplus}$	H_2O	20	+79 ± 2[b]		+256[b]	450[b]	−0.8404	1.6680
	H_2O	20		+60 ± 3[c]	+272[c]	490[c]		
	MeOH	6.4		+50 ± 3[c]	+282[c]	420[c]		
$MeCHNO_2^{\ominus}Na^{\oplus}$	H_2O	16	+79 ± 3[b]		+256[b]	660[b]		
	H_2O	30		+60 ± 4[c]	+272[c]	740[c]		
	MeOH	7		+56 ± 3[c]	+276[c]	450[c]		
$EtCHNO_2^{\ominus}Na^{\oplus}$	H_2O	30		+68 ± 4[c]	+268[c]	680[c]		
	MeOH	9.5		+62 ± 3[c]	+270[c]	550[c]		
$Me_2CNO_2^{\ominus}Na^{\oplus}$	H_2O	18	+88 ± 4[b]		+246[b]	660[b]		
	H_2O	30		+79 ± 4[c]	+254[c]	1100[c]		
	MeOH	15		+74 ± 4[c]	+258[c]	670[c]		
$HC(NO_2)_2^{\ominus}Na^{\oplus}$	H_2O	12		+20 ± 1[c]	+312[c]	40[c]		
	DMF	10		+21 ± 1[c]	+312[c]	80[c]		
	DMSO	25		+20 ± 3[c]	+312[c]	450[c]		
$HC(NO_2)_2^{\ominus}K^{\oplus}$	H_2O	14	+20 ± 0.5[b]		+314.5[b]	46[b]	+0.8579	1.0842
	DMF	11	+20 ± 0.5[b]		+314.5[b]	50[b]		
$MeC(NO_2)_2^{\ominus}Na^{\oplus}$	H_2O	25		+22.5 ± 1[c]	+310[c]	42[c]		
	MeOH	20		+22.5 ± 1[c]	+310[c]	165[c]		
	MeOH	10		+23 ± 1[c]	+310[c]	100[c]		
	MeOH	5		+22.5 ± 1[c]	+310[c]	80[c]		
$MeC(NO_2)_2^{\ominus}K^{\oplus}$	H_2O	5	+24 ± 1[b]		+310[b]	40[b]		
$MeC(NO_2)_2^{\ominus}NH_4^{\oplus}$	DMF	3	+18 ± 1[b]		+316[b]	56[b]		
	DMF	10	+26 ± 1[b]		+308[b]	87[b]		

Compound	Solvent		Shift (±)	δ	Δ		
$EtC(NO_2)_3^{-}Na^{+}$	H_2O	10	+30 ± 1[b]	+304[b]	250[b]	+0.8669	
$C(NO_2)_3^{-}Na^{+}$	H_2O	25	+28 ± 2[c]	+304[c]	80[c]		
	H_2O	30	+31 ± 1[c]	+302[c]	20[c]		
	H_2O	15	+31 ± 1[c]	+302[c]	19[c]		
	H_2O	7	+30 ± 1[c]	+302[c]	16[c]		
	H_2O	3.5	+31 ± 1[c]	+302[c]	13[c]		
	MeOH	30	+31 ± 1[c]	+302[c]	19[c]		
$C(NO_2)_3K^{+}$	H_2O	2	+27.0 ± 0.5[b]	+307.5[b]	25[b]		1.6730
	H_2O	3	+28 ± 1[c]	+304[c]	13[c]		
	MeOH	5	+28 ± 1[c]	+304[c]	11[c]		
	Me_2CO	10	+28 ± 1[c]	+304[c]	11[c]		
	DMF	30	+30 ± 1[c]	+302[c]	21[c]		
	DMSO	40	+28 ± 1[c]	+304[c]	85[c]		
$C(NO_2)_3Cs^{+}$	DMF	30	+28 ± 1[c]	+304[c]	17[c]		
$NCC(NO_2)_2^{-}Na^{+}$	H_2O	30	(NO_2) +28 ± 1[c]	+304[c]	48[c]		
			(CN) +101 ± 4[c]	+232[c]	450[c]		
$BrC(NO_2)_2^{-}K^{+}$	H_2O	4	+28 ± 1[c]	+304[c]	17[c]		
$ClC(NO_2)_2^{-}K^{+}$	H_2O	5	+30 ± 1[c]	+302[c]	17[c]		
$^{-}(NO_2)_2CC(NO_2)_2^{-}\,2K^{+}$	H_2O	3	+27 ± 2[c]	+306[c]	48[c]		
$MeNNO_2^{-}Na^{+}$	H_2O		+26 ± 1[d]	+306[d]	43[d]		
$^{-}O_2NNCH_2CH_2NNO_2^{-}\,2Na^{+}$	H_2O		+28 ± 1[d]	+304[d]	103[d]		

a Recalculated to Me_4N^{+} shift, 334.5 p.p.m. upfield from NO_3^{-}, 332.5 p.p.m. from $MeNO_2$.
b Ref. 85, relative to internal NO_3^{-}.
c Ref. 110, 111 and 79, relative to external $MeNO_2$.
d Ref. 86, relative to external $MeNO_2$.

the resonance of $M^{\oplus}CH(NO_2)_2^{\ominus}$. The shifts have been explained in terms of the average excitation energy approximation and changes in the π-orbital system of the anions [85]. They have been shown to follow, in the downfield direction, the increasing total mobile bond order rather than the decreasing π-charge density at the nitrogen atom. A comparison of the results $[85]^{\ominus}$ of SCF—MO calculations for the π-electron system in these ions with the corresponding nitrogen chemical shifts suggests that the anions derived from mono-nitroalkanes have structures close to that represented by the classical formula illustrated in Diagram 4.22, whilst the

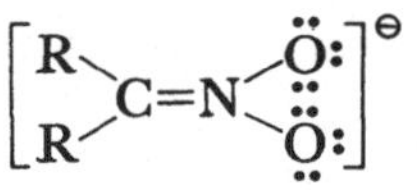

$$\left[\begin{array}{c} R \\ R \end{array} C=N \begin{array}{c} \ddot{O}: \\ \ddot{O}: \end{array}\right]^{\ominus}$$

Diagram 4.22

anions derived from geminal polynitroalkanes may be considered as nitro-carbanions where the NO_2 moiety is structurally closer to the true nitro group. The nitrogen NMR spectra of the latter indicate that the nitro groups in such polynitrocarbanions are equivalent [85, 87, 111] since only one narrow signal is observed for both ^{14}N and ^{15}N down to $-75°$. In the ^{15}N spectrum of $^{\ominus}CH(NO_2)_2$, the signal is always a doublet due to spin–spin coupling to ^1H (about 6 Hz) [87].

There is also a downfield shift for the $(X—NO_2)^{\ominus}$ ions with increasing electronegativity of the elements in $X = R_2C$, RN, O.

4.4.16   Nitroso Compounds and Nitrites

Molecules which contain the nitroso group give nitrogen NMR signals at very low fields (high frequencies). Their

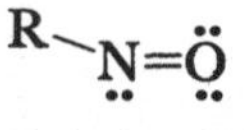

$$R \diagdown N=\ddot{O}$$

Diagram 4.23

chemical shifts are clearly outside the spectral range of all other organic compounds. Their ^{14}N resonance linewidths are usually large. It is possible to make a spectroscopic distinction in this group of molecules between the following:

	Screening-constant scale (p.p.m.)	Frequency scale (p.p.m.)
nitrosamines, R_2NNO, and nitrites, $R-O-NO$	ca. -150 to -200	$+480$ to $+530$
thionitrites, $R-S-NO$ and nitrosoalkanes, $R-NO$	ca. -350 to -450	$+680$ to $+780$
aromatic nitroso compounds	ca. -500	$+830$

The shifts for nitrites, RONO, distinguish them from the isomeric nitro compounds, RNO_2, as mentioned in Section 4.4.14. There is also a remarkable difference in nitrogen chemical shifts between the isomeric structures of nitroso compounds and oximes:

	Screening-constant scale (p.p.m.)	Frequency scale (p.p.m.)
nitroso compounds, R_2CH-NO	-400 to -500	$+730$ to $+830$
oximes, $R_2C=N-OH$	ca. $+30$	ca. $+300$

The nitrogen chemical shifts of the nitroso group show a roughly linear corrleation [91] with the energies of the lowest transitions observed in the electronic absorption spectra. The decreasing energy is accompanied by a downfield nitrogen chemical shift. The nitroso structure is a very peculiar example, theoretical calculations [94] indicate a dominant contribution (deshielding) to the nitrogen screening constant by the lowest-energy excitation connected with the presence of the lone electron pair at the nitrogen atom in the NO group. This is an exception which constitutes an explanation of the observed relationship between the shifts and the transitions. For other organic compounds, attempts to explain nitrogen chemical shifts in terms of experimental electronic transitions are unwarranted and actually have not been successful.

4.5   Medium Effects on Nitrogen Chemical Shifts

Nitrogen chemical shifts may provide an insight into molecular interactions in liquids. So far we have been interested mostly in changes in nitrogen chemical shifts resulting from structural changes within the molecule concerned now we consider the more subtle effects exerted by neighbouring molecules. The best way to follow intermolecular effects in liquids by nitrogen NMR methods is to refer the chemical

TABLE 4.30.   Nitrogen Chemical Shifts of some Nitroso Compounds

| Compound | Solvent or state | Nitrogen chemical shift (p.p.m.) | | ^{14}N Resonance half-height width (Hz) |
		Screening-constant scale referred to $MeNO_2$ or $NO_3^{\ominus}$.	Frequency scale[f] referred to $Me_4N^{\oplus}X^{\ominus}$	
Me_2NNO	neat	-158 ± 10[a]	$+492$[a]	355[a]
MeONO	neat	-188 ± 10[a]	$+522$[a]	265[a]
n-BuONO	neat	-189 ± 1[e]	$+522$[e]	
CF_3SNO ($-85°$)	neat	-342 ± 10[a]	$+676$[a]	1140[a]
EtSNO	neat	-412 ± 10[a]	$+746$[a]	455[a]
CF_3NO ($-90°$)	neat	-427 ± 10[a]	$+762$[a]	450[a]
C_3F_7NO	neat	-487 ± 10[a]	$+822$[a]	?
PhNO	ether	-489 ± 10[a]	$+824$[a]	1120[a]
	ether	-523[b]	$+856$[b]	
	acetone	-530 ± 4[c]	$+862$[c]	750[c]
$NO_2^{\ominus}$	H_2O	$-237 + 4$[d]	$+572$[d]	520[d]

[a] Ref. 91, direct ^{14}N measurements, originally referred to saturated aqueous $NO_2^{\ominus}$ shift (237 p.p.m. downfield from $NO_3^{\ominus}$, ref. 60).

[b] Ref. 23, direct ^{15}N measurements, originally referred to 70% HNO_3, recalculated relative to $MeNO_2$ through neat $MeNH_2$ shift (378 p.p.m. upfield from $MeNO_2$, Table 4.6).

[c] Ref. 121, direct ^{14}N measurements, referred to internal $MeNO_2$.

[d] Ref. 60, direct ^{14}N measurements, referred to internal $NO_3^{\ominus}$.

[e] Ref. 26, direct ^{15}N measurements, referred to anhydrous NH_3; recalculated with neat NH_3 shift ($+383$ p.p.m. upfield from $MeNO_2$, 48 p.p.m. from $NMe_4^{\oplus}$, Table 4.6).

[f] Recalculated to $Me_4N^{\oplus}$ shift, 332.5 p.p.m. upfield from $MeNO_2$ (Table 4.5), 334.5 p.p.m. upfield from $NO_3^{\ominus}$ [Table 4.6, footnote (a)].

shifts of a compound in the liquid phase to the shifts observed in the gaseous phase. The NMR spectrum of a gaseous compound seems to be insensitive, as far as the chemical shift is concerned, to a wide range of changes in pressure and temperature. This has been found [70, 89, 103] for NH_3, NMe_3 and acetonitrile. However, volume-susceptibility corrections should be made in relating the shifts in liquids to those of the gaseous phase. This is a potential source of error, as shown in the case of the NH_3 nitrogen chemical shifts [70] which have been recently corrected [89].

Investigations [70, 89] of the relatively simple molecular structures of NH_3 and NMe_3 in a large variety of solvents (Table 4.31) has led to the conclusion that changes in the

TABLE 4.31. ^{15}N Solvent Shifts for NH_3 and NMe_3
(Data from refs. 70 and 89)

Solvent	^{15}N shift (p.p.m.) at infinite dilution, frequency, scale[a] referred to corresponding gaseous compound[b]	
	NH_3	NMe_3
H_2O	+21.5	+11.1
Me_2O	+ 9.8	+ 4.0
Me_3N	+12.1	+ 4.2
Me_4C	+13.2	+ 4.2
$MeNH_2$	+14.4	+ 5.1
$MeOH$	+15.5	+ 9.2
$EtOH$	+18.7	+ 8.9
Me_2NH	+12.2	+ 5.0
$EtNH_2$	+15.7	+ 5.1
Et_2O	+12.8	+ 4.6
Et_3N	+14.5	+ 4.6
Et_2NH	+14.8	+ 5.0
NH_3	+18.0	+ 5.4
CCl_4	+18.0	+ 7.3

[a] This means that high frequency (low field) shifts are positive.

[b] Shift of NMe_3 (vapour) is 28.7 p.p.m. in this scale relative to NH_3 (vapour), ref. 103; shifts of neat liquids are underlined.

shifts are linear with those in the mole % concentrations of the solute and that at infinite dilution the shifts may be explained in terms of an additive scheme of interactions between the various sites in molecules of the solute and solvent. It is a very interesting approach which may reveal specific interactions. The 'infinite dilution shift' (referred to the gas phase) may be represented by a sum of contributions due to specific interactions

$$\delta_N \text{(inf. dil.)} = \sum_x C_x \sigma_A(x) + \sum_y C_y \sigma_B(y) \qquad (4.17)$$

where σ represents the contributions due to interactions of the NH_3 or NMe_3 nitrogen lone-pair electrons with

$$
\begin{array}{ll}
\text{solvent OH proton} & \sigma_A(OH) \\
\text{solvent NH proton} & \sigma_A(NH) \\
\text{solvent Me group} & \sigma_A(Me) \\
\text{solvent Et group} & \sigma_A(Et)
\end{array}
$$

(no interactions with basic centres are considered) or due to interactions of the methyl groups in NMe_3 or the H atoms in NH_3 with

$$
\begin{array}{ll}
\text{solvent oxygen lone-pair electrons} & \sigma_B(O) \\
\text{solvent nitrogen lone-pair electrons} & \sigma_B(N).
\end{array}
$$

The coefficients C_x and C_y express the ratios of the number of the specified sites in a solvent molecule to the total number of active sites in this molecule. Thus, for example, the infinite dilution shift for NMe_3 dissolved in $Et_2 NH$ will be

$$
\delta(NMe)_{Et_2NH} = \tfrac{2}{3}\sigma_A(Et) + \tfrac{1}{3}\sigma_A(NH) + \sigma_B(N) \qquad (4.18)
$$

The calculations give the following values [89] for the contributions (in p.p.m., frequency scale):

	NMe_3	NH_3
$\sigma_A(OH)$	11.9	25.2
$\sigma_A(NH)$	5.4	19.1
$\sigma_A(Et)$	4.7	16.5
$\sigma_A(Me)$	4.4	13.2
$\sigma_B(O)$	0	−3.3
$\sigma_B(N)$	0	−2.1

An interesting conclusion to be drawn from such calculations is that there are comparable effects arising from the interactions between the lone electron pair of the solute and the OH or NH protons of the solvent and interactions between the lone electron pair of the solute and the methyl or ethyl groups of the solvent. The former are of the hydrogen-bonding type but the latter must be of a different nature. It seems, therefore, that nitrogen chemical shifts cannot be used as a measure of hydrogen-bonding. Also of some interest is the downfield (high-frequency) shift for NMe_3 in CCl_4 which shows that considerable solvent–solute interactions take place and that donor-acceptor complexes may be formed between the amine and carbon tetrachloride. The same conclusion may be reached in the case of ammonia where, curiously enough, almost no effect on the nitrogen chemical shift is observed upon dilution with CCl_4 (Table 4.31). This may mean that either the NH_3 molecules remain in aggregates, even at infinite dilution, or two opposing effect must operate. One due to the breaking of interactions between NH_3 molecules (resulting in an upfield shift) and another involving interactions between NH_3 and CCl_4 (resulting in a downfield shift). The latter explanation seems to be more acceptable if a comparison is made with the results for trimethylamine.

TABLE 4.32.   Dilution Shifts of Nitrogen Resonance of some Amides

| Compound | Chemical shift (p.p.m.) (screening-constant scale) referred to external $NO_3^{\ominus}$ in 4.5M NH_4NO_3 in 3M HCl | | | |
	Neat	Inf. dil. acetone	Inf. dil. methanol	Inf. dil. dioxane
$HCONH_2$	$+265.5 \pm 0.4$[a]	$+274 \pm 2$[a]	$+268 \pm 1.5$[a]	$+274 \pm 0.5$[b]
$HCONHMe$	$+266.4 \pm 0.7$[a]	$+272 \pm 2$[a]	$+267 \pm 1.5$[a]	$+278 \pm 0.5$[b]
$HCONMe_2$	$+272.3 \pm 0.5$[a]	$+273 \pm 1.5$[a]	$+264 \pm 1.5$[a]	—
$HCONHEt$	$+247.1 \pm 0.7$[a]	$+253 \pm 2$[a]	$+247 \pm 1.5$[a]	—
$HCONEt_2$	$+246.5 \pm 0.5$[a]	$+247 \pm 1.5$[a]	$+237 \pm 1.5$[a]	—

[a] Ref. 59, ^{14}N double resonance data.
[b] Ref. 100, ^{14}N double resonance data.

High-field (low-frequency) dilution shifts for simple amides in dioxane solutions [100] (Table 4.32) have been interpreted in terms of the breaking of hydrogen-bonded associates of the amides upon dilution. A similar explanation has been given [99] for analogous observations of pyrrole derivatives diluted with CCl_4 or $CDCl_3$ (Table 4.33). However, one should be cautious in reaching conclusions since interactions other than hydrogen-bonding may strongly influence nitrogen chemical shifts, even in such recognized 'inert' solvents as carbon tetrachloride, as has been shown in the preceding discussion. This has also been suggested by the recent ^{15}N

TABLE 4.33. ^{14}N Dilution Shifts of some Pyrrole Derivatives (Ref. 99)

Infinite dilution shift (p.p.m.) (error ±0.5); screening constant scale referred to NO$_3^{\ominus}$ in 4.5M NH$_4$NO$_3$ in 3M HCl

Solvent	Pyrrole	2-Acetyl-pyrrole	2,5-Diacetyl-pyrrole	α-Methyl indole	β-Methyl indole	Indole	Carbazole
neat	+228.0	—	—	—	—	—	—
DMSO	+218.5	+215.5	+220.5	—	—	—	—
H$_2$O	+220.5	—	—	—	—	—	—
NEt$_3$	+221.0	+220.5	—	—	—	—	—
DMF	+222.0	+220.0	—	—	—	—	—
MeCN	+225.0	+223.0	+228.5	—	—	—	—
Me$_2$CO	+226.0	+223.5	+229.0	—	—	—	+264 ± 1
dioxane	+227.0	+222.5	+225.5	—	—	—	—
MeCOOH	+228.0	—	+228.0	—	—	—	—
benzene	+231.0	+226.0	+231.5	+246 ± 1	+256 ± 1	—	—
CCl$_4$	+232.5	+226.5	—	—	—	+254 ± 1.4	—
CDCl$_3$	—	+226.0	+232.0	—	—	—	—

double-resonance study [88] (Table 4.34) of solvent shifts in nitrogen NMR. Solvents which are generally considered as hydrogen-bond acceptors give downfield (high-frequency) nitrogen chemical shifts for aniline, methylamine and *N*-methylacetamide but a similar effect is observed for solutions in CCl_4. In the case of aniline, the downfield dilution shift in CCl_4 solutions is even larger than in dioxane solutions. Usually, solvent shifts in proton–donor solvents such as water and methanol follow the direction indicated by the nitrogen chemical shift of the corresponding conjugate acid of the solute. This has been found in amines (Tables 4.6, 4.9 and 4.31), aromatic heterocycles (Tables 4.22, 4.34), and nitriles (Table 4.18, ref. 32).

$$NMe_3 \text{ (neat } Me_4C) \xrightarrow[\text{shift}]{\text{downfield}} NMe_3 \text{ (MeOH, EtOH, } H_2O) \xrightarrow[\text{shift}]{\text{downfield}}$$

$$HNMe_3^{\oplus} \text{ (HCl}_{aq.})$$

$$\text{quinoline } (C_6H_{12}) \xrightarrow[\text{shift}]{\text{upfield}} \text{quinoline (MeOH)} \xrightarrow[\text{shift}]{\text{upfield}}$$

$$\text{quinolinium ion (HCl}_{aq.})$$

$$MeCN \text{ (neat)} \xrightarrow[\text{shift}]{\text{upfield}} MeCN (H_2O, MeOH) \xrightarrow[\text{shift}]{\text{upfield}}$$

$$MeCNH^{\oplus}(FSO_3H)$$

The nitrogen resonance signals of ammonium ions, $Me_x \overset{\oplus}{N}H_{4-x}$, in aqueous HCl have been shown [104] to move downfield, almost linearly with an increase in total $Cl^{\ominus}$ molarity. They are independent of the amine hydrochloride concentration (Table 4.35). The sensitivity of the shifts to the total $Cl^{\ominus}$ concentration has been found to increase with an increasing number of hydrogen atoms in the ion.

The entire effect on nitrogen chemical shifts of molecular interactions in liquids may be found from the gas–liquid resonance shift at the melting point temperature. Such nitrogen chemical shifts have been measured [103] for simple amines and acetonitrile (Table 4.36). The liquid-association shifts are downfield for all of the amines in Table 4.36 with the exception of acetonitrile. The shifts in the liquids depend almost linearly on temperature, and the corresponding temperature coefficients are closely related to the magnitudes of the corresponding liquid-association shifts (Table 4.36). The ratio of the latter shift to the temperature coefficient is fairly constant at -500 ± 100. For neat liquid NH_3 where an equilibrium between hydrogen-bonded and non-hydrogen-bonded molecular species should exist, and where no interactions other than those between N and H should be

TABLE 4.34. Solvent Shifts of ^{15}N Magnetic Resonance in Various Types of Organic Molecules (Data from ref. 88)

Compound	Solvent (concentration 0.1M)	^{15}N Chemical shifts (p.p.m.) frequency scale[a] relative to arbitrary references
aniline	cyclohexane-d_{12}	0
	CCl$_4$	+2.2
	dioxane-d_8	+0.6
	pyridine-d_5	+5.2
	DMF-d_7	+4.2
	DMSO-d_6	+7.7
MeNH$_2$	neat	−1.5
	cyclohexane-d_{12}	0
	CCl$_4$	+2.3
	DMF-d_7	+0.2
	DMSO-d_6	+2.2
MeNHCOMe	CCl$_4$ (1M)	0
	CCl$_4$ (0.25M)	+1.3
	DMSO (0.25M)	+1.5
	neat	+7.8
quinoline	cyclohexane-d_{12}	0
	CDCl$_3$	−10.5
	CD$_3$OD	−22.8
NMe$_3$	cyclohexane-d_{12}	0
	CDCl$_3$	+3.5
	CD$_3$OD	+4.3

a This means that high-frequency (lowfield) shifts are positive; double resonance ^{15}N data.

important, the temperature dependence of the nitrogen chemical shift has been tentatively explained [103] as being due to changes in the equilibrium with temperature. An adjustment of the parameters in equation

$$\delta(NH_3) = \frac{\delta(\text{H-bond})K_0 \exp\left(-\Delta H/RT\right)}{1 + K_0 \exp\left(-\Delta H/RT\right)} \tag{4.19}$$

where $\delta(NH_3)$ is the observed shift, $\delta(\text{H-bond})$ is the difference in nitrogen chemical shift between the 'monomeric' and hydrogen-bonded molecules, K_0 is a constant, and ΔH is

TABLE 4.35.   Variation of ^{15}N Shifts in Me_xNH_{4-x} with Total Chloride Concentration (5M in Amine Hydrochloride)

Ion	Total $Cl^{\ominus}$ molarity	^{15}N shift[a] frequency scale referred to NH_3 liquid
$NH_4^{\oplus}$	2	24
	6	26
	11	32
$MeNH_3^{\oplus}$	6	25
	17	30.5
$Me_2NH_2^{\oplus}$	6	27
	17	31
$Me_3NH^{\oplus}$	3	33
	15	35.5
$Me_4N^{\oplus}$	6	45
	17	45.5

a Ref. 104, direct ^{15}N measurements, read from plot with accuracy ± 0.5 p.p.m.

the enthalpy difference involved in hydrogen-bonding, gives the following values which reproduce satisfactorily the temperature dependence of the shift:

δ (H-bond) = 24.5 ± 0.7 p.p.m. to lower fields for the H-bonded species,

and $K_0 = 0.22 \pm 0.05$,

$H = -1.5 \pm 0.2$ kcal/mole (6280 joule/mole).

TABLE 4.36.   Liquid–Vapour ^{15}N Shifts for Some Simple Amines and Acetonitrile (Data from ref. 103)

Compound	Chemical shifts (p.p.m.), frequency scale[a] referred to gaseous NH_3			
	^{15}N shift vapour	^{15}N shift liquid[b]	Liquid association shift[c]	Temperature coefficient of ^{15}N shift (p.p.m./°C)
NH_3	0	+22.6 −77.7°	+22.6	-4.3×10^{-2}
$MeNH_2$	+14.5	+23.9 −93.5°	+ 9.4	-1.6×10^{-2}
Me_2NH	+26.1	+29.3 −96°	+ 3.2	-0.5×10^{-2}
Me_3N	+28.7	+35.6 −117°	+ 6.9	-2.0×10^{-2}
$MeCN$	+273.4	+262.1 −45.7°	−11.3	$+2.1 \times 10^{-2}$

a High-frequency (downfield) shifts are positive in this scale.
b Shift at melting point temperature; the latter given in parentheses.
c Difference between gas and liquid at melting point with correction for volume susceptibility of gas.

This corresponds to about 91% H-bonded nitrogen atoms in ammonia at its melting point and about 83% at its boiling point. However, simple amines, including NMe_3, show even larger liquid-association shifts in nitrogen NMR which must result, either partly or entirely, from the thermal perturbations of interactions other than hydrogen-bonding. This is another example of the considerable influence of such interactions which may be more marked in nitrogen NMR spectra than the effects of hydrogen-bonding.

4.6   Shift Reagents in Nitrogen NMR Spectroscopy

The use of various chelates of lanthanide ions for increasing relative chemical shifts of ^1H nuclei is now widespread. An application of such 'shift reagents' to nitrogen NMR, however, has not been attempted until very recently [106, 114]. The most important aspect of such an application seems to involve differences in the magnitude of the induced shifts for various types of organic compounds of nitrogen. The examples (Table 4.37) of the nitrogen chemical shifts induced by *tris*-dipivalo-

TABLE 4.37.   Effect of $Eu(DPM)_3$ and $Yb(DPM)_3$ on Nitrogen Chemical Shifts of Organic Molecules (Data from ref. 106)

Compound	Concentration (CCl_4 mole/l)	^{14}N Chemical shift (p.p.m.), screening-constant scale referred to internal $MeNO_2$	Screening-constant scale (p.p.m.), induced shift extrapolated to 1:1 solute to chelate molar ratio	
			$Yb(DPM)_3$	$Eu(DPM)_3$
pyrrolidine	3.86	$+339 \pm 2$	-410 ± 20	$+1500 \pm 200$
n-propylamine	2.34	$+355 \pm 1$	-340 ± 20	$+1600 \pm 200$
pyridine	3.19	$+58 \pm 1$	-425 ± 20	$+1500 \pm 200$
N-methylimidazole	3.38	$+218 \pm 1$ (NMe)	-10 ± 10	$+80 \pm 50$
		$+116 \pm 3$ (N:)	-205 ± 20	$+1400 \pm 300$
acetonitrile	3.08	$+130 \pm 1$	-135 ± 5	$+490 \pm 100$
dimethylformamide	2.22	$+277 \pm 1$	-30 ± 30	$+50 \pm 30$

methanates (DPM) of europium and ytterbium [106] reveal a remarkable differentiation between various structures. The largest effects are observed for molecules with 'nucleophilic' lone electron pairs (amines, pyridine and related compounds) but steric hindrance may also make an important contribution. If the lone electron pair on a nitrogen atom is involved in a delocalized π-electron system then the magnitude of the induced nitrogen chemical shift is drastically decreased, as can be seen from the comparison of the two types of nitrogen atoms in N-methylimidazole (Table 4.37) and the weak response to shift reagents in dimethylformamide. For acetonitrile where the nitrogen lone electron pair is a weak nucleophile, the induced shift is considerably reduced compared with amines. The nitrogen resonance of the nitro group is practically unaffected by shift reagents [114].

TABLE 4.38.   Comparison of Pyridine ^{14}N Shifts Induced by *tris*-(DPM)-Chelates of Lanthanides and Actinides (Data from ref. 114)

Chelate	Concentration of pyridine (CCl_4 mole/l)	Induced ^{14}N shift extrapolated to 1:1 chelate to solute molar ratio, screening-constant scale (p.p.m.)	Approximate signal-broadening[a] limit for chelate-to-solute ratio
La(DPM)$_3$	2.60	-380 ± 30	0.013
Ce(DPM)$_3$	2.52	-290 ± 50	0.010
Pr(DPM)$_3$	2.42	-450 ± 300	0.008
Nd(DPM)$_3$	2.74	-360 ± 300	0.008
Sm(DPM)$_3$	2.65	-540 ± 50	0.010
Eu(DPM)$_3$	2.05	$+1500 \pm 200$	0.029
Gd(DPM)$_3$	2.60	$+2900 \pm 500$	0.009
Dy(DPM)$_3$	2.80	$+4000 \pm 300$	0.030
Ho(DPM)$_3$	2.88	$+2100 \pm 200$	0.020
Er(DPM)$_3$	2.60	$+870 \pm 100$	0.020
Yb(DPM)$_3$	3.19	-425 ± 20	0.057
Th(DPM)$_3$	2.92	0 ± 100	0.011
U(DPM)$_3$	2.58	0 ± 300	0.004

[a] Approximate maximum chelate-to-solute ratio which is still useful for extrapolation of induced shift to 1 : 1 molar ratio; indirect measure of signal broadening by individual shift reagents.

A comparison of M(DPM)$_3$ shift reagents (Table 4.38) indicates that for upfield nitrogen shifts the best of them is Dy(DPM)$_3$ followed by the chelates of Ho and Eu. Among the downfield-shift reagents, the most convenient is Yb(DPM)$_3$ which gives the least signal broadening. One should note that Pr(DPM)$_3$ which is a commonly used shift reagent for ^{1}H, seems to be rather useless in nitrogen NMR because of the strong signal-broadening effect.

The induced shifts in nitrogen NMR spectra are highly characteristic of the molecular environment of the nitrogen nuclei and may be used for signal assignments and structural investigations.

References

1. W. G. PROCTOR and F. C. YU, *Phys. Rev.*, 77, 716 (1950).
2. W. G. PROCTOR and F. C. YU, *Phys. Rev.*, 77, 717 (1950).
3. Y. MASUDA and T. KANDA, *J. Phys. Soc. Japan*, 8, 432 (1953).
4. B. E. HOLDER and M. P. KLEIN, *J. Chem. Phys.*, 23, 1956 (1955).
5. R. A. OGG, Jr. and J. D. RAY, *J. Chem. Phys.*, 25, 1285 (1956).
6. R. A. OGG, Jr. and J. D. RAY, *J. Chem. Phys.*, 26, 1339 (1957).
7. J. D. RAY and R. A. OGG, Jr., *J. Chem. Phys.*, 26, 1452 (1957).
8. J. D. RAY, L. H. PIETTE, and D. P. HOLLIS, *J. Chem. Phys.*, 29, 1022 (1958).
9. B. M. SCHMIDT, L. C. BROWN, and D. H. WILLIAMS, *J. Mol. Spectrosc.*, 2, 539 (1958).
10. B. M. SCHMIDT, L. C. BROWN, and D. H. WILLIAMS, *J. Mol. Spectrosc.*, 2, 551 (1958).
11. B. M. SCHMIDT, L. C. BROWN, and D. H. WILLIAMS, *J. Mol. Spectrosc.*, 3, 30 (1959).
12. G. E. ENGLERT, *Z. Elektrochem.*, 65, 854 (1961).
13. J. D. BALDESCHWIELER and E. W. RANDALL, *Proc. Chem. Soc.*, 303 (1961).
14. E. B. BAKER, *J. Chem. Phys.*, 37, 911 (1962).
15. T. KANDA, Y. SAITO, and K. KAWAMURA, *Bull. Chem. Soc. Japan*, 35, 172 (1962).
16. J. D. BALDESCHWIELER, *J. Chem. Phys.*, 36, 152 (1962).
17. D. HERBISON-EVANS and R. E. RICHARDS, *Trans. Faraday Soc.*, 58, 845 (1962).
18. J. D. BALDESCHWIELER and E. W. RANDALL, *Chem. Rev.*, 63, 81 (1963).
19. M. R. BAKER and N. F. RAMSEY, *Phys. Rev.*, 133A, 1533 (1964); S. I. CHAN, M. R. BAKER, and N. F. RAMSEY, *Phys. Rev.*, 136A, 1224 (1964).
20. M. WITANOWSKI, T. URBANSKI, and L. STEFANIAK, *J. Amer. Chem. Soc.*, 86, 2568 (1964).
21. V. M. S. GIL and J. N. MURRELL, *Trans. Faraday Soc.*, 60, 248 (1964).
22. J. D. RAY, *J. Chem. Phys.*, 40, 3440 (1964).
23. J. B. LAMBERT, G. BINSCH, and J. D. ROBERTS, *Proc. Natl. Acad. Sci. US.*, 51, 735 (1964).
24. D. HERBISON-EVANS and R. E. RICHARDS, *Mol. Phys.*, 8, 19 (1964).
25. J. B. LAMBERT, B. W. ROBERTS, G. BINSCH, and J. D. ROBERTS, in *Nuclear Magnetic Resonance in Chemistry* (B. PESCE, Ed.), Academic Press, New York, London, 1965, p. 269.
26. J. B. LAMBERT and J. D. ROBERTS, *J. Amer. Chem. Soc.*, 87, 4087 (1965).

27. J. B. LAMBERT, W. L. OLIVER, and J. D. ROBERTS, *J. Amer. Chem. Soc.*, **87**, 5085 (1965).
28. B. W. ROBERTS, J. B. LAMBERT, and J. D. ROBERTS, *J. Amer. Chem. Soc.*, **87**, 5439 (1965).
29. N. LOGAN and W. L. JOLLY, *Inorg. Chem.*, **4**, 1508 (1965).
30. H. SAITÔ, K. NUKADA, H. KATO, T. YONEZAWA, and K. FUKUI, *Tetrahedron Letts.*, 111 (1965).
31. M. BOSE, N. DAS, and N. CHATTERJEE, *J. Mol. Spectrosc.*, **18**, 32 (1965).
32. A. LOEWENSTEIN and Y. MARGALIT, *J. Phys. Chem.*, **69**, 4152 (1965).
33. D. T. CLARK and J. D. ROBERTS, *J. Amer. Chem. Soc.*, **88**, 745 (1966).
34. J. A. HAPPE and M. MORALES, *J. Amer. Chem. Soc.*, **88**, 2077 (1966).
35. J. E. KENT and E. L. WAGNER, *J. Chem. Phys.*, **44**, 3530 (1966).
36. G. FRAENKEL, Y. ASAHI, H. BATIZ-HERNANDEZ, and R. A. BERNHEIM, *J. Chem. Phys.*, **44**, 4647 (1966).
37. P. HAMPSON and A. MATHIAS, *Mol. Phys.*, **11**, 541 (1966).
38. W. McFARLANE, *J. Chem. Soc.*, A, 1660 (1967).
39. M. WITANOWSKI and L. STEFANIAK, *J. Chem. Soc.*, B, 1061 (1967).
40. M. WITANOWSKI and H. JANUSZEWSKI, *J. Chem. Soc.*, B, 1063 (1967).
41. M. WITANOWSKI, L. STEFANIAK, and G. A. WEBB, *J. Chem. Soc.*, B, 1065 (1967).
42. K. HENSEN and K. P. MESSER, *Theoret. Chim. Acta*, **9**, 17 (1967).
43. P. HAMPSON and A. MATHIAS, *Chem. Comm.*, 371 (1967).
44. M. WITANOWSKI, *Tetrahedron*, **23**, 4299 (1967).
45. A. MATHIAS, *Mol. Phys.*, **12**, 381 (1967).
46. P. HAMPSON and A. MATHIAS, *Mol. Phys.*, **13**, 361 (1967).
47. C. F. PORANSKI, Jr. and W. B. MONIZ, *J. Phys. Chem.*, **71**, 1142 (1967).
48. F. J. WEIGERT and J. D. ROBERTS, *J. Amer. Chem. Soc.*, **89**, 2967 (1967).
49. T. K. WU, *J. Chem. Phys.*, **49**, 1139 (1968).
50. Y. VIGNOLLET, J. C. MAIRE, and M. WITANOWSKI, *Chem. Comm.*, 1187 (1968).
51. P. HAMPSON and A. MATHIAS, *Chem. Comm.*, 825 (1968).
52. L. O. ANDERSSON and J. MASON, *Chem. Comm.*, 99 (1968).
53. J. W. EMSLEY, *J. Chem. Soc.*, A, 1387 (1968).
54. W. McFARLANE and R. R. DEAN, *J. Chem. Soc.*, A, 1535 (1968).
55. P. HAMPSON and A. MATHIAS, *J. Chem. Soc.*, B, 673 (1968).
56. A. H. BERRIE, P. HAMPSON, S. W. LONGWORTH, and A. MATHIAS, *J. Chem. Soc.*, B, 1308 (1968).
57. L. T. EREMENKO, A. A. BORISENKO, S. I. PETROV, and V. F. ANDRONOV, *Izv. Akad. Nauk SSSR, Ser. Khim.*, 428 (1968).
58. L. T. EREMENKO and A. A. BORISENKO, *Izv. Akad. Nauk SSSR, Ser. Khim.*, 675 (1968).

59. H. KAMEI, *Bull. Chem. Soc. Japan*, 41, 1030 (1968).
60. M. WITANOWSKI, *J. Amer. Chem. Soc.*, 90, 5683 (1968).
61. E. W. RANDALL and J. J. ZUCKERMAN, *J. Amer. Chem. Soc.*, 90, 3167 (1968).
62. G. A. OLAH and T. E. KIOVSKY, *J. Amer. Chem. Soc.*, 90, 4666 (1968).
63. W. McFARLANE, in *Annual Reports of NMR Spectroscopy*, Vol. 1 (E. F. MOONEY, Ed.), Academic Press, New York, London, 1968, p. 135.
64. W. B. MONIZ and C. F. PORANSKI, Jr., *J. Phys. Chem.*, 73, 4145 (1969).
65. W. M. LITCHMAN, M. ALEI, and A. E. FLORIN, *J. Chem. Phys.*, 50, 1897 (1969).
66. T. K. WU, *J. Chem. Phys.*, 51, 3622 (1969).
67. W. BREMSER, J. I. KROSCHWITZ, and J. D. ROBERTS, *J. Amer. Chem. Soc.*, 91, 6189 (1969).
68. T. TOKUHIRO and G. FRAENKEL, *J. Amer. Chem. Soc.*, 91, 5005 (1969).
69. M. WITANOWSKI and H. JANUSZEWSKI, *Canad. J. Chem.*, 47, 1321 (1969).
70. W. M. LITCHMAN, M. ALEI, and A. E. FLORIN, *J. Amer. Chem. Soc.*, 91, 6574 (1969).
71. A. M. QURESHI, J. A. RIPMEESTER, and F. AUBKE, *Canad. J. Chem.*, 47, 4247 (1969).
72. M. R. BRAMWELL and E. W. RANDALL, *Chem. Comm.*, 250 (1969).
73. J. P. KINTZINGER, J. M. LEHN, and R. M. WILLIAMS, *Mol. Phys.*, 17, 135 (1969).
74. W. McFARLANE and D. H. WHIFFEN, *Mol. Phys.*, 17, 603 (1969).
75. R. J. CHUCK, D. G. GILLIES, and E. W. RANDALL, *Mol. Phys.*, 16, 121 (1969).
76. H. HENSEN and K. P. MESSER, *Chem. Ber.*, 102, 957 (1969).
77. E. F. MOONEY and P. H. WINSON, in *Annual Reports of NMR Spectroscopy*, Vol. 2 (E. F. MOONEY, Ed.), Academic Press, New York, London, 1969, p. 125.
78. W. N. LITCHMAN, M. ALEI, and A. E. FLORIN, *J. Chem. Phys.*, 50, 1031 (1969).
79. M. Y. MÄGI, E. T. LIPPMAA, T. I. PEHK, S. A. SHEVELEV, V. I. ERASHKO, and A. A. FAINZILBERG, *Izv. Akad. Nauk SSSR, Ser. Khim.*, 730 (1969).
80. M. Y. MÄGI, E. T. LIPPMAA, Y. O. PAST, V. I. ERASHKO, and S. A. SHEVELEV, *Izv. Akad. Nauk SSSR, Ser. Khim.*, 2089 (1969).
81. J. DABROWSKI, A. SKUP, and M. SONELSKI, *Org. Mag. Res.*, 1, 341 (1969).
82. W. BECKER and W. BECK, *Z. Naturforsch.*, 25b, 101 (1970).
83. W. BECKER, W. BECK, and R. RIECK, *Z. Naturforsch.*, 25b, 1332 (1970).
84. R. K. HARRIS, N. C. PYPER, R. E. RICHARDS, and G. W. SCHULZ, *Mol. Phys.*, 19, 145 (1970).

85. M. WITANOWSKI and S. A. SHEVELEV, *J. Mol. Spectrosc.*, **33**, 19 (1970).
86. M. Y. MÄGI, E. T. LIPPMAA, S. A. SHEVELEV, V. I. ERASHKO, and A. A. FAINZILBERG, *Izv. Akad. Nauk SSSR, Ser. Khim.*, 1450 (1970).
87. V. I. ERASHKO, S. A. SHEVELEV, A. A. FAINZILBERG, M. Y. MÄGI, and E. T. LIPPMAA, *Izv. Akad. Nauk SSSR, Ser. Khim.*, 958 (1970).
88. L. PAOLILLO and E. D. BECKER, *J. Mag. Res.*, **2**, 168 (1970).
89. M. ALEI, A. E. FLORIN, and W. M. LITCHMAN, *J. Amer. Chem. Soc.*, **92**, 4828 (1970).
90. K. F. CHEW, W. DERBYSHIRE, N. LOGAN, A. H. NORBURY, and A. I. P. SINHA, *Chem. Comm.*, 1708 (1970).
91. L. O. ANDERSSON, J. (BANUS) MASON, and W. van BRONSWIJK, *J. Chem. Soc., A*, 296 (1970).
92. J. (BANUS) MASON and W. van BRONSWIJK, *J. Chem. Soc., A*, 1763 (1970).
93. R. GRINTER and J. MASON, *J. Chem. Soc., A*, 2196 (1970).
94. F. AUBKE, F. G. HERRING, and A. M. QURESHI, *Canad. J. Chem.*, **48**, 3504 (1970).
95. H. KATÔ, H. KATO, and T. YONEZAWA, *Bull. Chem. Soc. Japan*, **43**, 1921 (1970).
96. H. BÖHLAND and E. MÜHLE, *Z. anorg. allg. Chemie*, **379**, 273 (1970).
97. F. W. WEHRLI, W. GIGER, and W. SIMON, *Helv. Chim. Acta*, **54**, 229 (1971).
98. K. M. MACKAY and S. R. STOBART, *Spectrochim. Acta*, **27A**, 923 (1971).
99. H. SAITÔ and K. NUKADA, *J. Amer. Chem. Soc.*, **93**, 1072 (1971).
100. H. SAITÔ, Y. TANAKA, and K. NUKADA, *J. Amer. Chem. Soc.*, **93**, 1077 (1971).
101. E. D. BECKER, R. B. BRADLEY, and T. AXENROD, *J. Mag. Res.*, **4**, 136 (1971).
102. E. D. BECKER, *J. Mag. Res.*, **4**, 142 (1971).
103. M. ALEI, Jr., A. E. FLORIN, W. M. LITCHMAN, and J. F. O'BRIEN, *J. Phys. Chem.*, **75**, 932 (1971).
104. M. ALEI, Jr., A. E. FLORIN, and W. M. LITCHMAN, *J. Phys. Chem.*, **75**, 1758 (1971).
105. J. M. BRIGGS, L. F. FARNELL, and E. W. RANDALL, *Chem. Comm.*, 680 (1971).
106. M. WITANOWSKI, L. STEFANIAK, H. JANUSZEWSKI, and Z. W. WOLKOWSKI, *Tetrahedron Letts.*, 1653 (1971).
107. M. WITANOWSKI, L. STEFANIAK, H. JANUSZEWSKI, and G. A. WEBB, *Tetrahedron*, **27**, 3129 (1971).
108. J. MASON and W. van BRONSWIJK, *J. Chem. Soc., A*, 791 (1971).
109. P. HAMPSON, A. MATHIAS, and R. WESTHEAD, *J. Chem. Soc., B*, 397 (1971).
110. E. T. LIPPMAA, M. Y. MÄGI, Y. O. PAST, S. A. SHEVELEV, V. I. ERASHKO, and A. A. FAINZILBERG, *Izv. Akad. Nauk SSSR, Ser. Khim.*, 1006 (1971).

111. E. T. LIPPMAA, M. Y. MÄGI, Y. O. PAST, S. A. SHEVELEV, V. I. ERASHKO, and A. A. FAINZILBERG, *Izv. Akad. Nauk SSSR, Ser. Khim.*, 1012 (1971).
112. E. W. RANDALL and D. G. GILLIES, in *Progress in Nuclear Magnetic Resonance Spectroscopy*, Vol. 6 (J. W. EMSLEY, J. FEENEY, and L. H. SUTCLIFFE Eds.), Pergamon Press, Oxford, 1971, p. 119.
113. E. D. BECKER, Private communication.
114. M. WITANOWSKI, L. STEFANIAK, H. JANUSZEWSKI, and Z. W. WOLKOWSKI, *Chem. Comm.*, 1573 (1971).
115. J. DABROWSKI, K. KAMIEŃSKA-TRELA, and A. J. SADLEJ, *Org. Mag. Res.*, 3, 589 (1971).
116. H. SAITÔ, Y. YOSHIZAWA, Y. TANAKA, and K. NUKADA, *Tetrahedron Letts.*, 3677 (1971).
117. R. L. LICHTER, in *Determination of Organic Structures by Physical Methods*, Vol. 4 (F. C. NACHOD and J. J. ZUCKERMAN Eds.), Academic Press, New York and London, 1971, p. 195.
118. M. Y. MÄGI, V. I. ERASHKO, S. A. SHEVELEV, and A. A. FAINZILBERG, *Eesti. NSV Tead. Akad. Toim., Keem.-Geol.*, 20, 297 (1971).
119. M. Y. MÄGI, *Eesti. NSV Tead. Akad. Toim., Keem.-Geol.*, 20, 364 (1971).
120. P. S. PREGOSIN, E. W. RANDALL, and A. I. WHITE, *Chem. Comm.*, 1602 (1971).
121. M. WITANOWSKI, L. STEFANIAK, H. JANUSZEWSKI, Z. GRABOWSKI, and G. A. WEBB, *Bull. Acad. Polon. Sci., Ser. Chim.*, 20, 917 (1972) and unpublished results.
122. L. STEFANIAK, unpublished results.
123. R. L. LICHTER and J. D. ROBERTS, *J. Amer. Chem. Soc.*, 93, 5218 (1971).
124. M. WITANOWSKI, L. STEFANIAK, S. PEKSA, and H. JANUSZEWSKI, *Bull. Acad. Polon. Sci., Ser. Chim.*, 20, 921 (1972), and unpublished results.
125. E. GROCHOWSKI and L. STEFANIAK, unpublished results.
126. M. WITANOWSKI, L. STEFANIAK, H. JANUSZEWSKI, Z. GRABOWSKI, and G. A. WEBB, *Tetrahedron*, 28, 637 (1972).
127. T. AXENROD, P. S. PREGOSIN, M. J. WIEDER, E. D. BECKER, R. B. BRADLEY, and G. W. A. MILNE, *J. Amer. Chem. Soc.*, 93, 6536 (1971).
128. M. WITANOWSKI, T. SALUVERE, L. STEFANIAK, H. JANUSZEWSKI, and G. A. WEBB, *Mol. Phys.*, 23, 1071 (1972).
129. T. AXENROD and M. J. WIEDER, private communication.
130. W. GIGER and P. SCHAUWECKER, *Helv. Chim. Acta*, 54, 2488 (1971).
131. M. WITANOWSKI and G. A. WEBB, in *Annual Review of NMR Spectroscopy*, Vol. 5 (E. F. MOONEY, Ed.), Academic Press, New York and London, 1972, p. 395.
132. H. PIOTROWSKA, M. WITANOWSKI, L. STEFANIAK, and H. JANUSZEWSKI, unpublished results.
133. W. BECK, W. BECKER, K. F. CHEW, W. DERBYSHIRE, N. LOGAN, D. M. REVITT, and D. B. SOWERBY, *J. Chem. Soc. Dalton*, 245 (1972).

CHAPTER 5

Correlations of Nitrogen Coupling Constants with Molecular Structure

Theodore Axenrod

Department of Chemistry, The City College of the City University of New York, New York, N.Y. 10031

5.1   Introduction

In recent years, as the commercial availability of ^{15}N-enriched compounds has grown, interest has increasingly focused on the measurement and interpretation of nitrogen spin–spin couplings as an aid in spectral analysis and structure elucidation. Although present theory is unable to predict accurate values of these couplings, a knowledge of the range of coupling constants found for different molecular systems has considerable potential for structure determination. This chapter is concerned with the correlation of nitrogen coupling constants with molecular structure. The coupling constant data presented here are largely derived from ^{15}N-enriched compounds, but in certain cases data for ^{14}N couplings are available. To facilitate comparisons the latter are converted to the corresponding ^{15}N values by multiplication by $| \gamma (^{15}N)/\gamma(^{14}N) | = 1.402$. The subject of nitrogen NMR has previously been reviewed by Witanowski and Webb [1], Randall and Gillies [10], Mooney and Winson [11] and by Lichter [12].

5.2   Spin–Spin Coupling

5.2.1   Theoretical Considerations

Since the coupling constant, J, reflects the energy of interaction of two coupled nuclei in a molecule, it is a molecular property that can reveal valuable structural and stereochemical information as well as provide insights into the nature and distribution of bonding electrons. The theory of spin–spin coupling is covered in Chapter 1.

In the Ramsay Theory [13] of indirect spin coupling, the coupling constant is expressed as the sum of nuclear spin interaction contributions which arise by three different mechanisms: the Fermi contact interaction, the interaction of valence orbital electronic currents with nuclear magnetic moments and the dipole–dipole interactions between nuclear magnetic moments and electronic magnetic moments. Contributions from the latter two mechanisms are frequently very small for light elements and it is generally accepted that the coupling is dominated by the Fermi contact interaction. In coupling between nuclei A and X the Fermi contact term will, in part, depend on the electron densities at each nucleus and since only s atomic states have finite density at the nuclei, a dependence of J on the atomic hybridizations is expected. From second-order perturbation theory the coupling constant, $^nJ(A-K)$, where n represents the number of intervening bonds between the coupled nuclei, has been approximated as being proportional to the percent s characters, S_A and S_X, in the bonding orbitals and inversely proportional to the mean triplet excitation energy, $^3\Delta E$. A recent account of the theory of nuclear spin–spin coupling is recommended [185].

$$^nJ(A-K) \propto \frac{S_A S_X}{^3\Delta E} \qquad (5.1)$$

If $^3\Delta E$ remains constant throughout a series of compounds the coupling constant can be a useful parameter with which bond hybridization can be investigated. This relationship has been widely exploited in the study of $^{13}C-H$ couplings, and a vast number of semi-empirical correlations have been proposed [14]. However, the assumptions involved in these derivations have not been free of criticism [10, 15] and in particular, the effect of increased nuclear charge has been shown to increase $^{13}C-H$ coupling constants in the absence of any rehybridization [16, 17].

Recently, by analogy with $^{13}C-H$ coupling and using the same assumptions, the hypothesis has been extended to include the dependence of the magnitude of the coupling between nitrogen and directly bonded hydrogen on the hybridization of nitrogen [18, 19].

5.2.2 Signs of Nitrogen Spin–Spin Interactions

When nitrogen is coupled to another nucleus interactions having unequal energies can arise from different combinations of their allowed spin states. This interaction may be positive or negative and by convention a positive sign is used to denote that state where the coupled nuclei having opposed spins is of lower energy than the state with aligned spins. From a fundamental viewpoint, it is desirable to know both the magnitude and the absolute sign of coupling constants. These allow comparisons with theoretical predictions [20, 21] and toward this end considerable data have been accumulated for a variety of couplings to nitrogen-14 and nitrogen-15. As shown in Section 3 of Chapter 1, all of the terms that contribute to the coupling are proportional to the product of the gyromagnetic ratios of the nuclei involved. In the case of nitrogen the gyromagnetic ratio is positive for ^{14}N and negative for ^{15}N. It is convenient to compare the reduced coupling constants, $^nK(AX)$, which are obtained by dividing the observed coupling constant by the product of the gyromagnetic ratios of the coupled nuclei [20]. The reduced coupling constant is a measure of the electronic interactions in the molecule and is independent of the specific properties of the nuclei.

$$^nK(A-K) = \frac{2\pi}{h\gamma_A \gamma_K} \, ^nJ(A-K) \qquad (5.2)$$

Directly bonded nuclei are predicted to have positive reduced couplings unless one of the nuclei, such as fluorine, has tightly bound s-valence electrons [20]. Single resonance experiments have been used to determine the relative signs of spin couplings but these are limited to spin systems exhibiting second-order features which are not usually encountered in heteronuclear couplings [22]. Double resonance techniques, which include selective decoupling [23], spin tickling [24], nuclear Overhauser effects [25] and localized saturation effects [26], are generally used to obtain relative signs of coupling constants. Absolute signs of couplings whose relative values are known can be related to the $^{13}C-H$ coupling which is predicted to be positive [50]. Liquid crystal solvent studies on methyl fluoride have confirmed that $J(^{13}C-H)$ is positive and $J(^{13}C-^{19}F)$ is negative, in agreement with theory [51].

Table 5.1 summarizes the magnitudes and absolute signs of a variety of couplings involving nitrogen in different functional groups and electronic environments. The entries include examples of spin coupling between other nuclei and ^{14}N and ^{15}N. Spin coupling between ^{14}N and ^1H nuclei is not usually observable because of the rapid quadrupolar relaxation of the ^{14}N nucleus which is discussed in Chapter 3. However, in

TABLE 5.1. Signs of Some Experimentally Determined Nitrogen Coupling Constants

Compound	J, (Hz)	$K \times 10^{-20}$ (cm^{-3})	Reference
N–H Coupling			
pyrrole-^{15}N	−96.5	+79.3	27, 28
2,5-di-t-butylpyrrole-^{15}N	−91.5	+75.2	29
HCO^{15}NH$_2$	−91.3[a],	+75.0[a],	19, 31, 32
	−86.9[b]	+71.4[b]	
HCO^{15}NHCOCH$_3$	−90.2	+74.1	30
(CF$_3$)$_2$P^{15}NH$_2$	−85.6	+70.4	43
CH$_3$^{15}NH$_2$	−65.0	+53.4	33
^{14}NH$_4^{\oplus}$Br	+52.8	+60.8	44, 45
N–C–H Coupling			
CH$_3$CH=^{15}N–OH	−15.9[a],	+13.1[a],	34
	+2.9[b]	−2.4[b]	
HCO^{15}NH$_2$	−14.5	+11.9	19, 31, 32
CH$_2$=^{15}N–OH	−13.9[a],	+11.4[a],	34, 36
	+2.7[b]	−2.2[b]	
quinoline-^{15}N	−11.1	+9.1	34, 37
pyrrole-^{15}N	−4.5	+3.7	27, 28
	−3.6(H$_{(1)}$)	+2.7	
	−1.7(H$_{(3)}$)	+1.4	151
	−1.4(H$_{(2)}$)	+1.1	
N–C–H Coupling			
quinolinium-^{15}N-ethiodide	−1.6	+1.3	37
HCO^{15}N(CH$_3$)$_2$	+1.1[a], +1.2[b]	−0.9, −1.0	32, 35
CH$_3$^{15}NO$_2$	+2.3	−1.9	225
CH$_3$^{15}NH$_2$	−1.0	+0.8	33
CH$_3$^{14}N≡C	−2.3	−2.6	38
[(CH$_3$)$_3$^{14}NCH=CH$_2$]$^{\oplus}$Br$^{\ominus}$	+3.5[c]	+4.0[c]	40, 41
N–C–C–H Coupling			
pyrrole-^{15}N	−5.4	+4.4	27, 28
2,5-di-tert-butylpyrrole-^{15}N	−5.2	+4.3	29
quinoline-^{15}N oxide	−5.0	+4.1	37
^{15}N-ethylquinolinium iodide	−4.3	+3.5	37
CH$_3$CH=^{15}N–OH	−4.2[a], −2.6[b]	+3.4[a], +2.1[b]	34
(CH$_3$)$_2$C=^{15}N–OH	−4.0[a], −2.2[b]	+3.2[a], +1.8[b]	34
quinoline-^{15}N	−1.4	+1.1	37
CH$_3$C≡^{15}N	−1.8	+1.5	42
[(CH$_3$)$_3$^{14}NCH=CH$_2$]$^{\oplus}$Br$^{\ominus}$	+5.6[d], +2.5[e]	+6.4[d], +2.9[e]	40, 41
CH$_3$CH$_2$^{14}N≡C	+5.6[d], +2.5[e]	+2.9	39
(CH$_3$)$_3$C^{14}N≡C	+2.0	+2.3	38
[(CH$_3$CH$_2$)$_4$^{14}N]$^{\oplus}$OH$^{\ominus}$	+1.8	+2.1	38

TABLE 5.1—*continued*

Compound	J (Hz)	$K \times 10^{-20}$ (cm^{-3})	Reference
Longer Range NH Coupling			
$^{15}NH_2$, H_6, Cl, H_5, H_3, H_4 (structure)	−1.98 (H_6)	+1.63	102
	−0.79 (H_3)	+0.65	102
	−0.36 (H_5)	+0.30	102
	+0.17 (H_4)	−0.14	102
$N-^{13}C$ *Coupling*			
$CH_3{}^{13}C\equiv{}^{15}N$	−17.5	+57.0	42
$^{13}CH_3-{}^{15}NH_2$	−4.5	+14.7	33
$^{13}CH_3{}^{15}NO_2$	−10.5	+34.3	225
$^{13}CH_3C\equiv{}^{15}N$	+3.0	−9.8	42
$^{13}CH_3{}^{14}N\equiv C$	+7.0	+34.7	38
$N-^{31}P$ *Coupling*			
$(CF_3)_2{}^{31}P^{15}NH_2$	+52.6	−106.8	43
$N-P-C-^{19}F$ *Coupling*			
$(CF_3)_2P^{15}NH_2$	−1.5	+1.3	43

[a] Coupling to proton *trans* to oxygen atom.
[b] Coupling to proton *cis* to oxygen atom.
[c] Geminal coupling to vinyl proton.
[d] *Trans* coupling.
[e] *Cis* coupling.

several instances such as isocyanides [46, 119], tetraalkyl-ammonium salts [47], nitramines [48, 49] and nitrates [2, 3, 9], where the electric field gradient at the ^{14}N nucleus is highly symmetrical, quadrupole-induced relaxation is evidently quite slow since the $^{14}N-H$ spin coupling is preserved. This situation has been used with considerable advantage in the determination of the signs of the spin couplings in these systems.

The data in Table 5.1 show that in accord with theory, all one-bond N—H couplings have positive reduced coupling constants. Similarly, all three-bond N—C—C—H couplings are also found to have positive reduced coupling constants. The situation is rather intriguing in the case of the two-bond N—C—H couplings. In saturated systems, such two-bond couplings are found to be quite small with typical values falling in the range 0–2 Hz. However, incorporation of a multiple bond to either the nitrogen or carbon produces a substantial increase in the magnitude of the coupling.

In oximes, of which formaldoxime (I) is representative, the two geminal ^{15}N—C—H couplings are found to be of different

$$
\begin{array}{cc}
\text{H}_A \diagdown & \\
\phantom{\text{H}_A}\text{C}=^{15}\text{N} & \\
\text{H}_B \diagup \qquad \text{OH} &
\end{array}
\qquad
\begin{array}{l}
^2K(\text{NH}_A) = +11.4 \\
^2K(\text{NH}_B) = -2.2
\end{array}
$$

(I)

magnitudes and also of opposite signs; $^2K(\text{N}-\text{H}_A)$ is positive whereas $^2K(\text{N}-\text{H}_B)$ is negative. A similar reversal of the signs of the N—C—H coupling with stereochemistry is not, however, found for the geometric isomers, cis-formanilide (II) and trans–formanilide (III). In both cases, the one- and two-bond reduced coupling constants have been shown to be positive [32, 52].

$$
\begin{array}{cc}
\text{(II)} & \text{(III)}
\end{array}
$$

The signs and magnitudes of two- and three-bond N—H couplings in oximes and quinoline derivatives have been thoroughly investigated in terms of solvent effects and stereochemistry [34, 37]. It is observed that protonation in aldoximes produces an algebraic decrease in two-bond reduced couplings to each of the geometrically different protons without any accompanying change in sign, whereas in quinoline, $^2K(\text{N}=\text{C}-\text{H})$, which is also positive, undergoes a similar algebraic decrease on protonation or quaternization finally becoming negative in the N-oxide in acid medium. For all three-bond couplings that have been investigated so far the reduced coupling constants are found to be positive and protonation brings about an algebraic increase in $^3K(\text{N}-\text{H})$. Directly bonded N—H couplings are also uniformly found to have positive reduced coupling constants, and evidence is available to show that protonation similarly leads to an algebraic enhancement in $^1K(\text{N}-\text{H})$. This alternating effect of

protonation on the reduced coupling constants for one- two-
and three-bond couplings has been suggested as a means of
determining the signs of spin couplings [34].

It is interesting to compare the reduced geminal coupling
constants for the structurally related N—CH$_3$ couplings in the
series methylamine (IV), N,N-dimethylformamide (V) and
methylisocyanide (VI). Geminal ^{13}C—C—H coupling has been
correlated with the ^{13}C hybridization [53, 54] and

$$CH_3-NH_2 \qquad \begin{matrix} CH_3 \\ \\ CH_3 \end{matrix} \!\!\! \overset{\oplus}{N}\!\!=\!\!C \!\!\! \begin{matrix} O^{\ominus} \\ \\ H \end{matrix} \qquad CH_3-\overset{\oplus}{N}\!\!\equiv\!\!C^{\ominus}:$$

$^2K(N\!-\!C\!-\!H) = +0.8$ $^2K(N\!-\!C\!-\!H) = -0.9, -1.0$ $^2K(N\!-\!C\!-\!H) = -2.6$

(IV)   (V)   (VI)

$J(^{13}$C—C—H) is found to become progressively more negative
as carbon passes through tetrahedral, trigonal and digonal
hybridization. In an entirely analogous manner, ^2K(N—C—H)
in the present series becomes more negative, undergoing a sign
inversion, as the hybridization of nitrogen changes from sp^3
to sp^2 to sp. The signs and magnitudes of the coupling
constants between all of the ring protons and ^{15}N in
2-chloroaniline-^{15}N have recently been reported [102]. The
three- and four-bond reduced couplings between the ^{15}N and
the *ortho* and *meta* protons are found to be positive, whereas
the five-bond coupling to the *para* proton is negative. If the
hybridization of the nitrogen is influenced by the nature of
ring substituents [78], the signs of these couplings may be
inverted in other aniline derivatives.

Although only limited data are available [42, 119], the
magnitude of N^{13}C coupling seems to be influenced by the
hybridization of both the carbon and the nitrogen and, in
agreement with theory, ^1K(N^{13}C) in acetonitrile is positive
[42].

Recently, the directly bonded N—^{31}P reduced coupling has
been found to be negative in (CF$_3$)$_2$^{31}P^{15}NH$_2$, whereas the
longer range N—P—C—F coupling has the opposite sign [43]. In
difluorodiazine, the relative signs of the one- and two-bond
N—F couplings are opposite [125]. Some additional data
dealing with the relative signs of ^{15}N couplings in organic
compounds have recently appeared [79, 224].

5.3 One-Bond ^{15}N–H Coupling

The coupling interaction between directly bonded ^{15}N–H is generally regarded as being dominated by the Fermi contact term. This postulate is strengthened by the demonstrated dependence of $^1J(^{15}$N–H$)$ on the amount of s character in the bond. Table 5.2 lists some illustrative ^{15}N–H coupling

TABLE 5.2.   Representative ^{15}N–H Coupling Constants as a Function of Nitrogen Hybridization

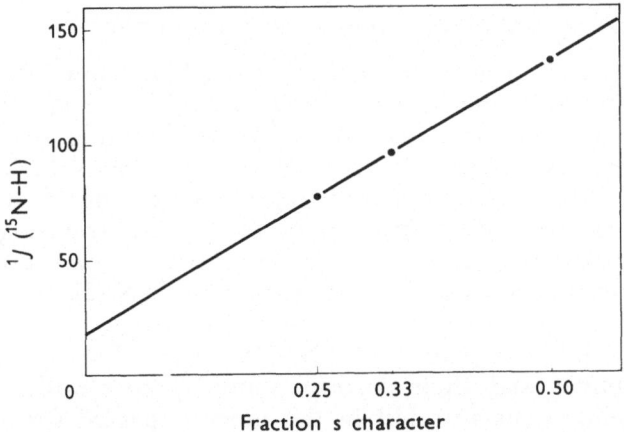

Compound	Solvent	Hybridization	$^1J(^{15}$N–H$)$, (Hz)
^{15}NH$_3^{\oplus}$	HFSO$_3$	sp^3	76.9
^{15}N$^{\oplus}$–H	HFSO$_3$	sp^2	96.0
C≡^{15}N$^{\oplus}$–H	HFSO$_3$–SbF$_5$–SO$_2$	sp	136.0

constants, determined under comparable solvent conditions, as a function of the s character in the bonding orbitals. Figure 5.1 shows a plot of these coupling constants against the fractions of s character in the respective bonds and as

Fig. 5.1   Plot of $^1J(^{15}$N–H$)$ against the fraction of s character in the N–H bond.

predicted the plot is linear. Using this approach, but confining their studies to sp^3 and sp^2 systems, pioneering work was carried on simultaneously by two independent groups of investigators. Binsch and collaborators [18] deduced the correlation

$$\% \, s = 0.43 \; {}^1J({}^{15}N\text{–}H) - 6 \qquad (5.3)$$

whereas Bourn and Randall [19] advanced a similar equation

$$\% \, s = 0.34 \; {}^1J({}^{15}N\text{–}H) \qquad (5.4)$$

having no intercept. Subsequently, Hogeveen [55] was able to show for the sp case of protonated propionitrile that ${}^1J({}^{15}N\text{–}H)$ was consistent with the above relationships.

Hybridization conclusions reached on the basis of these correlations do, however, require caution, not only because of the limitations of the approximations used in their derivation but also because solvent effects can profoundly influence one-bond ${}^{15}N\text{–}H$ couplings. The deviation of ${}^1J({}^{15}N\text{–}H)$ in diphenylketimine suggests that nuclear spin-electron orbital interaction contributions to the total coupling may not be negligible in systems where the nitrogen screening constant is highly anisotropic and this is supported by the abnormally low field resonance of the nitrogen in this compound [18]. Additionally, Paolillo and Becker [56] have observed that ${}^1J({}^{15}N\text{–}H)$ in aniline increases regularly with the hydrogen bonding ability of the solvent.

5.3.1 Aliphatic Amines and Ammonium Ions

Aliphatic amines are moderately strong bases and in solution rapid intermolecular proton exchange with traces of water present generally obscures the ${}^{15}N\text{–}H$ coupling [63]. This exchange has frequently proved to be very difficult to eliminate [56], although more aggressive drying efforts such as the use of sodium–potassium alloy are promising [64]. Removal of the nitrogen lone pair by protonation reduces the rate of exchange and the observation of ${}^{15}N\text{–}H$ coupling in ammonium ions presents no problem [65].

Table 5.3 compares ${}^1J({}^{15}N\text{–}H)$ values for a series of methylamines and their corresponding hydrochlorides. All of the coupling constants fall in the range expected for nominal sp^3 hybridization. Careful examination does, however, show that $^1J(^{15}N-H)$ increases with increasing methyl substitution both in the free amine and in the ammonium ion. Several possible explanations for this effect have been considered [58]. On the one hand, the trend can be satisfactorily accounted for by a small increase in the s character of the $^{15}N-H$ bond in going from $^{15}NH_3$ to $(CH_3)_2{}^{15}NH$, but

TABLE 5.3. One-Bond $^{15}N-H$ Coupling Constants in Ammonia, Some Aliphatic Amines and Ammonium Salts

Compound	Solvent	$^1J(^{15}N-H)$, (Hz)	Reference
$^{15}NH_3$	vapor	61.2	57, 58
		61.8	59
		64.0	60
$CH_3{}^{15}NH_2$	vapor	64.0	56
		64.5	58
		65.0	31
$(CH_3)_2{}^{15}NH$	vapor	67.0	58
$^{15}NH_4^{\oplus}Cl^{\ominus}$	H_2O	73.2	18
		73.3	61
		73.7	62, 82
$CH_3{}^{15}NH_3^{\oplus}Cl^{\ominus}$	H_2O	75.4	18
		75.6	61
$(CH_3)_2{}^{15}NH_2^{\oplus}Cl^{\ominus}$	H_2O	76.1	61
$(CH_3)_3{}^{15}NH^{\oplus}Cl^{\ominus}$	H_2O	76.1	61
$HO_2CCH_2{}^{15}NH_3^{\oplus}Cl^{\ominus}$	H_2O	77.0	18

available bond angle data [66] are not sufficiently precise to lend support to this postulate. On the other hand, either changes in the effective nuclear charge at nitrogen or variations in the electronic excitation energy can accommodate the observed data. At present the explanation for the increase in $^1J(^{15}N-H)$ with successive methyl substitution on nitrogen remains moot.

5.3.2 Ring-Substituted Aniline Derivatives

Substituent effects on the nuclear magnetic resonance parameters in ring-substituted aniline derivatives have received considerable attention. The solvent-dependent amino-proton chemical shifts have been correlated with Hammett substituent constants in acetonitrile [72], cyclohexane [73] and dimethylsulfoxide [74]. Calculated π-electron densities at nitrogen are in agreement with these chemical shifts [75].

The one-bond ^{15}N—H coupling constant in aniline is found to be intermediate between those expected for tetrahedral (sp^3) and trigonal (sp^2) nitrogen. The effect of substituents on $^1J(^{15}$N—H) in a series of ring-substituted aniline derivatives is shown in Table 5.4. Electron-withdrawing groups may be

TABLE 5.4.   One-Bond ^{15}N—H Coupling Constants of some Ring-Substituted Aniline Derivatives

G—⟨ring⟩—^{15}NH$_2$	$^1J(^{15}$N—H), (Hz)		
	DMSO[a]	CDCl$_3$[a]	Acetone[b]
2,4-(NO$_2$)$_2$	92.6		
4-NO$_2$	89.6	86.4	88.9
2,4,6-(Br)$_3$	87.4	85.5	87.8
4-NO$_2$-3,5-(CH$_3$)$_2$	87.0	83.2	
3-NO$_2$	86.2	83.0	83.0
2,4-(Br)$_2$	86.0	82.7	86.0
3-CF$_3$	85.1	81.0	
3-Br	85.3	80.5	
3-Cl	85.1	80.9	
3-I	84.4	80.4	
3-F	84.2	80.1	
4-I	84.0	79.7	84.1
4-Br	84.0	79.6	83.6
3,5-(CH$_3$O)$_2$	83.6	79.5	
4-Cl	83.7	78.9	
3-CH$_3$O	83.0	79.4	
H	82.6	78.6	81.9
3-CH$_3$	82.0	78.2	
3,5-(CH$_3$)$_2$	82.1	77.5	
4-F	81.6	77.8	
4-CH$_3$	81.4	76.5	79.6
4-CH$_3$O	79.4	75.6	
4-NH$_2$		74.9	
4-N(CH$_3$)$_2$	78.8	74.8	

[a] Ref. 78.
[b] Ref. 77.

seen to cause an increase in the magnitude of $^1J(^{15}$N—H). These observations have been interpreted in terms of a change in hybridization and a concomitant decrease in π-electron density at nitrogen [76–78]. The $^1J(^{15}$N—H) values correlate well with the Hammett substituent constants [76], as illustrated in Fig. 5.2.

That resonance delocalization of the nitrogen lone pair and not an inductive effect of the substituent is primarily

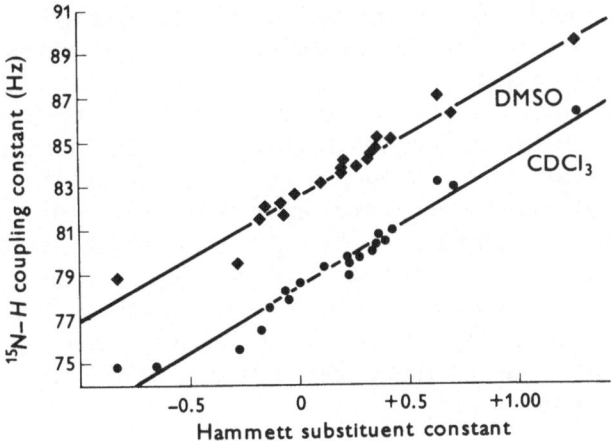

Fig. 5.2. Plot of $^1J(^{15}N-H)$ in aniline-^{15}N derivatives against the Hammett substituent constant.

responsible for the changes in $^1J(^{15}N-H)$ is indicated in the case of 4-nitro-3,5-dimethylaniline where steric inhibition of resonance results in diminished $^{15}N-H$ coupling [78]. This contention receives added confirmation from the fact that substituents have virtually no effect on $^1J(^{15}N-H)$ in compounds where the lone pair is either removed by protonation or the hybridization of the nitrogen is determined by its incorporation in an aromatic system [80]. Several such systems are shown in Table 5.5. The $^1J(^{15}N-H)$ values are

TABLE 5.5.   One-Bond $^{15}N-H$ Coupling Constants in some
Systems of Fixed Nitrogen Hybridization

System	G	$^1J(^{15}N-H)$, (Hz)	Solvent	Reference
G—⟨benzene⟩—$^{15}NH_3^{\oplus}$	H	76.0	H_2SO_4	80
	NO_2	76.0	H_2SO_4	80
G—⟨quinoline $^{15}N^{\oplus}$, H⟩	CH_3	96.0	$HFSO_3$	80, 81
	H	96.0	$HFSO_3$	80, 81
	NO_2	96.5	$HFSO_3$	80, 81
G—⟨indole ^{15}N, H⟩—CH_3	H	96.5	DMSO	80
	NO_2	96.9	DMSO	80

consistent with sp^3 hybridization of the nitrogen in the anilinium ions and sp^2 hybridization in the indole derivatives and protonated quinolines.

Evidence in favor of a contact contribution to the isotropic shifts induced in anilines by lanthanide chelates derives from the observation that with increasing concentration of the shift reagent, $^1J(^{15}N—H)$ decreases approaching the value in the sp^3-hybridized anilinium ion [4]. A pure pseudocontact interaction would be expected to have no effect on the $^{15}N—H$ coupling [85]. The magnitude of the effect parallels the basicity of the aniline [83] and is consistent with some degree of covalent bonding involving the nitrogen lone pair in the associated complex [84].

5.3.3   Solvent Effects and Hydrogen Bonding

The data of Paolillo and Becker [56] for aniline in various solvents are reproduced in Table 5.6. Although $^1J(^{15}N—H)$ increases and the nitrogen resonance occurs at lower field in solvents of increasing hydrogen-bonding ability, there is no simple correlation between these parameters and the solvent dielectric constant. These workers conclude that hydrogen-bonding, while important, must be considered together with other less well defined solvent interactions in interpreting the data.

TABLE 5.6.   Solvent Effects on $^1J(^{15}N—H)$ in Aniline[a]

Solvent	ϵ	$^1J(^{15}N—H)$, (Hz)
C_6D_{12}	2.02	78.0
CCl_4	2.22	78.0
$CDCl_3$	5.05	78.0
dioxane-d_8	2.20	80.6
pyridine-d_5	12.3	81.4
acetone	20.7	82.1
DMF-d_7	36.7	82.3
DMSO-d_6	48.9	82.3

[a] Ref. 56.

Axenrod and Wieder [86] have investigated the effect of solvents on $^1J(^{15}N—H)$ in a series of *ortho*-substituted anilines. Table 5.7 lists the $^1J(^{15}N—H)$ values measured in chloroform and dimethylsulfoxide solutions. Of particular interest here are the $\Delta^1J(^{15}N—H)$ values which represent the differences in the one-bond coupling constants in the two solvents. In contrast to the data in Table 5.4 where $\Delta^1 J(^{15}N-H)$ for *meta*- and *para*-substituted anilines falls between 3 and 4 Hz, it will be noted that as the hydrogen bonding ability of the *ortho* substituent increases, $\Delta^1 J(^{15}N-H)$ decreases reaching a value of 0.7 Hz in *ortho*-nitroanilines. Since it is generally acknowledged that *ortho*-nitroanilines are intramolecularly hydrogen

TABLE 5.7. Solvent Effects on One-Bond $^{15}N-H$ Coupling Consants of Some *Ortho*-Substituted Anilines

G ($^{15}NH_2$)	$^1J(^{15}N-H)$, (Hz)[a]		$\Delta^1 J(^{15}N-H)$, (Hz)
	CDCl$_3$	DMSO	
2-NO$_2$, 4-Cl	91.1	91.8	0.7
2-NO$_2$	90.3	91.0	0.7
2-COPh	88.1	89.3	1.2
2-Cl, 4-NO$_2$	89.2	90.5	1.3
2,4,6-(Br)$_3$	85.5	87.4	1.9
2-CF$_3$	83.6	86.5	2.9
2-Cl	81.5	84.3	2.9
2-Br	81.4	84.3	2.9
2-CH$_3$O	79.4	82.3	2.9
2-F	80.1	83.5	3.4
2-H	78.6	82.6	4.0

[a] Ref. 86.

bonded in chloroform and intermolecularly hydrogen bonded to the solvent in dimethylsulfoxide [87–89], the $\Delta^1 J(^{15}N-H)$ values may be taken as a measure of the strength of the intramolecular hydrogen bond.

In this connection, it is noteworthy that restricted rotation of the amino group should be favored by its participation in a strong hydrogen bond. In their study of base-pairing between purine and pyrimidines, Shoup, Miles and Becker [94] suggested restricted rotation of the amino group in 1-methyl-cytosine-7-^{15}N which has now been confirmed in the low temperature 220 MHz NMR spectrum [219]. Although the barrier to rotation may be influenced by hydrogen bonding and base-pairing, neither of these factors is essential to account for the experimental observations. Recently, several examples of this phenomenon have been found in *ortho*-nitro-anilines [81]. This is illustrated in the temperature-dependent

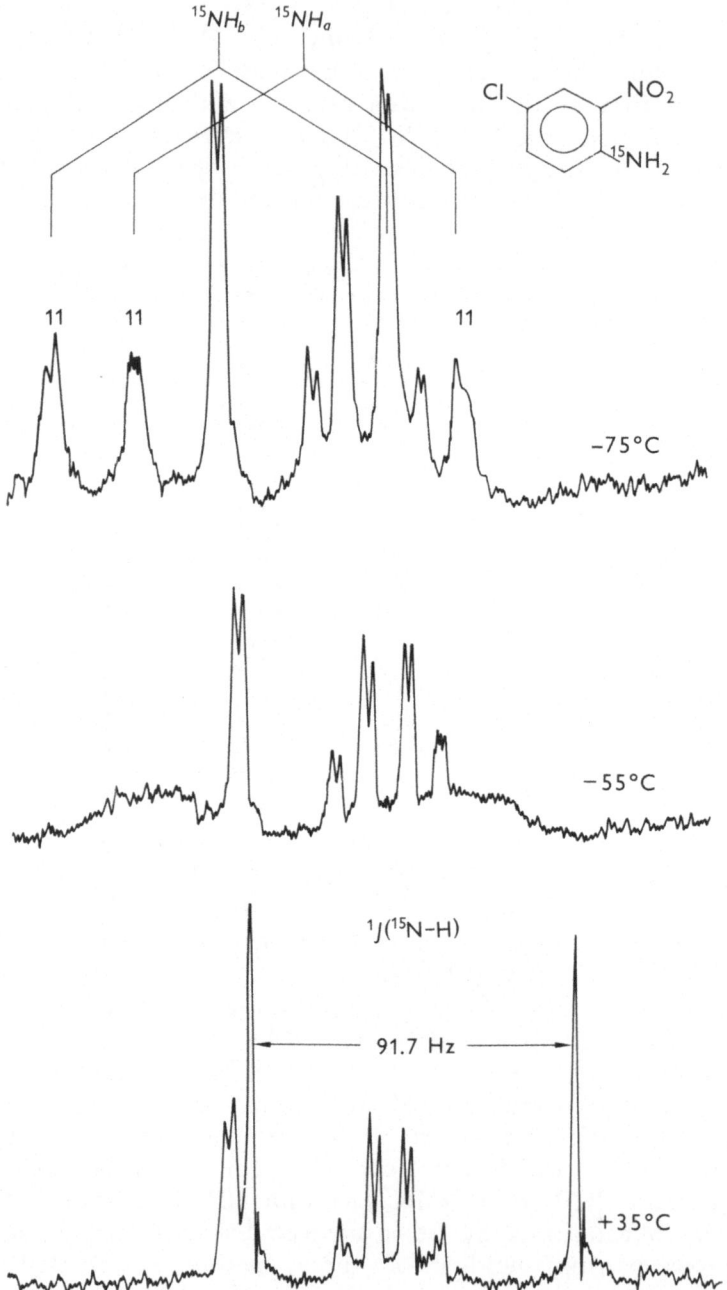

Fig. 5.3.   Temperature-dependent 60-MHz spectrum of 2-nitro-4-chloroaniline-
^{15}N in acetone.

NMR spectrum of 2-nitro-4-chloroaniline-^{15}N(VII) shown in Figure 5.3.

(VII)

At low temperature the two amino protons exhibit different chemical shifts of which, the upfield set is partially obscured by the aromatic proton resonances. Geminal $H_A NH_B$ coupling (2.2 Hz), similar to the situation in amides [5, 95, 96] and possibly reflecting the sp^2 hybridization of the nitrogen, is also found. A recent report concerned with the study of unlabeled anilines indicates similar observations [215].

5.3.4   N-Substituted Aniline Derivatives

Substituent effects on $^1J(^{15}N-H)$ in N-substituted aniline derivatives have not been extensively investigated in a systematic manner. However, some indications of the influence of the substituent attached to nitrogen on the one-bond $^{15}N-H$ coupling constant are shown in the compilation given in Table 5.8.

In line with previous observations in aliphatic amines [58], replacement of hydrogen by an alkyl group leads to an enhancement in $^1J(^{15}N-H)$. Thus, direct substitution of methyl on nitrogen changes $^1J(^{15}N-H)$ in the opposite direction to that found for *para*-methyl substitution in aniline. Moreover, the increase appears to be greatest for alkyl substituents containing electronegative groups as is evidenced by the large coupling in N-trifluroethylaniline compared with N-ethylaniline [91]. The coupling also shows a small dependence on stereochemistry [90]; the directly bonded proton coupling to the equatorial ^{15}N in *trans*-4-t-butylcyclohexylaniline (VIII) being smaller than the same coupling to the axial ^{15}N in *cis*-4-t-butylcyclohexylaniline (IX). Equally interesting is the unexpectedly low coupling in N-phenylhydroxylamine-^{15}N compared with phenylhydrazine-1-^{15}N. Relative to the parent aniline, replacement of a hydrogen on nitrogen by

TABLE 5.8.   One-Bond $^{15}N-H$ Coupling Constants of some
N-Substituted Aniline Derivatives

—$^{15}NH-G$

G	Solvent	$^1J(^{15}N-H)$, (Hz)	Reference
OH	DMSO	79.3	81
H	DMSO	82.6	78
trans-4-t-butylcyclohexyl	DMSO	86.3	90
CH_3	DMSO	87.0	91
$CH_2C(CH_3)_3$	DMSO	87.0	91
cis-4-t-butylcyclohexyl	DMSO	87.3	90
CH_2CH_3	DMSO	87.5	91
CH_2COPh	DMSO	88.2	91
CH_2CN	DMSO	88.5	91
CH_2Ph	DMSO	88.6	91
NH_2	DMSO	89.6	81
NHPh	DMF	90.5	81
CH_2CF_3	DMSO	90.2	91
	CDCl₃	91.4	92
	CCl₄	91.2	92
	CCl₄	92.8	92

an amino group leads to increased coupling, whereas introduction of a hydroxyl group reduces the magnitude of the
$^{15}N-H$ coupling [81]. The latter may indicate that contributions to the coupling from mechanisms other than the Fermi
contact interaction are important.

$^1J(^{15}N-H) = 86.3$ Hz $^1J(^{15}N-H) = 87.3$ Hz

(VIII)   (IX)

5.3.5   Imines and Hydrazones

As previously mentioned, ketimines exhibit anomalously low $^1J(^{15}N-H)$ values, probably because of contributions to the nuclear spin–spin coupling from electron orbital motion [18]. This apparent anomaly is removed when the imine is protonated [97]. *Cis-trans* isomerism about the carbon-nitrogen double bond has been observed in the low-temperature spectrum of sec-butylphenylketimine-^{15}N. The spectrum is sensitive to changes in temperature and amine concentration, but the presence of water or calcium oxide is without effect. Kinetic studies suggest that the exchange process could be best accounted for by the concerted shift of two protons in a bimolecular four-membered ring transition state (X) [98]. The reactants and products in such a degenerate process would, of course, be identical.

(X)

Spectroscopic evidence indicates that benzalaniline is not planar; the aniline ring being twisted from the plane of the azomethine linkage [105]. This has been confirmed for the crystal structure by X-ray analysis [106]. A consequence of this geometry is that substituents in the benzaldehyde ring are expected to have a more pronounced effect on the spectroscopic properties than similar substituents in the aniline ring. This accords with the findings for substituent effects on the chemical shift of the azomethine proton [105, 107]. Table 5.9 lists some one-bond $^{15}N-H$ coupling constants in several protonated benzalaniline derivatives. Noteworthy is the finding that $^1J(^{15}N-H)$ in these protonated benzalanilines is consistent with an sp^2-hybridized nitrogen but, analogous to the situation in anilinium and quinolinium ions [80] substituents have virtually no effect on the coupling.

For the related phenylhydrazones shown in Table 5.10, the situation is less than straightforward. The introduction of a 4-nitro group in phenylhydrazine-1-^{15}N leads to a greater enhancement in $^1J(^{15}N-H)$ than was the case in aniline [78]. However, in phenylhydrazones and 4-nitrophenylhydrazones of various substituted benzaldehydes and other carbonyl

TABLE 5.9.   One-Bond ^{15}N—H Coupling Constants of some Imines and Protonated Benzalanilines

Compound	Solvent	$^1J(^{15}$N—H$)$, (Hz)	Reference
$Ph_2C=^{15}NH$	C_5H_{12}	51.2	97
$Ph(sec\text{-}Bu)C=^{15}NH$	C_5H_{12}	50.6, 50.9[a]	98
$Ph_2C=^{15}NH_2^{\oplus}$ $Cl^{\ominus}$	SO_2	92.6	97
$H—N=C=O$	CCl_4	90[b]	99, 100

G	G'			
H	CH_3	$HFSO_3$	91.5	81
CH_3	H	$HFSO_3$	91.5	81
H	H	$HFSO_3$	91.5	81
H	Cl	$HFSO_3$	91.5	81
NO_2	H	$HFSO_3$	92.0	81
H	NO_2	$HFSO_3$	91.9	81

[a] Geometrical isomers present.
[b] Calculated from the value reported for $^1J(^{14}$N—H$)$.

compounds, the ^{15}N—H coupling constants differ by much less. The relationship of $^1J(^{15}$N—H$)$ to electronic structure requires further clarification.

5.3.6   Protonated Nitriles

Nitriles in strong acid media undergo protonation on nitrogen to form nitrilium ions for which the coupling constants listed in Table 5.11 provide persuasive argument in favor of sp hybridization at nitrogen [55, 103]. The interesting case of substituent effects on $^1J(^{15}$N—H$)$ and ^{15}N chemical shifts in protonated benzonitriles has recently been examined by Axenrod and Macchia [104]. Electron-withdrawing substituents have previously been shown to cause deshielding of the nitrogen in nitriles [108, 109], and in the protonated benzonitriles, a similar deshielding is observed for both the ^{15}N and ^{15}N-bound proton resonances [104]. The unexpected finding is that electron-withdrawing substituents in the aromatic ring bring about a decrease in the magnitude of $^1J(^{15}$N—H$)$ as shown in Table 5.11. The situation is contrary to the effect previously observed for electron-withdrawing

TABLE 5.10.   One-Bond $^{15}N-H$ Coupling Constants of some Phenylhydrazones and Related Derivatives

Compound	Solvent	$^1J(^{15}N-H)$, (Hz)	Reference
$^{15}NHNH_2$ (phenyl)	DMSO	89.6	81
O_2N-(phenyl)-$^{15}NHNH_2$	DMSO	99.2	81

(phenyl)-$^{15}NHN=G$

G			
$CH_3CH_2CH=$	DMSO	90.5, 89.5[a]	101
$PhCH=$	DMSO	92.7	92
$3-NO_2PhCH=$	DMSO	93.7	101
$4-NO_2PhCH=$	DMSO	93.8	101
$(CHO)_2C=$	CCl_4	96	102
$CH_3CO(CO_2Et)C=$	CCl_4	96.1, 94.7[a]	92
$OS=$	DMSO	99	81

O_2N-(phenyl)-$^{15}NHN=G$

$(CH_2)_2C=$	DMSO	94.3	101
$3-NO_2PhCH=$	DMSO	95.1	101
$4-CH_3OPhCH=$	DMSO	94.9	101

[a] Geometrical isomers present.

substituents in anilines [78], and it stands in marked contrast to the substituent effect on $^1J(^{13}C-H)$ in the isoelectronic phenylacetylenes. Electron-withdrawal in phenylacetylenes is accompanied by decreased shielding of the acetylenic proton [110], but $^1J(^{13}C-H)$ increases from 251.7 Hz in the unsubstituted compound to 254.3 Hz in the p-nitro derivative [104]. The direction of the latter changes are also in accord with the $^1J(^{13}C-H)$ data reported for a series of haloacetylenes [111].

5.3.7   Amides, Ureas, Thioamides and Thioureas

Delocalization of the lone pair of electrons on nitrogen in amides and related compounds produces substantial double bond character in the central C–N bond and the concomitant relatively high barrier to internal rotation results in the

TABLE 5.11. One-Bond ^{15}N–H Coupling Constants of some Protonated Nitriles

Compound	Solvent	$^1J(^{15}N–H)$, (Hz)	Reference
H–C≡^{15}N$^{\oplus}$–H	$FSO_3H–SbF_5–SO_2$	134	99
CH_3–C≡^{15}N$^{\oplus}$–H	$FSO_3H–SbF_5–SO_2$	136	99
CH_3CH_2–C≡^{15}N$^{\oplus}$–H	$HF–BF_3$	130	55

G

4-CH_3	$FSO_3H–SbF_5–SO_2$	136.2	104
H	$FSO_3H–SbF_5–SO_2$	136.2	104
2-Br	$FSO_3H–SbF_5–SO_2$	135.1	104
4-CH_3O [a]	$FSO_3H–SbF_5–SO_2$	134.7	104
4-NO_2	$FSO_3H–SbF_5–SO_2$	132.6	104
4-NO_2, 2-Cl	$FSO_3H–SbF_5–SO_2$	132.5	104
3,5-$(CH_3O)_2$ [a]	$FSO_2H–SbF_5–SO_2$	131.8	104
3,5-$(CH_3O)_3$ [b]	$FSO_3H–SbF_5–SO_2$	131.6	104

[a] O-protonated species.
[b] Ring-protonated species.

non-equivalence of the substituents on nitrogen even when $G_{(cis)} = G_{(trans)}$ (XI) [134]. In formamide an essentially planar amide skeleton is indicated by X-ray crystal data [135] and by microwave studies [136]. To the extent that the dipolar resonance structure (XI) is important and the magnitude of the one-bond ^{15}N–H coupling depends on the hybridization at nitrogen, $^1J(^{15}N–H)$ in amides should be in the range expected for sp^2 nitrogen.

(XI)

The data for a variety of amides, ureas, thioamides and thioureas, summarized in Table 5.12, are consistent with this view; the $^1J(^{15}N–H)$ values almost all fall within the range 88–92 Hz. It is to be noted, however, that the coupling shows a stereochemical dependence. The shorter bond between the ^{15}N and the hydrogen *trans* to the carbonyl oxygen is generally associated with stronger coupling than the longer bond to the amide hydrogen having the *cis* orientation.

TABLE 5.12.   One-Bond ^{15}N—H Coupling Constants of some Amides, Ureas, Thioamides and Thioureas

		Solvent	$^1J(^{15}N-H)$, (Hz) cis	trans	Reference
G_{cis}	G_{trans}				
H	H	diglyme	87.1	89.3	221
H	H	butanone	87.7	90.0	221
H	H	none	88.0	92.0	112
H	H	CH_3CN	87.5	90.1	95
H	H	acetone	88.0	91.1	112
H	H	acetone	86.4	89.7	19
H	H	H_2O	95.4	91.9	112
CH_3	H	none		92.6	19
Ph	H	$CDCl_3$		91.2	52
H	Ph	$CDCl_3$	83.0		52
n-Bu	H	none		92.2	113
H	n-Bu	none	89.8		113
H	$COCH_3$	CF_3CO_2H	90.2		30

G_{cis}	G_{trans}	X	Solvent	cis	trans	Reference
H	H	O	acetone	87.6	89.2	5
H	H	O	DMSO	87	89	5
H	H	O	H_2O	88.4	90.9	5
H	H	O	H_2O		89	18
H	H	S	$CDCl_3$	91	94	114
n-Bu	H	O	none		92.0	113
H	Ph	S	$CDCl_3$	90		114
Ph	H	S	$CDCl_3$		92	114
H	Ph	O	acetone	90.0		18
H	Ph	O	$CDCl_3$	89.9		92
H	CH_2CO_2H	O	DMSO	94.5		18

G		Solvent	cis	trans	Reference
3-CF_3		DMSO	88.7	89.1	81
3-Br		DMSO	88.2	88.8	81
$3,5$-$(CH_3)_2$		DMSO	88.0	88.8	81
2-NO_2		DMSO	88.4	89.6	81

TABLE 5.12—*continued*

Compound	Solvent	$^1J(^{15}N-H)$, (Hz) cis	trans	Reference
2-CF$_3$	DMSO	88.7	89.7	81
2-Br	DMSO	88.2	90.0	81
4-F	DMSO	88.2	88.6	91
4-I	DMSO	88.1	89.2	91

$$G_1 HN-\underset{\underset{C}{\overset{\|}{}}}{\overset{X}{}}-^{15}NHG_2$$

G$_1$	G$_2$	X	Solvent	cis	trans	Reference
H	H	O	DMSO	89		18
Ph	PH	O	DMSO	89.7		92
Ph	CH$_3$	S	Ethanol	91.2		18
Ph	Ph	S	DMSO	89.9		92

The one-bond $^{15}N-H$ couplings show little variation with structure but pronounced changes are observed when the solvent is varied. In aqueous solution, the relative magnitudes of the two $^{15}N-H$ coupling constants in formamide are reversed from those observed in the neat liquid [112], whereas temperature variation is found to be significant only for the coupling between the ^{15}N atom and the directly bonded proton *trans* to the carbonyl group [221]. This may reflect variations in the contributions of the several resonance forms, changes due to different hydrogen bonded species, or a combination of these effects. In formamide, hydrogen-bonding at the hydrogen *trans* to the carbonyl oxygen is favored over the *cis* hydrogen, and this may account for the formation of hydrogen-bonded chains by N-monosubstituted amides preferentially involving the hydrogen *trans* to oxygen [95]. In benzamides, the two $^{15}N-H$ couplings differ only very little and they show no trend which can be attributed to an electronic effect arising from ring substituents. This is surprising in view of the observation that the ^{15}N resonances is substituted benzamides do show a small but regular downfield shift induced by electronegative substituents [109]. The latter shifts correlate with Hammett σ values supporting the contention that π-electron density changes contribute to the observed effect.

5.3.8 Amino Groups Attached to Elements of Groups IV and V

A number of reports concerning the determination of $^{15}N–H$ couplings in amino groups bound to elements other than carbon have appeared. These coupling constants are summarized in Table 5.13. The measured values have been utilized

TABLE 5.13. One-Bond $^{15}N–H$ Coupling Constants of some Amino Groups Attached to Elements Other than Carbon

Compound	$^1J(^{15}N–H)$, (Hz)	Reference
$[(CH_3)_3Si]_2{}^{15}NH$	66.5	115
$(CF_3)_2As^{15}NH_2$	73.4	116
$(CH_3)_3Sn^{15}NHC_6H_5$	73.8	115
$(CH_3)_3Si^{15}NHC_6H_5$	76.0	115
$(CH_3)_3Ge^{15}NHC_6H_5$	77.1	115
$[(CF_3)(CH_3)P]_2{}^{15}NH$	77.4, 78.9[a]	116
$(CF_3)(CH_3)P^{15}NH_2$	78.9	116
$[(CF_3)_2As]_2{}^{15}NH$	79.0	116
$CF_3S^{15}NH_2$	80.6	116
$(CF_3)_2P^{15}NH_2$	85.6	116
$PF_3(^{15}NH_2)_2$	87.5	116
$(CF_3S)_2{}^{15}NH$	99.1	116

[a] Diastereoisomers.

as a structural probe to elucidate the stereochemical situation at nitrogen and to assess the importance of multiple bonding in the N–X bonds of these compounds. On the assumption of a dominant Fermi contact interaction, the results suggest trigonal planar geometry at nitrogen in $PF_3(NH_2)_2$, $(CF_3)_2PNH_2$ and $(CF_3S)_2NH$, and tetrahedral geometry in aminophosphines and CF_3SNH_2 [116]. These data are consistent with $(2p-3d)-\pi$-bonding in those N–P and N–S compounds having strongly electronegative substituents at phosphorus and sulfur. On the other hand, $^1J(^{15}N–H)$ in aminoarsines accords with an sp^3-hybridized nitrogen in these compounds which may be attributed to the less effective overlap possible between the $2p$ nitrogen orbital and the $4d$ arsenic orbitals [116].

The relationship between $^1J(^{15}N–H)$ and the stereochemical situation at nitrogen to the question of $(p-d)-\pi$-bonding in ^{15}N-substituted trimethylsilyl-, germyl-, and -stannylanilines has been examined [137]. The magnitude of the $^{15}N–H$ couplings are all less than $J(^{15}N-H)$ in aniline itself and this has been interpreted in terms of a more pyramidal nitrogen and the absence of $(p$-$d)$-π-bonding in these compounds [115]. A word of caution is perhaps in order in view of the fact that the nitrogen atom in related silylamines has been shown to be planar [138] and that the effect on $^1J(^{15}N-H)$ of substituents directly on nitrogen is less than well-understood [58, 70].

5.4 Coupling Between Nitrogen and Nuclei other than Hydrogen

5.4.1 $^{15}N-^{13}C$ Coupling

In their early investigation of ^{15}N couplings Binsch and coworkers [18] extended their studies to include the couplings between directly bonded nitrogen-15 and carbon-13. Again, assuming the dominance of the Fermi contact term these investigators were able to correlate a variety of $^{15}N-^{13}C$ couplings with the product of the s-characters of the nitrogen and carbon orbitals forming the σ-bond. The empirical equation

$$\%S_N\%S_C = 80^1J(^{15}N-^{13}C) \tag{5.5}$$

gives reasonable values for hybridization parameters at the two centers, but large deviations are noted particularly for those compounds in which the nitrogen atoms are highly deshielded. The latter phenomenon seems to be associated with low-lying excited states and strongly indicates that contributions to the total coupling mechanism from the orbital terms are not negligible. As additional data, summarized in Table 5.14, show the results provide little

TABLE 5.14. Some One- and Two-Bond $^{15}N-^{13}C$ Coupling Constants

Compound	$J(^{15}N\underline{-}^{13}C)$ (Hz)[a]	Reference
One-Bond Coupling		
$^{13}CH_3{}^{15}N=CHC_6H_5$ *(trans)*	<3	18
$(CH_3{}^{13}CH_2)_4{}^{15}N^{\oplus}$	4.0	117
O_2N〈benzene ring〉$^{13}C_\alpha(H)\!-\!^{15}N\!-\!^{13}C_\beta H(CH_3)_2$ (O)	3.1 (C_α),[c] 5.5 (C_β)[c] 4.9 (C_α),[d] 5.9 (C_β)[d]	226 226

TABLE 5.14—*continued*

Compound	$J(^{15}N\underline{^{13}C})$, (Hz)[a]	Reference
$^{13}CH_3{}^{15}NH_2$	4.5, 7	33, 18
$CH_3{}^{13}CH(^{15}NH_3^{\oplus})CO_2^{\ominus}$	5.6	118
$(^{13}CH_3)_4{}^{15}N^{\oplus}$	5.8	117
$^{13}C{\equiv}^{15}N^{\ominus}$	5.9	120
$C_6H_5{}^{13}CH{=}^{15}NCH_3$ (*trans*)	7.1	97
$(CH_3)_3C^{15}N{\equiv}^{13}C$	7.2	119
$CH_3CH_2{}^{15}N{\equiv}^{13}C$	7.3	119
$^{13}CH_3{}^{15}NH_3^{\oplus}$	<8	18
$CH_3{}^{15}N{\equiv}^{13}C$	9.1	38
$(CH_3)_3{}^{13}C^{15}N{\equiv}C$	9.4	119
$CH_3{}^{13}CH_2{}^{15}N{\equiv}C$	10.2	119
$^{13}CH_3{}^{15}N{\equiv}C$	10.6	38

	9.9 $^{15}N{-}^{13}CH_2$	217
	12.9 $^{15}N{-}^{13}CO$	217

Compound		
$CH_3{}^{13}CO^{15}NHC_6H_5$	13.1	97
$^{13}CH_3{}^{15}N{=}C{=}S$	13.4	18
$H^{13}CO^{15}N(CH_3)_2$	13.4	118
$CH_3{}^{13}CO^{15}NH_2$	<15	18
$^{13}CH_3{}^{15}NH(CS)NHC_6H_5$	<15	18
$(CH_3)_3C^{13}C{\equiv}^{15}N$	15.0	120
$(CH_3)_2CH^{13}C{\equiv}^{15}N$	15.4	120
$CH_3CH_2{}^{13}C{\equiv}^{15}N$	16.4	120
$CH_3{}^{13}C{\equiv}^{15}N$	17.5	18

Two-Bond Coupling

Compound		
$^{13}CH_3C{\equiv}^{15}N$	3.0	42
$CH_3{}^{13}CH_2CO^{15}NH_2$	6.6	12
$^{13}CH_3CO^{15}NHC_6H_5$	9.3	12
$^{13}CH_3CO^{15}NH_2$	9.5	118

	11.4	217

Compound		
$(^{13}CH_3)_2N^{15}NH_2$	<1	118
$(^{13}CH_3)_2N^{15}NO$	7.5 (*anti*), 1.4 (*syn*)	118

[a] In cases where the coupling constants were reported for the ^{14}N isotopomer the corresponding ^{15}N value was obtained by multiplication by $|\gamma(^{15}N)/\gamma(^{14}N)| = 1.402$.

[b] R is a sugar moiety.

[c] *Trans*-isomer.

[d] *Cis*-isomer.

encouragement for attempts to relate $^1J(^{15}N-^{13}C)$ to carbon and nitrogen s-characters.

Successive α-methyl substitution in acetonitrile [143] and the corresponding isonitrile [38, 119] leads to a regular decrease in $^1J(^{15}N-^{13}C)$ in the cyano and isocyano groups, respectively. The former trend has been successfully predicted by INDO molecular orbital calculations [120], but the method fails to give agreement with the experimentally established sign of the coupling [42].

Proton decoupling and spectrum accumulation [118] in the case of pyridine-^{15}N and pulsed Fourier techniques [121] in the case of quinoline-^{15}N have recently made available the

TABLE 5.15.   Some Carbon–Nitrogen Coupling Constants[a] in 2-Pyridone-^{15}N, Pyridine-^{15}N and Quinoline-^{15}N

$^{15}N-^{13}C$ Coupling	CDCl$_3$[b]	CH$_3$OH[c]	CH$_3$OH/HCl[c]	CH$_3$OH[d]	H$_2$SO$_4$[d]
C-2	11.2	0.7	12.0	2.4	15.9
C-3	<0.5	2.6	2.1	2.6	1.0
C-4	10.5	3.8	5.3	3.7	4.6
C-5	5.2			~0	~0
C-6	2.5			~0	~0
C-7				3.9	2.7
C-8				8.4	~1.0
C-9				1.5	13.8
C-10				2.0	<1.0

[a] In hertz.
[b] Ref. 7.
[c] Ref. 118.
[d] Ref. 121.

values of all of the $^{15}N-^{13}C$ couplings in these two nitrogen heterocycles [7]. The values of these $^{15}N-^{13}C$ couplings for the free bases and their protonated forms are compared in Table 5.15 with those found in 2-pyridone-^{15}N. In the nitrogen-containing rings of these bases, the magnitudes of the couplings increase with the number of bonds intervening between the coupled nuclei, whereas protonation at nitrogen causes a marked increase in the coupling to the α-carbon atoms. Whether or not an inversion in the sign of $^1J(^{15}N-^{13}C)$ accompanies protonation is unknown. Noteworthy in quinoline is the relatively large ^{15}N–^{13}C-8 coupling which presumably stems from a geometrical dependence on the nitrogen lone-pair orientation. The magnitude of this coupling is sensitive to solvent hydrogen bonding and is dramatically reduced on protonation [121].

Comparatively few data are available regarding coupling between ^{15}N and ^{13}C nuclei which are separated from each other by more than one bond. In view of the absence of ^{15}N–N–^{13}C coupling in 1,1-dimethylhydrazine-2-^{15}N, and the appreciable ^{15}N–N–^{13}C and ^{15}N–C(O)–^{13}C coupling in dimethylnitrosamine-^{15}N and amides, it has been suggested that an intervening π-system is a likely requisite for the transmission of spin–spin interaction [118]. Recently, the one- and two-bond ^{13}C–^{15}N couplings in the N,N-phthaloyl derivative of 6-amino-^{15}N, 6-deoxy-1,2 : 3,5-di-O-isopropyl-idene-α-D-glucofuranose have been reported [217]. It is interesting that the directly bonded couplings, ^{15}N–^{13}CH$_2$ and ^{15}N–^{13}C(O) of 9.9 Hz and 12.9 Hz, respectively, differ very little from the 11.4 Hz coupling found for the two-bond ^{15}N–C(O)–^{13}C situation.

5.4.2 ^{14}N–^{19}F, ^{15}N–^{19}F, ^{15}N–^{15}N and ^{15}N–^{31}P Couplings

Relaxation of the nitrogen-14 nucleus by interaction of its electric quadrupole moment with fluctuating electric field gradients is generally so efficient that spin coupling to other nuclei is totally obscured. However, in a number of instances not readily predictable from gross structural considerations, scalar ^{14}N–^{19}F coupling constants have been measured in molecules where the electric field gradient is either zero by symmetry or where the gradient is fortuitously small. The corresponding ^{15}N–^{19}F absolute values, obtained by multiplication by $|\gamma(^{15}$N$)/\gamma(^{14}$N$)| = 1.402$, along with other such couplings determined using either enriched materials or measured from the ^{15}N satellites in the ^{19}F spectrum are summarized in Table 5.16.

One-bond ^{15}N–^{19}F couplings are comparatively large and trends paralleling the hybridization at nitrogen can be seen, but the significance of the role played by the Fermi contact term in these couplings is uncertain although without doubt of lesser importance than in ^{15}N–H couplings. The couplings in trifluoroammonia [125] or its N-oxide [124] differ very little from those observed in the *cis-* and *trans-*difluorodiazines [125]. In the latter pair, larger coupling is found in the *cis*-isomer which has a slightly shorter N—N bond, longer N—F bonds, and is thermodynamically more stable than the *trans* [161]. Comparison of these values with that observed for the

TABLE 5.16. Absolute values of some Couplings of Nitrogen-15 to Elements Other Than Hydrogen

Compound	J,(Hz)[a]	Reference
^{15}N–^{19}F *Coupling*		
$^{19}F^{15}NO_2$	157.7	148, 149
$^{19}F_4{}^{15}N_2$	164	172
$CF_3O^{15}N^{19}F_2$	176.3, 164	122, 123
$^{19}F_3{}^{15}NO$	190	124, 147, 148
^{19}F–^{15}N=^{15}N–^{19}F (*trans*)	190	125
^{19}F–^{15}N=^{15}N–^{19}F (*cis*)	203	125
$^{19}F_3C^{15}NO_2$	214, 217	148, 125, 150
$FCO^{14}N^{19}F_2$	221	126
$[^{15}N^{19}F_4]^{\oplus}SbF_6{}^{\ominus}$	323	127
$[^{15}N^{19}F_4]^{\oplus}AsF_6{}^{\ominus}$	328	128
$[^{15}N_2{}^{19}F]^{\oplus}AsF_6{}^{\ominus}$	459	133
^{15}N–C–^{19}F *Coupling*		
$^{19}FC(^{15}NO_2)_3$	13.7	129
$^{19}F_3C^{15}N(O)$=NCF_3	20.3	141
$^{19}F_3{}^{15}NO_2$	21.3	141

	52.3	130, 131

	52.6	130, 131

^{15}N–N–^{19}F *Coupling*		
^{19}F–N=^{15}N–F (*trans*)	102	125
^{19}F–N=^{15}N–F (*cis*)	52	125
Three-Bond ^{15}N–^{19}F *Coupling*		
$(C^{19}F_3)_2P^{15}NH_2$	1.5	116
$Ph^{15}NHN$=$C^{19}FCHO$	11	132
$C^{19}F_3CO^{15}NH$–R^b	<0.6	217

TABLE 5.16—*continued*

Compound	J, (Hz)[a]	Reference
$^{15}N-^{15}N$ *Coupling*		
$Ph^{15}N=^{15}N(O)Ph$	13.7	18
$(PhCH_2)_2 {}^{15}N-^{15}NO$	19	139[c]

	7	140

$^{15}N-^{31}P$ *Coupling*		
$(CF_3)_2 {}^{31}P^{15}NH_2$	52.6	43

[a] In cases where the coupling constants were reported for the ^{14}N isotopomer the corresponding ^{15}N value was obtained by multiplication by $|\gamma(^{15}N)/\gamma(^{14}N)| = 1.402$.

[b] R is a sugar moiety.

product resulting from the reaction of *cis*-difluorodiazine with arsenic pentafluoride indicates a substantial increase in *s*-character. This is taken as evidence in favor of an ionic structure for the fluorodiazonium hexafluoroarsenate (XII),

$$[F-N\equiv N]^{\oplus} AsF_6^{\ominus}$$

(XII)

rather than a covalent complex between the reactants [133]. Indeed, this is the largest coupling involving nitrogen that has been reported to date. However, attention should be drawn to the fact that a similar large enhancement in $^1J(^{15}N-^{19}F)$ is found in the tetrahedral perfluoroammonium cation [127, 128].

An interesting but as yet unexplained finding is the apparent substituent effect evidenced by the one-third decrease in the magnitude of $^2J(^{15}N-^{19}F)$ in trinitrofluoromethane [129] compared with the similar coupling in trifluoronitromethane [141]. Substantial coupling between nitrogen and α-fluorine atoms in pyridines is found [130, 131] and the geometric dependence of the $^{15}N-N-^{19}F$ coupling, possibly reflecting a lone-pair effect, in the *cis*- and *trans*-difluorodiazines [125] is noteworthy. The three-bond coupling between ^{15}N and ^{19}F in the trifluoroacetamido group is reported to be 0.6 Hz [217].

A discussion of $^{15}N-^{15}N$ couplings is limited by the existence of relatively few data. Couplings between directly bonded ^{15}N nuclei in *trans*-azoxybenzene [6, 18] and dibenzylnitrosamine [139] are 13.7 Hz and 19 Hz, respectively. The 7 Hz coupling observed in the ^{15}N spectrum of the thiotrithiazyl cation has been assigned to coupling across the $^{15}N-S-^{15}N$ fragment confirming the cyclic structure (XIII) for this ion [140, 142].

(XIII)

Thus far, the 52.6 Hz coupling in $(CF_3)_2PNH_2$ is the only reported value of coupling between nitrogen and phosphorus [43].

5.5   Two-Bond $^{15}N-C-H$ Coupling

5.5.1 $^{15}N-C-H$ Coupling Across a Tetracoordinate Carbon Atom

Geminal coupling between nitrogen-15 and hydrogen nuclei separated by a saturated carbon atom is small with values generally falling in the range 0–2 Hz. A summary of such couplings is shown in Table 5.17. The magnitude of this coupling shows little variation with structure although compared with other functional groups and paralleling the s-character at nitrogen a slight enhancement is noted in isonitriles [38], N-alkylnitrilium ions [103] and oxaziridines [152].

A significant and highly interesting observation is the stereochemical dependence of the geminal $^{15}N-C-H$ coupling found in molecules which owe their configurational stability to non-inversion at a pyramidal nitrogen center. In the aziridine [151] (XIV), oxaziridine [152] (XV) and tetrahydro-1,3-oxazine [145] (XVI) the magnitude of the geminal coupling strongly depends on the orientation of the lone-pair of electrons on nitrogen. The largest coupling is invariably found between the nitrogen and that hydrogen which is *cis* to

TABLE 5.17. Some Two-Bond $^{15}N-C-H$ Coupling Across An Intervening Tetracoordinate Carbon Atom

Compound	$^2J(^{15}N-H)$, (Hz)	Reference
$CH_3{}^{15}N=CHPh$	0.6[a]	18
$CH_3{}^{15}NH_3^{\oplus}\ Cl^{\ominus}$	0.8	18
$CH_3{}^{15}NHCHO$	0.8[b]	19
$CH_3{}^{15}NHC(S)NHPh$	0.9	18
$(CH_3)_4{}^{15}N^{\oplus}I^{\ominus}$	0.9	9
$CH_3{}^{15}NH_2$	1.0	18, 33
$(CH_3)_2{}^{15}NCHO$	1.1[c], 1.2[b]	35

	0 $(H_{(2)}, CH_3)$	
	1.1 $(H_{(3)})$	145
	1.5 $(H_{(1)})$	

Compound	$^2J(^{15}N-H)$, (Hz)	Reference
N-methylphthalimide-^{15}N	1.4	18
$(CH_3)_2CCH_2CH_2{}^{15}N\equiv C$	2.0	146
$CH_3CH_2{}^{15}N\equiv C$	2.7	38
$PhCH_2{}^{15}N\equiv C$	3.2	9
$CH_3{}^{15}N\equiv C$	3.2	38
$[CH_3{}^{15}N\equiv C-CH_3]^{\oplus}$	3.8	103

	5.4[d], 0[a]	152

[a] *Trans*-isomer.
[b] Coupling to methyl group *cis* to carbonyl.
[c] Coupling to methyl group *trans* to carbonyl.
[d] *Cis*-isomer.

the long-pair [153]. For example, in the tetrahydro-1,3-oxazine, coupling to the C-4 equatorial proton, but not to either of the axial protons at C-2 and C-4, is observed [145].

(XIV)   (XV)

(XVI)

A method for the assignment of configurations in oxaziridines based on the coupling to the C-3 proton in the *cis*-isomer and the absence of such coupling in the *trans*-isomer has been proposed [152].

5.5.2 ^{15}N—C—H Coupling Across a Tricoordinate Carbon Atom

In contrast to ^{15}N—C—H coupling across a saturated carbon atom, the presence of a π-system between the carbon and nitrogen atoms or a multiple bond to the carbon alone leads to enhancement of the geminal coupling and is frequently accompanied by the existence of stable geometric isomers. A selected compilation of ^{15}N—C—H couplings across an intervening tricoordinate carbon atom is given in Table 5.18.

In formamides coupling in the range 14–16 Hz is found between the nitrogen atom and the formyl proton. Dilution with water substantially increases this coupling in formamide-^{15}N [112], but surprisingly a similar effect is not observed in N-methylformamide [19, 52]. Hindered rotation and proton exchange in formamide have been investigated with the aid of ^{14}N decoupling and the solvent effects on these processes are attributed mainly to hydrogen bonding to the carbonyl oxygen [171]. Both dielectric and hydrogen bonding effects on the stabilization of the planar resonance form appear to contribute to these solvent-induced changes, but in a manner that is not clearly understood [19, 112, 158]. In N-monosubstituted formamides coupling to that formyl proton which is *cis* to the adjacent amido-hydrogen is slightly but consistently larger than the coupling in the opposite configuration [52, 113].

The ^{15}N—C—H coupling in several derivatives of N'-phenyl-N,N-dimethylformamidine^{15}-N(XVII) has been investigated by Bose and Kugajevsky [92]. In the parent compound $^2J(^{15}NCH)$ is found to be 8.4 Hz, which is essentially

TABLE 5.18. Some Selected Two-Bond ^{15}N—C—H Couplings Across an Intervening Tricoordinate Carbon Atom

Compound	Solvent	$^2J(^{15}\text{N}-\text{H})$, (Hz)	Reference
HCO^{15}NH$_2$	acetone	16.4	112
	none	19.0	112
	H$_2$O	23.3	112
HCO^{15}NHCH$_3$	none	15.6[a]	52
	H$_2$O	14.6	19
HCO^{15}NHPh	CDCl$_3$	16.3[a], 15.0[b]	52
HCO^{15}N(CH$_3$)$_2$	none	15.6	35
CH$_2$=^{15}N—OH	H$_2$O	13.8[c], 2.6[d]	34
CH$_3$CH=^{15}N—OH	H$_2$O	15.9[c], 2.9[d]	34
PhCH=^{15}N—OH	CDCl$_3$	2.6[d]	154

| | CDCl$_3$ | 14.4 | 155 |

| | none | 4.36 | 28 |

| | CDCl$_3$ | 14.2 | 154, 155 |
| | H$_2$SO$_4$ | 4.2 | 154, 155 |

| | CCl$_4$ | 11.1 | 34 |
| | D$_2$SO$_4$—CF$_3$CO$_2$D | 2.0 | 37 |

Compound	Solvent	$^2J(^{15}\text{N}-\text{H})$, (Hz)	Reference
CH$_3$CH=^{15}NNHC$_2$H$_5$		14.5[c], 4.5[d]	118
PhCH=^{15}NPh	CDCl$_3$	3.8[d]	92
PhCH=^{15}NCH$_3$		3.9[d]	18
(CH$_3$)$_2$NCH=^{15}NPh	CCl$_4$	2.4	92
CH$_2$=CH^{15}N≡C	CCl$_4$	3.1	146
[CH$_2$=CH^{15}N(CH$_3$)$_3$]$^{\oplus}$ Br$^{\ominus}$	D$_2$O	5.1	40, 156
PhN=CH^{15}N(CH$_3$)$_2$	CCl$_4$	8.4	92
HC(^{15}NO$_2$)$_2^{\ominus}$ Na$^{\oplus}$	DMF	6	157

[a] Coupling to formyl proton *cis* to NH.
[b] Coupling to formyl proton *trans* to NH.
[c] *Anti*-isomer.
[d] *Syn*-isomer.

unchanged by *p*-methoxy substitution, whereas *p*-nitro substitution reduces the coupling to 7.5 Hz. These coupling constants apparently represent average values due to free rotation about the central C—N bond, although under different solvent conditions separate resonances for distinct geometric isomers have been detected [159]. In the analog of (XVII) having the isotopic label at the anilinonitrogen the

$$\text{C}_6\text{H}_5-\text{N}=\text{C}\overset{^{15}\text{N(CH}_3)_2}{\underset{\text{H}}{\diagdown}}$$

(XVII)

coupling across the $^{15}\text{N}=\text{C}-\text{H}$ moiety is reported [92] to be 2.4 Hz, but under the circumstances it is difficult to assess the relation of the magnitude of this coupling to the geometry of the azomethine linkage.

Kintzinger and Lehn [154] have uncovered a very striking effect of the orientation of the nitrogen lone-pair on $^{15}\text{N}-\text{H}$ coupling constants in the $^{15}\text{N}=\text{C}-\text{H}$ group. In oximes measured $^{15}\text{N}=\text{C}-\text{H}$ coupling constants are *ca.* 3 Hz and *ca.* 16 Hz for the *syn*-(XVIII) and *anti*-(XIX) isomers, respectively.

$$\overset{\text{R}}{\underset{\text{H}}{\diagup}}\text{C}={}^{15}\text{N}\diagdown\text{OH} \qquad \overset{\text{R}}{\underset{\text{H}}{\diagup}}\text{C}={}^{15}\text{N}\diagup\text{OH}$$

(XVIII)   (XIX)

Coupling comparable with that found in *anti*-oximes is also observed in cyclic systems such as pyridine [118], quinoline [34], 5-phenylisothiazole [155], and isoxazole [154] which similarly have the nitrogen lone-pair *cis* to the adjacent hydrogen. Confirmation of the importance of the lone-pair orientation is further derived from the fact that removal of the nitrogen lone-pair by protonation or quaternization [37, 160] results in the reduction of $^2J(^{15}\text{N}-\text{H})$ to the value found in *syn*-oximes [154] and pyrrole [28]. A theoretical interpretation of the effect of the lone-pair on geminal N=C—H coupling has appeared [216].

The ethylhydrazone of acetaldehyde has been shown to exist in *syn*- and *anti*-forms for which $^{15}\text{N}=\text{C}-\text{H}$ coupling constants of 4.5 Hz and 14.5 Hz respectively, have been reported [18]. It may be inferred from the $^{15}\text{N}=\text{C}-\text{H}$

coupling of 3.8 Hz in benzalaniline [92], 3.9 Hz in benzal-methylamine [18] and 3.5 Hz in phenylhydrazone of benzaldehyde [81] that these Schiff bases exist predominately in the *syn*-configuration.

5.5.3 Other Two-Bond $^{15}N-H$ Couplings

Two-bond couplings between nitrogen and hydrogen attached to a dicoordinate carbon atom have been reported for hydrogen cyanide-^{15}N [162] and its ^{15}N-protonated conjugate acid [103].. For the latter, $^{2}J(^{15}N-H)$ was found to be 19.0 Hz which is twice the magnitude of the $^{15}N-C-H$ coupling in the unprotonated form.

Geminal coupling across a heteroatom is rarely encountered because hydrogen exchange is usually facile. However, for O-protonated conjugate acid of dibenzylnitrosamine-^{15}N a value of 2.6 Hz was found for the $^{15}N-O-H$ coupling [163].

5.6 Three-Bond N–H Coupling

5.6.1 $^{14}N-C-C-H$ and $^{15}N-C-C-H$ Coupling Across Two Tetracoordinate Carbon Atoms

The majority of available data regarding coupling between nitrogen and hydrogen nuclei removed from each other by two saturated carbon atoms comes from investigation of quaternary ammonium-^{14}N salts [164]. This coupling which is of opposite sign and larger than the two-bond geminal coupling is highly dependent on the field symmetry situation at the nitrogen center [207]. In ordinary amines, for example, such coupling is not observed and even small structural changes in ammonium salts profoundly affect the quadrupolar relaxation of the ^{14}N nucleus with attendant consequences on the ability to detect coupling [166, 167]. Successive replacement of the ethyl groups by methyl groups in the tetraethylammonium ion results in a regular increase in the $^{15}N-C-C-H$ coupling from 2.5 Hz to 3.0 Hz in the trimethylethylammonium ion [164, 165]. A similar effect which may reflect bond angle changes is found in the 0.7 Hz coupling to the β-hydrogens in N,N-diethylaniline-^{15}N [18] compared with 3.0 Hz in N-ethylaniline-^{15}N [81].

With the aid of proton double- and triple-resonance techniques, Terui and coworkers [168] were able to demonstrate an angular dependence of the vicinal ^{14}N—C—C—H coupling in the rigid bicyclic systems, dibenzobicyclo [2.2.2] octa-2,5-dien-7-yltrimethyl-ammonium bromide (XX) and the 2-*exo*-(XXIa) and 2-*endo*-hydroxybornan-3-*endo*-yltrimethylam-monium bromides (XXIb). A dihedral angle of 0° between

(XX)

(XXIa) (R_1 = OH, R_2 = H)
(XXIb) (R_1 = H, R_2 = OH)

the interacting nuclei lead to maximum coupling which corres-ponds to a $^3J(^{15}N$—H) value of 3.8 Hz. Without experimental confirmation it was inferred that $^3J(^{15}N$—H) would also be at a maximum with a dihedral angle in the vicinity of 180°. This prediction is provisionally verified by the angular dependence of the vicinal ^{15}N—H coupling reported for amino acids [169]. Conformational analysis of the temperature-variable coupling constants in 3,3-dimethylbutyl isocyanide and related compounds provides additional support for the suggestion that couplings in these systems conform to a Karplus-type curve [146].

5.6.2 ^{15}N—C—C—H Coupling Across One or More Tri-coordinate Carbon Atoms

Data for several three-bond ^{15}N—C—C—H couplings in oximes [34], imines [81] and hydrazones [81, 118] are listed in Tables 5.19. The proton spectra of these carbonyl derivatives reveal the presence of *syn*- and *anti*-isomers and configura-tional dependence of this coupling is clearly identifiable. For example, in acetonoxime-^{15}N (XXII) coupling to *trans*-CH$_3$ group (4.0 Hz) is approximately twice as large as the coupling to the *cis*-CH$_3$ group (2.2 Hz) [34]. Invariably, larger coupling is found between the ^{15}N nucleus and that hydro-gen-three bonds removed which lies closest in space to the nitrogen lone-pair. On protonation, small but unequal in-creases are noted for both $^3J(^{15}N$—H) values pointing to identical signs for these couplings [34].

TABLE 5.19. Geometric Dependence of
Three-Bond $^{15}N{=}C{-}C{-}H$ Coupling[a]

$$G_1{-}^{15}N{=}C\underset{G_3}{\overset{CH_2{-}G_2}{\Big\langle}}$$

G_1	G_2	G_3	\multicolumn{2}{c}{$^3J(^{15}N{-}H)$, (Hz)}	Reference
			Syn	Anti	
OH	H	H	2.6	4.2	34
OH	H	CH_3	2.2 (3.2)[b]	4.0 (4.5)[b]	34
OH	H	t-Bu	1.8	—	34
OCH_3	H	CH_3	0.4	~0	170
Ph	H	$(CH_3)_2N$[c]	1.1		92
Ph	H	CH_3	1.8	3.4	81
Ph	H	t-Bu	1.8	—	81
PhNH	H	CH_3	1.8	3.6	81
PhNH	H	t-Bu	1.8	—	81
PhNH	Ph	H	2.2	3.6	81
EtNH	H	H	2.3	5.1	118

[a] Syn- and anti-isomers are defined as the configurations in which the substituents, G_1 and G_3, are oriented trans and cis to each other, respectively.

[b] Protonated form.

[c] Configuration uncertain.

Similarly, in thioacetanilide-^{15}N $^3J(^{15}N{-}H)$ is found to be 3 Hz in the configuration with the phenyl and methyl groups trans to each other compared with 2 Hz in the other rotational isomer and in thioacetamide [114]. The comparable couplings of 0.7 Hz and 1.3 Hz, respectively, in acetanilide-^{15}N and acetamide-^{15}N have been reported, but stereospecificity was not observed [5, 18].

$$HO \overset{\displaystyle \bigcirc}{\underset{}{\diagdown}} {}^{15}N{=}C \overset{CH_3 \ (trans)}{\underset{CH_3 \ (cis)}{\Big\langle}}$$

(XXII)

Coupling across the $^{15}N{-}C{=}C{-}H$ fragment also appears to be dependent on the spatial arrangement of the coupled nuclei. Comparison of the 3.5 Hz coupling between the ^{15}N atom and the vinyl proton in the anilino-^{15}N-fumarate (XXIII) with the 1.7 Hz value observed in the anilino-^{15}N-maleate (XXIV) indicates stronger coupling in the trans-configuration [92]. Similar values are found in other systems which have a trans $^{15}N{-}C{=}C{-}H$ geometry [144, 173–175].

(XXIII)   (XXIV)

A value of 1.9 Hz consistent with *cis*-coupling is found in the dimedone derivative (XXV). *Trans*-coupling in vinyl isocyanide [146] and trimethylvinylammonium bromide [40, 71] is also found to be larger than this *cis*-coupling.

(XXV)

Other three-bond $^{15}N–H$ couplings across unsaturated carbon atoms are known. In pyrroles [27–29, 69], pyrimidines [177] and pyrazine derivatives [160] values generally fall in the range 4–6 Hz. The coupling in quinoline increases from 1.4 Hz in the free base to 4.5 Hz in acid medium [37]. In aniline-^{15}N derivatives [78] and other ^{15}N-nitro labeled aromatic compounds [179, 180] coupling between the ^{15}N and the *ortho* protons is observed. Values fall in the range 2–3 Hz, but show no well-defined variation with structure.

5.6.3 $^{15}N–N–C–H$ and $^{15}N–O–C–H$ Coupling

The temperature-dependent NMR spectra of compounds containing a nitroso group attached to an atom with electrons available for π-bonding reveal an equilibrium between *syn*-(XXVIa) and *anti*-(XXVIb) isomers resulting from restricted

G = Ar, CH$_2$R, NHR, OR

(XXVIa)   (XXVIb)

TABLE 5.20. Some Selected Three-Bond $^{15}N–H$ Coupling Constants

Compound	Type of Coupling	$^3J(^{15}N–H)$, (Hz)	Reference
$CH_3O^{15}NO_2$	$^{15}N–O–C–H$	4.2	9
$PhCH_2O^{15}N=O$	$^{15}N–O–C–H$	0 (syn)	184
		2.4 (anti)	
$(CH_3)_2N^{15}NO_2$	$^{15}N–N–C–H$	2.3	9
$CH_3CH_2{}^{15}N\equiv C$	$^{15}N–C–C–H$	2.8[a]	38
$CH_3CH={}^{15}NNHCH_2CH_3$	$^{15}N–N–C–H$	2.5	118
$Ph^{15}NHN=CHPh$	$^{15}N–N–C–H$	6.6	92
$CH_3C\equiv{}^{15}N$	$^{15}N–C–C–H$	1.80	42
$Ph^{15}N(CH_2CH_3)_2$	$^{15}N–C–C–H$	0.7	18
$CH_3C(^{15}NO_2)_3$	$^{15}N–C–C–H$	$\geqslant 2.5$[a]	3
$CH_3N(PhCH_2)^{15}N=O$	$^{15}N–N–C–H$	0.8 (syn)[b]	181
		2.2 (anti)[b]	181
$CH_3N(OCH_3)^{15}N=O$	$^{15}N–N–C–H$	0.7 (syn)[b]	183
		2.7 (anti)[b]	183
$(CH_3)_2NN(PhCH_2)^{15}N=O$	$^{15}N–N–C–H$	0 (syn)	182
		2.2 (anti)	182
$[(CH_3)_3{}^{15}N^{\oplus}CH=CH_2]Br^{\ominus}$	$^{15}N–C–C–H$	7.9 (trans)[a]	156, 40
		3.6 (cis)[a]	156, 40
$CH_3CO^{15}NH_2$	$^{15}N–C–C–H$	1.3	18
$CH_3CS^{15}NHPh$	$^{15}N–C–C–H$	2 (syn)	114
		3 (anti)	114
$CH_3CH(CO_2^{\ominus})^{15}NH_3^{\oplus}$	$^{15}N–C–C–H$	3.1	118
(aniline structure: $^{15}NH_2$ on benzene, ortho H, para NO_2)	$^{15}N–C–C–H$	1.8	78
(quinoline with ^{15}N, H at position 3)	$^{15}N–C–C–H$	1.3	34
		4.5[c]	34
(quinolinium N-methyl structure, $^{15}N^{\oplus}$, CH_3)	$^{15}N–C–C–H$	2.9[a]	160
($^{15}NO_2$ quinoline N-oxide structure, $O^{\ominus}$)	$^{15}N–C–C–H$	2.7	179
(pyrrole structure, ^{15}N, H)	$^{15}N–C–C–H$	5.4	28

[a] Calculated from the reported value of $^3J(^{14}N–H)$.
[b] Coupling to CH_3 group.
[c] Protonated form.

rotation of the nitroso group. In nitrosamines [181], nitrosohydrazines [182] and nitrosohydroxylamines [183] $^3J(^{15}N-N-C-H)$ shows a marked dependence on the orientation of the nitroso oxygen with respect to the α-hydrogen. Values in the range 2–3 Hz for the $^{15}N-N-C-H$ coupling in the *anti*-isomer compared with less than 1 Hz in the *syn*-isomer have permitted the assignment of configurations in these compounds. The identical geometric dependence has been observed in the $^{15}N-O-C-H$ fragment and the method has been extended to assign configurations in alkyl nitrites-^{15}N [184].

A few other three-bond couplings across heteroatoms are known. The $^{15}N-N=C-H$ coupling in benzaldehyde phenylhydrazone is 6.6 Hz for the *anti*-isomer [92]. One would expect smaller coupling in the *syn*-configuration, but this is not known. In dimethylnitramine-^{15}N and methyl nitrate-^{15}N the $^{15}N-N-C-H$ and $^{15}N-O-C-H$ couplings are 2.3 Hz and 4.2 Hz, respectively [2,3,9]. The $^{15}N-N-C-H$ couplings in 2-methyl-5-phenyltetrazole-^{15}N is reported to be 1.7 Hz [209].

5.7   Applications to Structure Elucidation

5.7.1   Tautomerism

The first and highly perceptive application of nitrogen-15 substitution to the study of keto-enol equilibria in Schiff bases was made by Dudek and Dudek [173]. The method has been elaborated in succeeding papers [174, 175, 186, 187, 208, 218]. The imine-amine equilibrium illustrated by the tautomeric forms, (XXVIIa) and (XXVIIb), is typical. Changes in

(XXVIIa)   (XXVIIb)

$^{15}N-H$ coupling due to temperature or solvent variation are attributed to rapid proton exchange between oxygen and nitrogen sites. The observed coupling constant, J(obsd.), is taken as the weighted average of $^1J(OH)$ and $^1J(^{15}NH)$, where P_O and P_N represent the mole fractions of

$$J(\text{obsd.}) = {}^1J(OH)P_O + {}^1J(^{15}NH)P_N \qquad (5.6)$$

the two tautomeric species. If $^1J(OH)$ is assigned a value of zero and $^1J(^{15}NH)$ in the pure amino-form (XXVIIa) is assumed to be ~89 Hz, then the equilibrium composition can be determined. In this manner, N-methyl-2-hydroxy-1-naphthaldehyde imine-^{15}N has been shown to exist predominately in the ketoamine tautomer (XXVIII) [173], whereas an N-phenyl substituent destabilizes this form as evidenced by the preponderance of the enolimine tautomer (XXIX) [175] in N-phenyl-2-hydroxy-1-naphthaldehyde imine-^{15}N.

(XXVIII)   (XXIX)

This approach has been used to estimate the azo-hydrazo tautomeric equilibria in arylazo-^{15}N-derivatives of phenols and naphthols [188, 189]. In the case of 1-phenylazo-2-napthol in chloroform solution, the results indicate that the hydrazoform (XXXa) is favored and at low temperature is present almost

(XXXa)   (XXXb)

exclusively [188]. This conclusion has been independently confirmed by Berrie and coworkers [190] who used the difference in ^{14}N chemical shifts of the azo- and hydrazo-forms to estimate the equilibrium composition. The latter investigators also found that in a non-polar solvent such as hexane the azo-tautomer dominates [190].

Agreement is found in the conclusions of two groups [189, 191] of workers who used ^{15}N-labeling studies to distinguish among the several tautomeric possibilities for 3-methyl-1-phenyl-4-phenylazo-2-pyrazolin-5-one-^{15}N. In chloroform solution, the ^{15}N-bound proton resonance occurs at very low field and the observed $^1J(^{15}N-H)$ value of 96.5 Hz remains invariant with temperature. Both facts are interpreted in terms of the intramolecularly hydrogen-bonded hydrazo-tautomer (XXXI). Analogous studies with 3-anilino-^{15}N-1-phenyl-2-pyrazolin-5-one provide evidence for the existence of the keto-tautomer (XXXII) only [93].

(XXXI)   (XXXII)

The 1,2,3-butanetrione-2-phenylhydrazone-1-^{15}N resulting from the coupling of acetoacetaldehyde with benzenediazonium-^{15}N tetrafluoroborate exists as an equilibrium mixture of the two internally hydrogen-bonded hydrazo-forms (XXXIIIa and XXXIIIb) [192] which agrees with an earlier report [193] concerning the structure of mesoxaldialdehyde-2-phenylhydrazone-1-^{15}N.

(XXXIIIa)   (XXXIIIb)

Evidence based on $^{15}N-H$ coupling has been presented in favor of the enamine-thione tautomer of the methyl dithioate

(XXXIV) [144] and a preliminary report indicates that a similar preference exists in the thiophene derivative (XXXV) [194].

(XXXIV) (XXXV)

The enamine-keto tautomer of the betaine (XXXVI) has been established as the predominant species in solution [195].

(XXXVI)

An interesting report concerning the tautomeric equilibria in 2-picolyl-^{15}N-ketones has appeared [196]. In addition to the keto (XXXVIIa) and enol (XXXVIIb) forms stable at low temperature, a new species is formed at elevated temperatures. Hückel MO calculations are consistent with the pyridinium structure (XXXVIIc) which has been postulated for this new tautomer, although ^{15}N—H coupling could not be observed. In line with the marked upfield nitrogen shift observed in the pyridinium-^{15}N ion [8, 118], and pyridone-^{14}N tautomers [67, 197], a determination of the ^{15}N chemical shift in the latter species would be of interest.

(XXXVIIa) (XXXVIIb) (XXXVIIc)

Nitrogen-15 substitution has resolved the question of the structure of formazans in favor of a tautomeric equilibrium rather than a resonance hybrid structure. Published data for symmetrical sugar formazans [198] and 1,3,5-triphenyl-formazan-[15]N [199] indicate an averaging of the [15]H—N coupling over two chemically equivalent nitrogen sites as exemplified by the equilibrium between (XXXVIIIa) and (XXXVIIIb). The observed $^1J(^{15}N{-}H)$ and position of the tautomeric equilibrium are a function of substituents in the ^{15}N-anilino ring [199]. Similar conclusions have been reached in the study of unsymmetrical sugar osazoneformazans [200, 201].

(XXXVIIIa)   (XXXVIIIb)

In contrast to other reports [222, 223], analysis of the proton NMR spectra of the coupling and reduction products derived from ^{15}N-labeled phenyldiazonium ion indicates the absence of the rearrangement (XXXIXa) to (XXXIXb) [220].

(XXXIXa)   (XXXIXb)

5.7.2   Sites of Protonation

In strong acid media, virtually all molecules undergo protonation, and the structure of conjugate acids of [15]N- enriched compounds can frequently be conclusively defined by the presence or absence of [15]N—H coupling in their NMR spectra. For example, the ~136 Hz [15]N—H coupling in nitrilium ions confirms the expected protonation on nitrogen and the linear structure for this ion [55, 103].

Low-temperature NMR studies of acetamide-[15]N [202] and dibenzylnitrosamine-[15]N [163] in anhydrous fluorsulfuric acid confirm earlier reports [203, 204] that these cations exist in the O-protonated form in this medium. Restricted rotation persists in these cations as evidenced by the configurational dependence of the [15]N—N—C—H coupling in the nitrosamine [181] and by the unequal ^{15}N–H couplings of 92.5 Hz and 96.5 Hz found in acetamide [68, 202]. In aqueous acid media, N-protonation of acetamide has been proposed but hydrogen exchange precludes the detection of ^{15}N–H coupling [202].

Olah and White [205] have studied the protonation of doubly-labeled urea-^{15}N in FSO_3H–SbF_5–SO_2. The low temperature spectrum indicates the formation of the diprotonated species (XL) with non-equivalent nitrogen atoms. The two

(XL)

observed ^{15}N–H coupling constants of 76.6 Hz and 96.8 Hz are consistent with the formulated sp^3 and sp^2 nitrogen nuclei. Heteronuclear decoupling demonstrates that the 4.6 Hz doublet due to the hydroxyl proton arises through coupling to the -^{15}NH$_3^{\oplus}$ nitrogen only. Phthalimide-^{15}N has similarly been shown to undergo protonation on both carbonyl oxygen atoms and the $^1J(^{15}$N–H) of 105 Hz reported [206] for this diprotonated imide is the largest value known for coupling between hydrogen and a tricoordinate nitrogen atom.

5.7.3 Molecules of Biological Interest

The geminal ^{15}N–H coupling constant in amino acids has been found to increase with pH [210]. Thus, changes in this coupling, rather than ^{14}N quadrupolar effects can account for the pH-dependence of the linewidths of the α-proton resonances in unlabeled amino acids. A complementary report has described the angular-dependence of the vicinal ^{15}N–H coupling in amino acids [169]. Exchange rates for the peptide protons in some oligoglycines-^{15}N are slower in aqueous urea than in water and are markedly affected by their neighboring groups [211].

Simple purines and pyrimidines interact to form specific base-paired complexes as shown by the hydrogen bonding shifts that result on admixture [94]. The observation of

[15]N—H coupling in the amino group of 1-methylcytosine-7-[15]N (XLI) in both the free base and its complex with 9-ethylquanine rules out any change of tautomeric form on pair formation.

(XLI)

Proton and nitrogen-15 magnetic resonance spectra of several derivatives of pyrimidine-^{15}N have been studied [7, 176, 178, 212]. Protonation of triply-^{15}N-labeled 1-methylcytosine [212] has been shown to occur at N-3 in agreement with studies on material labeled in the amino group [176]. Hindered rotation of the amino group has been found in the hydrohalide salts, and in the methylamino derivative this gives rise to geometric isomers [176]. Long-range [15]N—H couplings have been measured [212] in uracil-[15]N$_2$, 2,4-dichloro-pyrimidine-[15]N$_2$, 2-4-dimethoxypyrimidine-^{15}N, and 1-methyl-4-methoxy-2-pyrimidone-[15]N$_2$. The complexities of the ^{15}N spectra resulting from proton couplings have prevented the determination of $J($[15]N—[15]N$)$ in these compounds [97].

An interesting application of ^{15}N substitution has been made by Happe and Morales [213] in their study of cation binding in adenosine-[15]N$_5$ triphosphate (XLII). The ^{15}N spectrum taken in D$_2$O shows five well-separated resonances, of which the 6-amino and N-9 resonances are singlets, N-7 is a doublet (J = 10 Hz) due to coupling with H-8, and N-1 and N-3 are each doublets arising from coupling (J = 16 Hz) to H-2. Evidently, [15]N—C—[15]N couplings are quite small. Addition of Zn$^{2\oplus}$ ions produces shifts in the 6-amino, N-7 and N-9 resonances whereas no significant shift in any nitrogen resonance is observed on addition of Mg$^{2\oplus}$ ions. These observations have been interpreted in terms of interactions at the 6-amino and N-7 positions in the case of Zn$^{2\oplus}$, and the complete absence of any binding to the adenine ring in the case of Mg$^{2\oplus}$. Complex formation, particularly if the nitrogen lone-pairs are involved, would be expected to have significant influence on the ^{15}N—C—H couplings, but the effect of the metal ions on these couplings has not been reported.

(XLII)

Lichter [12] cites some unpublished work of Hunter who has found that the ^{15}N resonance in complexes of ethylene-diaminetetra-acetic acid is a function of the metal ion, its concentration and the pH. These variables can produce compensatory effects on the ^{15}N shifts suggesting that an investigation of ^{15}N—C—H couplings in metal complexes of adenosing triphosphate might be worthwhile.

Some preliminary work aimed at defining the applicability of measurements of ^{15}N magnetic resonance parameters to the structural and conformational analysis of amino sugars, which are of biomedical interest, has appeared [214, 217].

5.8   Conclusions

An attempt has been made to give a comprehensive account of the corelations between nitrogen coupling constants and molecular structure. The broad spectrum of ^{15}N NMR applications to the solution of structural and theoretical problems should be apparent. Advances in signal enhancement techniques and the commercial development of versatile multinuclei spectrometers promise to accelerate the marked progress that the past few years have witnessed. Biological systems and other complex molecules which have thus far resisted ^{15}N-NMR study provide a formidable and exciting challenge for the future. Evidence is now available to indicate that the effort required to develop methods for the routine investigation of ^{15}N at the natural abundance level will bring a just return. Continued advances in this area may confidently be expected.

References

1. M. WITANOWSKI and G. A. WEBB, in *Annual Review of N.M.R. Spectroscopy*, Vol. 5 (E. F. MOONEY, Ed.), Academic Press, London, 1972, p. 395.
2. C. M. SHEPPERD, T. SCHAEFFER, B. W. GOODWIN, and J. T. RAA, *Can. J. Chem.*, 49, 3158 (1971).
3. J. P. KINTZINGER, J. M. LEHN, and R. L. WILLIAMS, *Mol. Phys.*, 17, 135 (1969).
4. T. AXENROD and F. MACCHIA, unpublished results.
5. M. LILER, *J. Mag. Resonance*, 5, 333 (1971).
6. J. B. LAMBERT, G. BINSCH, and J. D. ROBERTS, *Proc. Nat. Acad. Sci. U.S.*, 51, 735 (1964).
7. T. J. BATTERHAM and C. BIGUM, *Org. Mag. Resonance*, in press.
8. M. WITANOWSKI, *J. Amer. Chem. Soc.*, 90, 5683 (1968).
9. E. D. BECKER, R. B. BRADLEY, and T. AXENROD, *J. Mag. Resonance*, 4, 136 (1971).
10. E. W. RANDALL and D. G. GILLIES, in *Progress in Nuclear Magnetic Resonance Spectroscopy*, Vol. 6 (J. W. EMSLEY, J. FEENEY, and L. H. SUTCLIFFE, Eds.), Pergamon Press, Oxford, 1971, p. 119.
11. E. F. MOONEY and P. H. WINSON, in *Annual Review of N.M.R. Spectroscopy*, Vol. 2 (E. F. MOONEY, Ed.), Academic Press, London, 1969, p. 125.
12. R. L. LICHTER, in *Determination of Organic Structures by Physical Methods*, Vol. 4 (J. J. ZUCKERMAN and F. C. NACHOD, Eds.), Academic Press, New York, London, 1972, p. 195.
13. J. W. EMSLEY, J. FEENEY, and L. H. SUTCLIFFE, in *High Resolution Nuclear Magnetic Resonance Spectroscopy*, Vol. 1, Pergamon Press, Oxford, 1965, pp. 103 ff.
14. W. McFARLANE, *Quart. Rev.*, 187 (1969).
15. G. E. MACIEL, J. W. McIVER, Jr., N. S. OSTLUND, and J. A. POPLE, *J. Amer. Chem. Soc.*, 92, 1 (1970).
16. D. M. GRANT and W. M. LITCHMAN, *J. Amer. Chem. Soc.*, 87, 3994 (1965).
17. T. L. BROWN and J. C. PUCKETT, *J. Chem. Phys.*, 44, 2238 (1966).
18. G. BINSCH, J. B. LAMBERT, B. W. ROBERTS, and J. D. ROBERTS, *J. Amer. Chem. Soc.*, 86, 5564 (1964).
19. A. J. R. BOURN and E. W. RANDALL, *Mol. Phys.*, 8, 567 (1964).
20. J. A. POPLE and D. P. SANTRY, *Mol. Phys.*, 8, 1 (1964).
21. Y. KATO, M. MIURA, and A. SAIKA, *Mol. Phys.*, 13, 491 (1967).
22. H. S. GUTOWSKY, C. H. HOLM, A. SAIKA, and G. A. WILLIAMS, *J. Amer. Chem. Soc.*, 79, 4596 (1957).
23. R. FREEMAN, K. A. McLAUCHLAN, J. I. MUSHER, and K. G. R. PACHLER, *Mod. Phys.*, 5, 321 (1962).
24. R. FREEMAN and W. A. ANDERSON, *J. Chem. Phys.*, 37, 2053 (1962).
25. K. KUHLMANN and J. D. BALDESCHWEILER, *J. Amer. Chem. Soc.*, 85, 1010 (1963).

26. R. FREEMAN and B. GESTBLOM, *J. Chem. Phys.*, **48**, 5008 (1968).
27. E. RAHKAMAA, *Mol. Phys.*, **19**, 727 (1970).
28. E. RAHKAMAA, *Z. Naturforsch*, **24A**, 2004 (1969).
29. D. GAGNAIRE, R. RAMASSEUL, and A. RASSAT, *Bull. Soc. Chim. Fr.*, 415 (1970).
30. G. A. KALABIN, V. F. BYSTROV, E. V. SHEPELEV, S. N. SHVEDOVA, O. B. LEBEDEV, and L. I. KHMEL'NITSKII, *Izv. Akad. Nauk. SSSR. Ser. Khim.*, 2627 (1969).
31. R. J. CHUCK, D. G. GILLIES, and E. W. RANDALL, *Mol. Phys.*, **16**, 121 (1969).
32. A. J. R. BOURN, D. G. GILLIES, and E. W. RANDALL in *Nuclear Magnetic Resonance in Chemistry* (B. PESCE, Ed.), Academic Press Inc., New York, N.Y., 1965, p. 277.
33. L. PAOLILLO and E. D. BECKER, *J. Mag. Resonance*, **3**, 200 (1970).
34. D. CREPAUX, J. M. LEHN, and R. R. DEAN, *Mol. Phys.*, **16**, 225 (1969).
35. A. J. R. BOURN and E. W. RANDALL, *J. Mol. Spectrosc.*, **13**, 29, (1964).
36. D. CREPAUX and J. M. LEHN, *Mol. Phys.*, **14**, 547 (1968).
37. K. TORI, M. OHTSURU, K. AONO, Y. KAWAZOE, and M. OHNISHI, *J. Amer. Chem. Soc.*, **89**, 2765 (1967).
38. W. McFARLANE, *J. Chem. Soc. (A)*, 1660 (1967).
39. J. P. MAHER, *J. Chem. Soc. (A)*, 1855 (1966).
40. M. OHTSURU, K. TORI, J. M. LEHN, and R. SEHER, *J. Amer. Chem. Soc.*, **91**, 1187 (1969).
41. W. McFARLANE and R. R. DEAN, *J. Chem. Soc. (A)*, 1187 (1968).
42. W. McFARLANE, *Mol. Phys.*, **10**, 603 (1966).
43. A. H. COWLEY, J. R. SCHWEIGER, and S. L. MANATT, *Chem. Commun.*, 1491 (1970).
44. W. McFARLANE, *J. Chem. Soc. (A)*, 1535 (1968).
45. R. B. JOHANNESEN, *J. Chem. Phys.*, **48**, 1414 (1968).
46. I. D. KUNTZ, Jr., P. von R. SCHLEYER, and A. ALLERHAND, *J. Chem. Phys.*, **35**, 1533 (1961).
47. E. GRUNWALD, A. LOEWENSTEIN, and S. MEIBOOM, *J. Chem. Phys.*, **27**, 630 (1957).
48. A. H. LAMBERTON, I. O. SUTHERLAND, J. E. THORPE, and H. M. YUSUF, *J. Chem. Soc. (B)*, 6 (1968).
49. P. HAMPSON and A. MATHIAS, *Chem. Commun.*, 825 (1968).
50. A. D. BUCKINGHAM and K. A. McLAUCHLAN, *Proc. Chem. Soc.*, 144 (1963).
51. R. A. BERNHEIM and B. J. LAVERY, *J. Amer. Chem. Soc.*, **88**, 1279 (1967).
52. A. J. R. BOURN, D. G. GILLIES, and E. W. RANDALL, *Tetrahedron*, **20**, 1811 (1964).
53. K. A. McLAUCHLAN and T. SCHAEFER, *Can. J. Chem.*, **44**, 321 (1966).
54. E. SACKMANN and H. DREESKAMP, *Spectrochim. Acta*, **21**, 2005 (1965).

55. H. HOGEVEEN, *Rec. Trav. Chim.*, **86**, 1288 (1967).
56. L. PAOLILLO and E. D. BECKER, *J. Mag. Resonance*, **2**, 168 (1970).
57. R. A. BERNHEIM and H. BATIZ-HERNANDEZ, *J. Chem. Phys.*, **40**, 3446 (1964).
58. M. ALEI, Jr., A. E. FLORIN, W. LITCHMAN, and J. F. O'BRIEN, *J. Phys. Chem.*, **75**, 932 (1971).
59. W. LITCHMAN, M. ALEI, Jr., and A. E. FLORIN, *J. Chem. Phys.*, **50**, 1897 (1969).
60. R. A. OGG and J. D. RAY, *J. Chem. Phys.*, **26**, 1515 (1957).
61. M. ALEI, Jr., A. E. FLORIN, and W. M. LITCHMAN, *J. Phys. Chem.*, **75**, 1758 (1971).
62. J. D. BALDESCHWEILER, *J. Chem. Phys.*, **36**, 152 (1962).
63. C. S. SPRINGER and D. W. MEEK, *J. Phys. Chem.*, **70**, 481 (1966).
64. K. L. HENOLD, *Chem. Commun.*, 1340 (1970).
65. D. W. MEEK and C. S. SPRINGER, *Inorg. Chem.*, **5**, 445 (1966).
66. *Tables of Interatomic Distances and Configurations in Molecules and Ions*, The Chemical Society, London, 1958.
67. M. WITANOWSKI, L. STEFANIAK, H. JANUSZEWSKI, and G. A. WEBB, *Tetrahedron*, **27**, 3129 (1971).
68. D. M. BROUWER and J. A. van DOORN, *Tetrahedron Letters*, 3339 (1971).
69. E. RAHKAMAA, *Z. Naturforsch.*, **26A**, 1187 (1971).
70. E. W. RANDALL and J. J. ZUCKERMAN, *Chem. Commun.*, 732 (1966).
71. M. OHTSURU and K. TORI, *Chem. Commun.*, 750 (1966).
72. L. K. DYALL, *Aust. J. Chem.*, **17**, 419 (1963).
73. Y. YONEMOTO, W. F. REYNOLDS, H. M. HUTTON, and T. SCHAEFFER, *Can. J. Chem.*, **43**, 2688 (1965).
74. B. M. LYNCH, B. C. MACDONALD, and J. K. G. WEBB, *Tetrahedron*, **24**, 3595 (1968).
75. B. M. LYNCH, *Tetrahedron Letts.*, 1357 (1969).
76. T. AXENROD, P. S. PREGOSIN, M. J. WIEDER, and G. W. A. MILNE, *J. Amer. Chem. Soc.*, **91**, 3681 (1969).
77. M. R. BRAMWELL and E. W. RANDALL, *Chem. Commun.*, 250 (1969).
78. T. AXENROD, P. S. PREGOSIN, M. J. WIEDER, E. D. BECKER, R. B. BRADLEY, and G. W. A. MILNE, *J. Amer. Chem. Soc.*, **93**, 6536 (1971).
79. V. V. NEGREBETSKII, A. V. KESSENIKH, S. S. NOVIKOV, L. I. KHMELINITII, A. S. PRIKHODIKO, and O. V. LEBEDEV, *Izv. Akad. Nauk. SSSR, Ser. Khim.*, 2163 (1971).
80. T. AXENROD, M. J. WIEDER, G. BERTI, and P. L. BARILI, *J. Amer. Chem. Soc.*, **92**, 6066 (1970).
81. T. AXENROD and M. J. WIEDER, unpublished results.
82. G. FRAENKEL, Y. ASAHI, H. BATIZ-HERNANDEZ, and R. A. BERNHEIM, *J. Chem. Phys.*, **44**, 4647 (1966).
83. L. ERNST and A. MANNSCHRECK, *Tetrahedron Letts.*, 3023 (1971).
84. C. BEAUTE, Z. W. WOLKOWSKI, and N. THOAI, *Tetrahedron Letts.*, 817 (1971).

85. J. K. M. SANDERS and D. H. WILLIAMS, *J. Amer. Chem. Soc.*, **93**, 641 (1971).
86. T. AXENROD and M. J. WIEDER, *J. Amer. Chem. Soc.*, **93**, 3541 (1971).
87. I. D. RAE, *Chem. Commun.*, 519 (1966).
88. I. D. RAE, *Aust. J. Chem.*, **20**, 1173 (1967).
89. V. BEKAREK, J. KAVALEK, J. SOCHA, and S. ANDRYSEK, *Chem. Commun.*, 630 (1968).
90. T. AXENROD, G. BERTI, and G. BELLUCCI, unpublished results.
91. T. AXENROD and P. S. PREGOSIN, unpublished results.
92. A. K. BOSE and I. KUGAJEVSKY, *Tetrahedron*, **23**, 1489 (1967).
93. G. J. LESTINA, G. P. HAPP, D. P. MAIER, and T. H. REGAN, *J. Org. Chem.*, **33**, 3336 (1968).
94. R. R. SHOUP, H. T. MILES, and E. D. BECKER, *Biochem. Biophys. Res. Commun.*, **23**, 194 (1966).
95. R. D. GREEN, *Can. J. Chem.*, **47**, 2407 (1969).
96. H. KAMEI, *Bull. Chem. Soc., Japan*, **38**, 1212 (1965).
97. J. B. LAMBERT, B. W. ROBERTS, G. BINSCH, and J. D. ROBERTS, in *Nuclear Magnetic Resonance in Chemistry* (B. PESCE, Ed.), Academic Press Inc., New York, N.Y., 1965, p. 269.
98. J. B. LAMBERT, W. L. OLIVER, and J. D. ROBERTS, *J. Amer. Chem. Soc.*, **87**, 5085 (1965).
99. J. NELSON, R. SPRATT, and S. M. NELSON, *J. Chem. Soc. (A)*, 583 (1970).
100. K. M. MACKAY and S. R. STOBART, *Spectrochim. Acta*, **27A**, 923 (1971).
101. T. AXENROD, G. BERTI, and P. L. BARILI, unpublished results.
102. R. WASYLISHEN and T. SCHAEFFER, *Can. J. Chem.*, **49**, 3627 (1971).
103. G. A. OLAH and T. E. KIOVSKY, *J. Amer. Chem. Soc.*, **90**, 4666 (1968).
104. T. AXENROD and F. MACCHIA, unpublished results.
105. K. TABEI and E. SAITOU, *Bull. Chem. Soc. Japan*, **42**, 1440 (1969) and references therein.
106. H. B. BÜRGI and J. D. DUNITZ, *Helv. Chim. Acta*, **53**, 1747 (1970); **54**, 1255 (1971).
107. V. BEKAREK, J. KLICNAR, F. KRISTEK, and M. VECERA, *Collect. Czech. Chem. Commun.*, **33**, 994 (1968).
108. M. WITANOWSKI, *Tetrahedron*, **23**, 4299 (1967).
109. P. S. PREGOSIN, E. W. RANDALL, and A. I. WHITE, *J. Chem. Soc. (B)*, 513 (1972).
110. C. D. COOK and S. S. DANYLUK, *Tetrahedron*, **19**, 177 (1963).
111. L. LUNAZZI, D. MACCIANTELLI, and F. TADDEI, *Mol. Phys.*, **19**, 137 (1970).
112. B. SUNNERS, L. H. PIETTE, and W. G. SCHNEIDER, *Can. J. Chem.* **38**, 681 (1960).
113. M. T. ROGERS and L. A. LAPLANCHE, *J. Phys. Chem.*, **69**, 3648 (1965).
114. W. WALTER, H. P. KUBERSKY, E. SCHAUMANN, and K. J. REUBKE, *Ann. Chem.*, **719**, 210 (1968).
115. E. W. RANDÂLL and J. J. ZUCKERMAN, *J. Amer. Chem. Soc.*, **90**, 3167 (1968).

116. A. H. COWLEY and J. R. SCHWEIGER, *Chem. Commun.*, 1492 (1970).
117. E. BULLOCK, D. G. TUCK, and E. J. WOODHOUSE, *J. Chem. Phys.*, **38**, 2318 (1963).
118. R. L. LICHTER and J. D. ROBERTS, *J. Amer. Chem. Soc.*, **93**, 5218 (1970).
119. I. MORISHIMA, A. MIZUNO, T. YONEZAWA, and K. GOTO, *Chem. Commun.*, 1321 (1970).
120. G. E. MACIEL, J. W. McIVER, Jr., N. S. OSTLUND, and J. A. POPLE, *J. Amer. Chem. Soc.*, **92**, 11 (1970).
121. P. S. PREGOSIN, E. W. RANDALL, and A. I. WHITE, *J. Chem. Soc.*, Perkin Trans II, 1 (1972).
122. W. H. HALE, Jr. and S. M. WILLIAMSON, *Inorg. Chem.*, **4**, 1343 (1965).
123. J. M. SCHREEVE, L. C. DUNCAN, and G. H. CADY, *Inorg. Chem.*, **4**, 1516 (1965).
124. W. B. FOX, J. S. MACKENZIE, E. R. McCARTHY, J. R. HOLMES, R. F. STAHL, and R. JUURICK, *Inorg. Chem.*, **7**, 2064 (1968).
125. J. H. NOGGLE, J. D. BALDESCHWEILER, and C. B. COLBURN, *J. Chem. Phys.*, **37**, 182 (1962).
126. W. B. FOX, G. FRANZ, and L. R. ANDERSON, *Inorg. Chem.* **7**, 383 (1968).
127. W. E. TOLBERG, R. T. REWICK, R. S. STRINGHAM, and M. E. HILL, *Inorg. Chem.*, **6**, 1156 (1967).
128. K. O. CHRISTIE, J. P. GUERTIN, A. E. PAVLATH, and W. SAWODNY, *Inorg. Chem.*, **6**, 533 (1967).
129. V. GRAKAUSKAS and K. BAUN, *J. Org. Chem.*, **33**, 3080 (1968).
130. A. V. CUNLIFFE and R. K. HARRIS, *Mol. Phys.*, **15**, 413 (1968).
131. R. K. HARRIS and N. C. PYPER, R. E. RICHARDS, and G. W. SCHULZ, *Mol. Phys.*, **19**, 145 (1970).
132. C. REICHARDT, personal communication.
133. D. MOY and A. R. YOUNG, *J. Amer. Chem. Soc.*, **87**, 1889 (1965).
134. For a review of restricted rotation in amides, see T. H. SIDDALL and W. E. STEWART, in *Progress in Nuclear Magnetic Resonance Spectroscopy*, Vol. 5 (J. W. EMSLEY, J. FEENEY, and L. H. SUTCLIFFE, Eds.) Pergamon Press, Oxford, 1969, p. 33.
135. J. LADELL and B. POST, *Acta Cryst.*, **7**, 559 (1954).
136. C. C. COSTAIN and J. M. DOWLING, *J. Chem. Phys.*, **32**, 158 (1960).
137. E. W. RANDALL, J. J. ELLNER, and J. J. ZUCKERMAN, *J. Amer. Chem. Soc.*, **88**, 622 (1966).
138. C. GLIDEWELL, D. W. H. RANKIN, A. G. ROBIETTE, and G. M. SHELDRICK, *J. Mol. Struct.*, **4**, 215 (1969).
139. T. AXENROD and E. D. BECKER, unpublished results.
140. N. LOGAN and W. L. JOLLY, *Inorg. Chem.*, **4**, 1508 (1965).
141. R. FIELDS, J. LEE, and D. J. MOWTHORPE, *Trans. Faraday Soc.*, **65**, 2278 (1969).
142. P. FRIEDMAN, *Inorg. Chem.*, **8**, 692 (1969)
143. G. A. GRAY, Ph.D. Dissertation, University of Calif., Davis, 1967.

144. G. LeCOUSTUMER and Y. MOLLIER, *Bull. Soc. Chim. Fr.*, 2244 (1970).
145. F. G. RIDDELL and J. M. LEHN, *J. Chem. Soc. (B)*, 1224 (1968).
146. A. A. BOTHNER-BY and R. H. COX, *J. Phys. Chem.*, 73, 1830 (1969).
147. N. BARTLETT, J. PASSMORE, and E. J. WELLS, *Chem. Commun.*, 213 (1966).
148. A. M. QURESHI, J. A. RIPMEESTER, and F. AUBKE, *Can. J. Chem.*, 47, 4247 (1969).
149. R. A. OGG and J. D. RAY, *J. Chem. Phys.*, 25, 797 (1956).
150. E. L. MUETTERTIES and W. D. PHILLIPS, *J. Amer. Chem. Soc.*, 81, 1084 (1959).
151. M. OHTSURU and K. TORI, *Tetrahedron Letts.*, 4043 (1970).
152. D. M. JERINA, D. R. BOYD, L. PAOLILLO, and E. D. BECKER, *Tetrahedron Letts*, 1483 (1970).
153. H. PAULSEN and W. GREVE, *Chem. Ber.*, 103, 486 (1970).
154. J. P. KINTZINGER and J. M. LEHN, *Chem. Commun.*, 660 (1967).
155. J. P. KINTZINGER and J. M. LEHN, *Mol. Phys.*, 14, 133 (1968).
156. J. M. LEHN and R. SEHER, *Chem. Commun.*, 847 (1966).
157. V. I. ERASHKO, S. A. SHEVELEV, A. A. FAINZIL'BERG, M.Ya. MÄGI, and E. T. LIPPMAA, *Izv. Akad, Nauk. SSSR. Ser. Khim.*, 958 (1970).
158. J. C. WOODBREY and M. T. ROGERS, *J. Amer. Chem. Soc.*, 84, 13 (1962).
159. D. J. BERTELLI and J. T. GERIG, *Tetrahedron Letts.*, 2481 (1967).
160. T. GOTO, M. ISOBE, M. OHTSURU, and K. TORI, *Tetrahedron Letts.*, 1511 (1968).
161. R. K. BOHN and S. H. BAUER, *Inorg. Chem.*, 6, 309 (1967).
162. G. BINSCH and J. D. ROBERTS, *J. Phys. Chem.*, 72, 4310 (1968).
163. T. AXENROD, *Spectrosc. Letts.*, 3, 263 (1970).
164. P. G. GASSMAN and D. C. HECKERT, *J. Org. Chem.*, 30, 2859 (1965).
165. E. W. RANDALL and D. SHAW, *Spectrochim. Acta*, 23A, 1235 (1967).
166. J. M. LEHN and M. FRANCK-NEUMANN, *J. Chem. Phys.*, 43, 1421 (1965).
167. J. F. BIELLMANN and H. CALLOT, *Bull. Soc. Chim. Fr.*, 397 (1967).
168. Y. TERUI, K. AONO, and K. TORI, *J. Amer. Chem. Soc.*, 90, 1069 (1968).
169. R. L LICHTER and J. D. ROBERTS, *J. Org. Chem.*, 35, 2806 (1970).
170. T. YONEZAWA, I. MORISHIMA, K. FUKUTA, and Y. OHMORI, *J. Mol. Spectrosc.*, 31, 341 (1969).
171. H. KAMEI, *Bull. Chem. Soc. Japan*, 41, 2269 (1968).
172. R. ETTINGER and C. B. COLBURN, *Inorg. Chem.*, 2, 1311 (1963).
173. G. O. DUDEK and E. P. DUDEK, *J. Amer. Chem. Soc.*, 86, 4283 (1964).
174. G. O. DUDEK and E. P. DUDEK, *Tetrahedron*, 23, 3245 (1967).

175. G. O. DUDEK and E. P. DUDEK, *J. Amer. Chem. Soc.*, **88**, 2407 (1966).
176. E. D. BECKER, H. T. MILES, and R. B. BRADLEY, *J. Amer. Chem. Soc.*, **87**, 5575 (1965).
177. B. W. ROBERTS, J. B. LAMBERT, and J. D. ROBERTS, *J. Amer. Chem. Soc.*, **87**, 5439 (1965).
178. H. T. MILES, R. B. BRADLEY, and E. D. BECKER, *Science*, **142**, 1569 (1963).
179. Y. KAWAZOE, M. ARAKI, S. SAWAKI, and M. OHNISHI, *Chem. Pharm. Bull. (Tokyo)*, **18**, 381 (1970).
180. Y. KAWAZOE, M. OHNISHI, and N. KATAOKA, *Chem. Pharm. Bull. (Tokyo)*, **13**, 396 (1965).
181. T. AXENROD, P. S. PREGOSIN, and G. W. A. MILNE, *Chem. Commun.*, 702 (1968).
182. T. AXENROD, P. S. PREGOSIN, and G. W. A. MILNE, *Tetrahedron Letts.*, 5293 (1968).
183. T. AXENROD, M. J. WIEDER, and G. W. A. MILNE, *Tetrahedron Letts.*, 401 (1969).
184. T. AXENROD, M. J. WIEDER, and G. W. A. MILNE, *Tetrahedron Letts.*, 1397 (1969).
185, J. N. MURRELL, in *Progress in Nuclear Magnetic Resonance Spectroscopy*, Vol. 6 (J. W. EMSLEY, J. FEENEY, and L. H. SUTCLIFFE, Eds.), Pergamon Press, Oxford, 1971, p. 7.
186. R. A. COBURN and G. O. DUDEK, *J. Phys. Chem.*, **72**, 1177 (1968).
187. S. M. BLOOM and G. O. DUDEK, *J. Org. Chem.*, **36**, 235 (1971).
188. V. BEKAREK, K. ROTHSCHEIN, P. VETESNIK, and M. VECERA, *Tetrahedron Letts.*, 3711 (1968).
189. V. BEKAREK, J. DOBAS, J. SOCHA, P. VETESNIK, and M. VECERA, *Collect. Czech. Chem. Commun.*, **35**, 1406 (1970).
190. A. H. BERRIE, P. HAMPSON, S. W. LONGWORTH, and A. MATHIAS, *J. Chem. Soc. (B)*, 1308 (1968).
191. G. J. LESTINA and T. H. REGAN, *J. Org. Chem.*, **34**, 1686 (1969).
192. C. REICHARDT and W. GRAHN, *Tetrahedron*, 27, 3745 (1971).
193. C. REICHARDT and W. GRAHN, *Chem. Ber.*, **103**, 1065 (1970).
194. V. S. BOGDANOV, M. A. KALIK, Ya. L. DANYUSHEVSKII, and Ya. L. GOLDFARB, *Izv. Akad. Nauk. SSSR. Ser. Khim.*, 2783 (1967).
195. M-L. BLANCHARD, H. STRZELECKA, G. J. MARTIN, M. SIMALTY, and R. FUGNITTO, *Bull. Soc. Chim. Fr.*, 2677 (1967).
196. G. KLOSE and E. UHLEMANN, *Tetrahedron*, 22, 1373 (1966).
197. P. HAMPSON and A. MATHIAS, *Chem. Commun.*, 371 (1967).
198. L. MESTER, A. STEPHEN, and J. PARELLO, *Tetrahedron Letts.*, 4119 (1968).
199. P. B. FISCHER, B. L. KAUL, and H. ZOLLINGER, *Helv. Chim. Acta*, **51**, 1449 (1968).
200. L. MESTER and G. VASS, *Tetrahedron Letts.*, 3847 (1969).
201. L. MESTER, G. VASS, A. STEPHEN and J. PARELLO, *Tetrahedron Letts.*, 4053 (1968).
202. M. LILER, *Chem. Commun.*, 115 (1971).
203. S. J. KUHN and J. S. McINTYRE, *Can. J. Chem.*, **44**, 105 (1966).

204. T. BIRCHALL and R. J. GILLESPIE, *Can. J. Chem.*, **41**, 2642 (1963).
205. G. A. OLAH and A. M. WHITE, *J. Amer. Chem. Soc.*, **90**, 6087 (1968).
206. G. A. OLAH and R. H. SCHLOSBERG, *J. Amer. Chem. Soc.*, **90**, 6464 (1968).
207. R. A. OGG, Jr. and J. D. RAY, *J. Chem. Phys.*, **26**, 1339, 1340 (1957).
208. G. O. DUDEK and E. P. DUDEK, *Chem. Commun.*, 465 (1965).
209. R. R. FRASER and K. E. HAGUE, *Can. J. Chem.*, **46**, 2855 (1968).
210. R. L. LICHTER and J. D. ROBERTS, *Spectrochim. Acta*, **26A**, 1813 (1970).
211. C. A. SWENSON and L. KOOB, *J. Phys. Chem.*, **74**, 3376 (1970).
212. B. W. ROBERTS, J. B. LAMBERT, and J. D. ROBERTS, *J. Amer. Chem. Soc.*, **87**, 5439 (1965).
213. J. A. HAPPE and H. A. MORALES, *J. Amer. Chem. Soc.*, **88**, 2077 (1966).
214. B. COXON, *Carbohyd. Res.*, **11**, 153 (1969).
215. M. L. FILLEUX-BLANCHARD, J. FIEUX, and J. C. HALLE, *Chem. Commun.*, 851 (1971).
216. V. M. S. GIL and S. J. S. FORMOSINHO-SIMÕES, *Mol. Phys.*, **25**, 639 (1968).
217. B. COXON and L. F. JOHNSON, *Carbohyd. Res.*, **20**, 105 (1971).
218. G. DUDEK and E. P. DUDEK, *J. Chem. Soc. (B)*, 1356 (1971).
219. R. R. SHOUP, E. D. BECKER, and H. T. MILES, *Biochem. Biophys. Res. Commun.*, **43**, 1350 (1971).
220. A. K. BOSE and I. KUGAJEVSKY, *J. Amer. Chem. Soc.*, **88**, 2325 (1966).
221. T. DRAKENBERG and S. FORSEN, *J. Phys. Chem.*, **74**, 1 (1970).
222. J. M. INSOLE and E. S. LEWIS, *J. Amer. Chem. Soc.*, **85**, 122 (1963); **86**, 32, 34 (1964).
223. E. S. LEWIS and P. G. KOTCHER, *Tetrahedron*, **25**, 4873 (1969).
224. V. V. NEGREBETSKII, V. S. BOGDANOV, and A. V. KESSENIKH, *Zh. Struct. Khim.*, **12**, (4) 716 (1971).
225. E. D. BECKER and R. B. BRADLEY, personal communication.
226. W. B. JENNINGS, D. R. BOYD, E. D. BECKER, R. B. BRADLEY, and D. M. JERINA, personal communication.

CHAPTER 6

Applications of ^{14}N NMR Data in the Study of Inorganic Molecules

N. Logan

*Department of Chemistry, University of Nottingham,
Nottingham, England*

6.1   Introduction

The foregoing chapters have indicated that in spite of inherent difficulties, ^{14}N NMR spectroscopy has proved to be of considerable value in the study of nitrogen compounds. For example, the last few years have witnessed an extensive and rewarding application of the technique in organic chemistry (Chapter 4). Three recent general reviews [1-3] of nitrogen nuclear magnetic resonance including both ^{14}N and ^{15}N, show that less attention has been paid to the study of inorganic molecules and reasons for this neglect are readily apparent. A high proportion of inorganic compounds are solids of generally low solubility, yet the lower limit of concentration for observance of suitably intense ^{14}N signals by direct continuous wave measurement is as high as *ca.* 1M. Even if a solvent can be found which shows little or no interaction with a nitrogen-containing inorganic solute, this level of solubility is frequently difficult or impossible to achieve. Valuable information can result from ^{14}N NMR studies of solids (Section 6.4) but thus far, few studies of this type have been reported. Secondly, the paramagnetism of transition-metal and other compounds leads to the additional complication of spectra by shifting of resonance positions and broadening of signals. The latter may in some instances be extensive enough to render signals unobservable; however, a number of paramagnetic transition metal complexes have been successfully studied (Section 6.3). Extensive broadening of signals may also result from quadrupole interactions where the ^{14}N nucleus finds itself in a particularly unsymmetrical electronic environment (Chapter 3). Such broadening effects can result in erroneous structural conclusions. Furthermore, spin–spin splitting of ^{14}N signals due to coupling with other magnetic nuclei e.g. ^{1}H is obscured in all but the few species where the nitrogen electronic environment possesses a high degree of symmetry e.g. $NH_4^{\oplus}$. Notwithstanding these intrinsic disadvantages, sufficient ^{14}N NMR spectra have now been accumulated to demonstrate that the technique is worthy of the attention of Inorganic Chemists interested in nitrogen-containing molecules. It is therefore timely to review the ^{14}N NMR data which have now accumulated for inorganic compounds. Before embarking on such a survey, a few general remarks on the recording of ^{14}N spectra and the measurement of chemical shifts for inorganic nitrogen species are appropriate. The majority of ^{14}N spectra have been recorded by direct, continuous wave, single resonance techniques on spectrometers operating at the ^{14}N resonant frequency in magnetic fields of 1.0, 1.45 or 2.35 T. The earlier studies at 1.0 and 1.45 T on variable frequency wide-line spectrometers usually employed 15-mm o.d. sample tubes necessitating a minimum sample volume of *ca.* 3 ml of concentrated solution ($>$1M). More recent investigations at 2.35 T, e.g. using the Varian HA 100 spectrometer, have involved the use of 5- and 8-mm o.d. tubes. Smaller sample volumes can thus be tolerated but the requirement for high concentration of sample is unchanged.

Owing to the generally high reactivity of many inorganic nitrogen compounds, and to solvent effects, no universal, internal standard is available as reference substance for their ^{14}N resonances. The compound most commonly used as external reference has been NH_4NO_3. This compound is highly soluble in water (118.3 g/100 ml at 273K; 14.8M) producing two sharp and strong resonances separated by 355 ± 0.5 p.p.m. In neutral aqueous solution, $NH_4^{\oplus}$ gives rise to a single sharp peak due to rapid proton exchange with H_2O molecules. This exchange may be retarded by acidification, whereupon the expected 1 : 4 : 6 : 4 : 1 quintet is observed. The positions of these resonances are neither temperature- nor concentration-dependent; however, the chemical shift of the nitrate ion is pH-dependent, moving to higher field in acidified solution, and of course addition of base results in liberation of ammonia. It is of interest in this context to recall that, although ^{14}N has been a relatively neglected nucleus, the ^{14}N resonances of NH_4NO_3 featured in one of the first reports of the phenomenon of chemical shift [4]. The presence of two resonances provides a check on chemical shifts and the field sweep linearity and since the signals saturate at quite different r.f. power levels, they are suitable for recording spectra under different conditions. Because of the high solubility in H_2O, narrow-bore tubing may be used to contain the saturated solution as a concentric external standard. Alternatively, the aqueous NH_4NO_3 is contained in a separate tube and the shift is measured by the substitution method. The two techniques give identical shifts within experimental error [5] and errors due to susceptibility differences arising from the use of such an external standard are usually considered to be insignificant in comparison to the errors involved in measurement of line positions and the large shifts observed in nitrogen NMR.

In some instances [6–9], shifts have been referred to $NH_4^{\oplus}$ but are most frequently quoted relative to $NO_3^{\ominus}$. The spectrum of NH_4NO_3 is illustrated in Fig. 6.1.

Other species which have found some application as standards for ^{14}N NMR studies are nitromethane [11], which gives a resonance for all practical purposes coincident with that of $NO_3^{\ominus}$ and is recommended as a primary standard for organic compounds, tetranitromethane [11], acetonitrile [11, 12], dimethylformamide [11] and saturated aqueous nitrite ion [13, 14, 78], but due to possible interaction with inorganic solutes and/or appropriate solvents, none of these is a subitable candidate for general use as an internal standard.

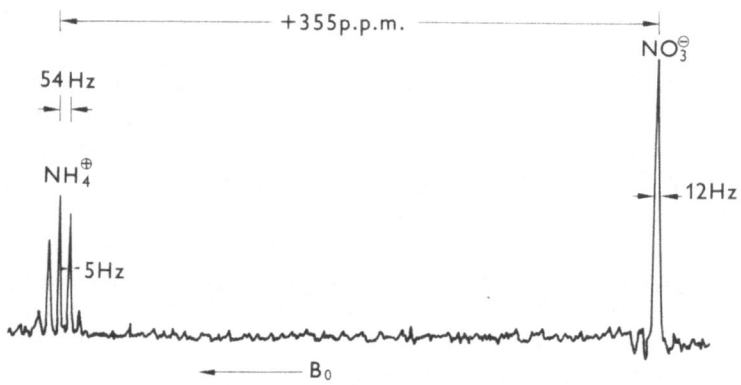

Fig. 6.1. ^{14}N NMR spectrum at 7.226 MHz of aqueous, acidified ammonium nitrate.

As mentioned in Chapter 2, ^{14}N chemical shifts can also be obtained indirectly from proton spectra by the double irradiation INDOR technique. The advantages of this method have been convincingly demonstrated for several classes of organic compounds containing protons directly coupled to nitrogen [15–20]. In this technique, first suggested by Baker [21], a suitable line in the proton spectrum is selected and the recorder pen is set on top of it. This fixes the main magnetic field B. A frequency sweep of appropriate range is then carried out in the neighbourhood of the ^{14}N frequency. As lines in the spectrum of ^{14}N are traversed, the recorder pen makes excursions whose size depends on the intensity of the line irradiated. This pen response may be recorded on an auxiliary moving chart and the resulting trace closely resembles the ^{14}N spectrum which would be obtained by direct observation, but is often superior to it. Reference 15 includes a detailed description of ^{14}N chemical shift measurement using this technique. Since proton resonances are used, the indirect method is more sensitive than the direct one and ^{14}N chemical shifts can be obtained readily from the dilute solutions used in proton work. Also, the ^{14}N irradiating frequency can normally be measured quite precisely, so that the resulting ^{14}N chemical shift can be determined more accurately than by the direct method. There are a few instances in which ^{14}N quadrupole relaxation and proton exchange effects have caused the method to fail, but it is applicable in principle to all compounds containing ^{1}H, ^{19}F or any other sensitive non-quadrupolar nucleus coupled to ^{14}N. The technique does not appear to have been applied as yet to inorganic ^{14}N–H or ^{14}N–F species with the exception of the ammonium ion [22], the partially deuterated ammonium ion [23] and protonated hydrogen cyanide [24] and is obviously worthy of wider exploitation. The INDOR technique has however been used in recent measurements of ^{15}N chemical shifts in penta-ammine complexes of cobalt(III) [118]. The *trans* ammine protons in corresponding ^{14}N complexes have a chemical shift different from that of the *cis* ammine protons [119, 120]. It is therefore reasonable to expect that the *trans* and *cis* nitrogens would also have different chemical shifts, however, the ^{14}N resonances are broad and the different signals cannot be resolved (see Table 6.7). Cobalt(III) complexes with ^{15}NH$_3$ as ligand can be prepared without much difficulty and the ^{1}H NMR spectrum shows clearly resolved ^{15}N coupling. The ^{15}N chemical shifts of $[Co(^{15}NH_3)_5X]^{n\oplus}$ complexes have been measured with respect to $[Co(^{15}NH_3)_6]^{3\oplus}$ by the double resonance technique. *Cis* and *trans* nitrogen atoms give distinct resonances upfield from the hexa-ammine complex but in contrast to the ^{59}Co chemical shifts [121, 122], the ^{15}N shifts show no simple correlation with ligand field splittings.

The ^{14}N chemical shifts so far reported for diamagnetic inorganic compounds span a range of approximately 800 p.p.m., the extremes of this range being presently represented by the compounds $[Ru(NH_3)_6]Cl_2$ ($\delta_{NO_3^\ominus}$ = +455 p.p.m.) and NOBr ($\delta_{NO_3^\ominus}$ = −341 p.p.m.). It is noteworthy that the extensive compilation of ^{14}N chemical shifts in organic compounds by Herbison-Evans and Richards [25] includes results for a significant number of inorganic species. It is also a very useful guide to the general location of nitrogen resonance signals for various classes of molecules. In this chapter the screening constant scale, discussed in Chapter 4, is used for reporting chemical shifts.

6.2 Elucidation of Electronic and Molecular Structure in Diamagnetic Inorganic Compounds

6.2.1   Compounds of Main Group Elements

6.2.1.1 Isocyanates and Cyanates

^{14}N NMR spectroscopy should ideally provide a simple means of distinguishing between the alternative bonding modes of an ambivalent group such as NCO. It is well known that this group shows a marked preference for N-bonding (isocyanate) [26], O-bonded (cyanate) species, e.g. PhOCN [27], being of comparative rarity. This behaviour can be readily understood if most of the negative charge in the cyanate ion resides on the nitrogen atom, a feature verified by point charge calculations [28]. ^{14}N chemical shifts for several isocyanates of main group elements and two organic cyanates [29] are listed in Table 6.1. In this instance the shifts are recorded relative to

TABLE 6.1. ^{14}N Chemical Shifts[d] of some Isocyanates and Cyanates [29]

Compound	Solvent	^{14}N shift (p.p.m.)[a]	Linewidth (Hz)
H_3GeNCO	benzene	$+64 \pm 1$	45
	neat	$+62 \pm 1$	45
MeNCO	benzene	$+61 \pm 1$	35
	neat	$+61 \pm 1$	35
HNCO[b]	cyclohexane	$+53 \pm 2$	20
	diethylether	$+51 \pm 2$	35
Me_3SiNCO	neat	$+53 \pm 1$	38
Et_3GeNCO	neat	$+52 \pm 1$	65
Pr_3^nGeNCO	benzene	$+50 \pm 1$	50
Pr^nNCO	neat	$+46 \pm 1$	70
EtNCO	neat	$+43 \pm 1$	37
PhNCO	neat	$+30 \pm 3$	50
$P(NCO)_3$	benzene	$+22 \pm 4$	88
EtOCN	diethylether	-78 ± 1	45
PhOCN[c]	neat	-92	

a Referred to external $NO_3^{\ominus}$; recalculated to external $NCO^{\ominus}$.

b Doublet, see Section 6.2.1(i) and Fig. 6.2.

c Ref. 56.

d The chemical shifts reported in this table and all subsequent tables in this chapter are given on the *screening-constant scale*.

aqueous $K^{\oplus}NCO^{\ominus}$ for which $\delta_{NO_3^{\ominus}} = +300\pm 1$ p.p.m., line-width 15 Hz, since a simple method for distinguishing between isocyanate and cyanate bonding is particularly obvious when the data are presented in this way. Isocyanates are seen to exhibit shifts to high field of the free ion and conversely, ethyl and phenyl cyanates give rise to large low-field shifts. The ^{14}N resonance of HNCO is a doublet with $J(N-H) = 69 \pm 5$ Hz, line-width of each component $= 20 \pm 4$ Hz, this representing one of the few examples where spin–spin coupling between ^{14}N and an attached proton is clearly resolved [30]. The ^{1}H resonance of HNCO has also been studied [31] and yields a value of $J_{N-H} = 64 \pm 1$ Hz. Figure 6.2 illustrates the ^{14}N spectrum of HNCO in relation to that of NH_4NO_3. HNCO is evidently not

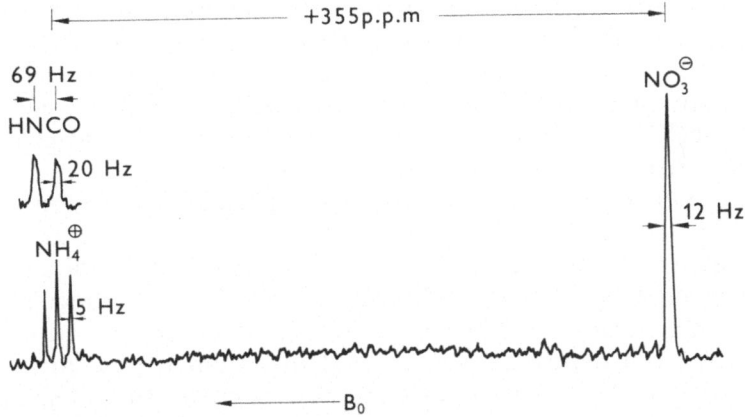

Fig. 6.2. [14]N NMR spectra at 7.226 MHz of isocyanic acid in cyclohexane and aqueous, acidified ammonium nitrate as external reference.

in equilibrium with its enol isomer, cyanic acid which would be expected to show a resonance in the region of -50 to -100 p.p.m. relative to $NCO^{\ominus}$. If the intermolecular exchange process

$$H-N=C=O \rightleftharpoons N\equiv C-O-H$$

occurs, the resonance observed should be a singlet with a smaller positive, or perhaps a negative, chemical shift. In fact, the isomer HOCN is reported to be present to the extent of about 2% in the rapidly cooled product of the reaction of HCl and NaNCO, but decomposes rapidly on warming to room temperature [32]. The coupling constant is close to the value of 68 ± 1 Hz obtained for pyridinium ion [33], where the nitrogen atom is trigonally sp^2-hybridized. This value is higher than the coupling constant observed in the case of ammonium ion, $J(N-H) = 54 \pm 0.5$ Hz, and ammonia, $J(N-H) = 46 \pm 2$ Hz, as expected if the magnitude of the coupling constant is largely determined by the Fermi contact term. An increase in s-character in the N–H bond should produce a corresponding increase in $J(N-H)$. A plot of $J(N-H)$ versus the degree of s-character, similar to that obtained by Shoolery [34] for ^{13}C–H coupling constants, affords a straight line [5]. It is reasonable to assume from the value of $J(N-H)$ that the nitrogen atom in HNCO is sp^2-hybridized. The experimental bond angle is 128° [35].

On the basis of vibrational spectroscopic studies [36], Griffiths suggested that the product of the reaction of GeH$_3$Br and AgNCO contains a significant concentration of the cyanate, H$_3$GeOCN, together with the isocyanate isomer. However, ^{14}N NMR data lends no support to this contention since a reaction product prepared in this way, termed H$_3$GeNCO in Table 6.1, gives a single resonance with the largest positive chemical shift of any of the compounds examined. Therefore, even if H$_3$GeNCO and H$_3$GeOCN are involved in an exchange process, the concentration of the latter must be very low [37].

Detailed interpretation of ^{14}N chemical shift variations even within a closely related series of compounds, such as those listed in Table 6.1, is not an easy matter and will not be attempted here for the following reasons. It is now recognized that the nuclear shielding arises from a rather complex interplay of electron densities, energies and stoichiometries as discussed in Chapter 1. In order to attempt any rationalization of shifts in the face of so many parameters, rather drastic assumptions must be made so far as the variation in their relative magnitudes with changes in molecular structure is concerned.

6.2.1.2 Thiocyanates and Isothiocyanates

^{14}N chemical shifts of several isothiocyanates of the lighter Group IV elements have been reported [30] (Table 6.2). A number of analogues of the isocyanates listed in Table 6.1. are also included in this study. The shifts show all the compounds to be isothiocyanates (i.e. N-bonded), since thiocyanates (S-bonded) show resonances to much lower field relative to aqueous NO$_3^{\ominus}$ (cf. EtSCN, $+99 \pm 3$; SCN$^{\ominus}$, $+166 \pm 1$; EtNCS, $+271 \pm 2$ p.p.m. [25]—see also ref. 76 and Table 6.11). It was hoped that the ^{14}N chemical shifts might reflect the varying degrees of $(p \rightarrow d)$ π-bonding inferred from other structural and spectroscopic studies (see ref. 30 for details of these) on this group of compounds. However, such effects are not evident (Table 6.2) from the shifts, the most striking feature of which is the difference of about 20 p.p.m. between the H_3MNCS and R_3MNCS (M = Si, Ge) species. This constitutes a large proportion of the rather narrow total range of values observed for covalently bound isothiocyanate groups. It is considered that the small change from C to Si in H_3MNCS may reflect a balance of changes in σ and π effects on the nitrogen shielding while the near identity of shifts in Si and Ge compounds is taken to suggest that any π effects are similar for these two elements. The pattern of shifts is similar to that found in the corresponding isocyanates (Table 6.1) where the H—MNCO to

TABLE 6.2. [14]N Chemical Shifts of some Isothiocyanates [30]

Compound	Solvent	[14]N shift (p.p.m.)[a]	Linewidth (Hz)
H_3CNCS	neat	+285 ± 3	40
H_3GeNCS	benzene	+280 ± 3	20
H_3SiNCS	benzene	+279 ± 3	15
HNCS	benzene	+265 ± 3	25
	diethylether	+265 ± 3	25
$Ph_2Si(NCS)_2$	benzene	+265 ± 5	ca. 200
PhNCS	neat	+265 ± 5	110
$Pr_3{}^nGeNCS$	benzene	+263 ± 3	25
Me_3SiNCS	benzene	+260 ± 3	10
Et_3GeNCS	benzene	+260 ± 3	10

[a] Referred to external $NO_3^{\ominus}$.

R—MNCO change again results in a downfield shift, ca. 10 p.p.m. in this case for Ge or C derivatives. H_3SiNCO is too unstable to allow measurements.

In contrast to HNCO, HNCS gives rise to only a fairly narrow single line ^{14}N resonance [30] and the proton spectrum [31] shows only a single broad resonance, even though the quadrupole coupling constant is lower than that for HNCO [38]. A rapid inter-molecular exchange of protons might be envisaged since, in contrast to the situation in cyanate ion [28], nitrogen and sulphur are indeed calculated to carry closely similar negative charges [28] in the thiocyanate ion. However, the ^{14}N chemical shift (Table 6.2) clearly indicates the predominance of the N-bonded tautomer, HNCS, in benzene and diethylether (see also ref. 76).

6.2.1.3 *Azides*

A further linear triatomic inorganic group for which the ^{14}N NMR technique should be potentially informative from a structural viewpoint is the N_3 (azide) unit. For the $N_3^{\ominus}$ ion, two signals of 1 : 2 intensity ratio are anticipated whilst an RN_3 molecule should give rise to three distinct ^{14}N resonances. Figure 6.3 and Table 6.3. show that these expectations are borne out in practice [39]. Earlier studies on azide ion [40, 41] and methyl [33] and ethyl [33, 40] azides have been extended [39] to included hydrazoic acid, HN_3, a further group of organic azides and certain organoarsenic(III) azides prepared as possible precursors of As—N ring com-

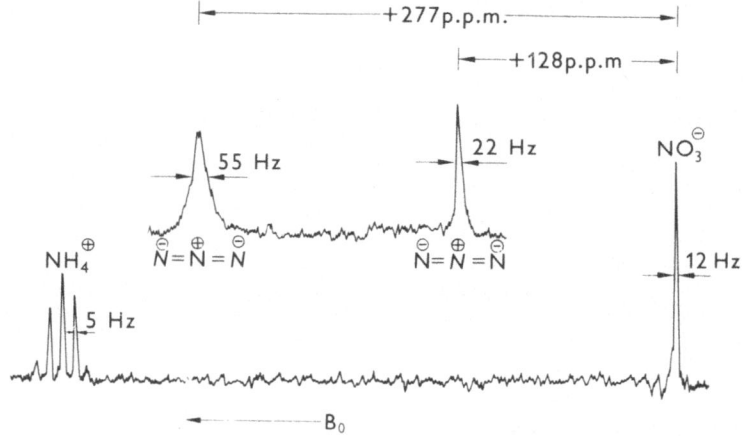

Fig. 6.3. ^{14}N NMR spectra at 7.226 MHz of aqueous azide and aqueous, acidified ammonium nitrate as external reference.

pounds. In no case has $^{14}N-{}^{14}N$ spin–spin coupling been observed although it has been estimated that this coupling constant in the azide ion must be less than 30 Hz [42]. In contrast to the situation in HNCO [30] (Table 6.1) no splitting of the N_a signal (Table 6.3) into a doublet as a result of $^{14}N-{}^1H$ coupling is observed for a solution of HN_3 in diethyl ether. However, the three-line ^{14}N spectrum of HN_3 closely resembles the spectra of MeN_3 and EtN_3 (Table 6.3). The assignment of the individual resonances of covalent RN_3 compounds in Table 6.3 to N_a, N_b and N_c has been made on the basis of the arguments given by Witanowski [33]. Two resonances only are observed for HN_3 in 'wet' ether or acidified aqueous solutions of $N_3^{\ominus}$, the high field shift appearing to lower field than that in aqueous $N_3^{\ominus}$.

TABLE 6.3. ^{14}N Chemical Shifts of some Covalent Azides, $R-\overset{\ominus}{N_a}-\overset{\oplus}{N_b}\equiv N_c$ [39]

Compound	Solvent	^{14}N shifts (p.p.m.)[a]		
		N_a	N_b	N_c
$Na^{\oplus}N_3^{\ominus}$	H_2O	$+277^b$ (55)	+128 (22)	+277 (55)
$Na^{\oplus}N_3^{\ominus}$	H_2SO_4	$+277^b$	+129 (23)	$+277^b$
		to +245		to +245
HN_3	$Et_2O + H_2O$	$+240^b$ (100)	+129 (30)	$+240^b$
HN_3	Et_2O	+300 (100)	+129 (24)	+165 (100)
MeN_3^c	$MeNO_2$	+320 (101)	+128 (17)	+170 (19)
EtN_3^c	$MeNO_2$	+305 (122)	+129 (22)	+167 (28)
PhN_3	C_6H_6	+286 (1200)	+135 (180)	+194 (55)
$PhCON_3$	C_6H_6	+322 (430)	+140 (300)	+237 (73)
$p\text{-}MeC_6H_4SO_2N_3$	liquid	+304 (410)	+152 (250)	+209 (120)
$p\text{-}NO_2C_6H_4N_3$	C_6H_6	+310 (270)	+139 (75)	+211 (45)
Me_2AsN_3	liquid	$+253^b$ (80)	+132 (37)	$+253^b$
Et_2AsN_3	liquid	$+251^b$ (85)	+133 (40)	$+251^b$
	CCl_4		+130 (23)	
Ph_2AsN_3	liquid		+134 (94)	
	CCl_4		+132 (37)	+188 (75)
$PhEtAsN_3$	liquid		+134 (100)	
	CCl_4		+131 (30)	
$PhClAsN_3$	liquid		+134 (43)	+170 (86)
$PhBrAsN_3$	liquid		+134 (87)	+170 (200)

a Referred to external $NO_3^{\ominus}$; linewidths in Hz at half peak height in parentheses.
b Resonance of intensity 2 relative to the N_b signal.
c Data from ref. 33.

Proton exchange phenomena are cited in explanation in both cases. As in the case of RNCO and RNCS [33], a correlation between decreasing N_a shift and decreasing electronegativity of R is also postulated [33] for the organic azides, RN_3 (Table 6.3). No similar dependence on the nature of the attached organic group has been recognized for the positions of the N_b and N_c signals.

None of the disubstituted azidoarsines shows three resonances, although each gives rise to the relatively sharp resonance at ca. +130 p.p.m. observed for all the azides in Table 6.3 and assigned to N_b. Unfortunately, this resonance is structurally uninformative but is the only signal observed for the compound $PhEtAsN_3$, even when diluted with carbon tetrachloride to minimize the possibility of quadrupolar broadening of the already broader N_a and N_c lines due to the high viscosity of the pure liquid. However, such dilution would diminish the intensity of already weak signals. Two resonances assignable to N_b and N_c of covalent azide have been observed for the diphenyl (CCl_4 solution only), chlorophenyl and bromophenyl derivatives. The N_a resonance is almost invariably the broadest of the three signals for a covalent azide and is presumably indistinguishable from the noise level for the above compounds. In the case of the dimethyl and diethyl compounds a different phenomenon is encountered. When run as pure liquids, two resonances are also observed, but the high field signal has twice the intensity of the N_b line and is situated approximately at the mean position of the N_a and N_c resonances observed in RN_3 compounds (Table 6.3). The implied equivalence of N_a and N_b atoms in Me_2AsN_3 and Et_2AsN_3 is rationalized in terms of azide bridging or azide exchange between R_2As moieties and other evidence is cited in support of these proposals [39]. The occurrence of the N_c resonance at $ca.$ +170 p.p.m. in Ph_2AsN_3, $PhClAsN_3$ and $PhBrAsN_3$ is attributed to greater steric hindrance of arsenic atoms by phenyl or substituted phenyl groups, resulting in a diminished tendency towards molecular association or exchange of N_3 groups. Spectra of Et_2AsN_3 have also been recorded in CCl_4 solution in an attempt to ascertain whether association or exchange is still indicated. Unfortunately, the consequent dilution of the sample allows only the single, sharp N_b resonance to be discerned [39].

6.2.1.4 *Nitrogen Oxides, Oxo-ions and Oxo-acids.*

Inspection of Table 6.4 shows that ^{14}N NMR parameters have now been reported for most of the principal nitrogen-oxygen species. The nitric oxide ^{14}N resonance does not appear to have been reported but the spectrum of liquid nitrous oxide, recorded at 193K, shows the two expected signals. Since the azide ion and nitrous oxide are isoelectronic, and are thus expected to have similar electronic distributions, the resonance positions should be similar in the two species (see Table 6.3) and are assigned on that basis [43].

It is particularly noteworthy that the results observed for liquid N_2O_3 provide a direct demonstration of the nitroso-nitro structure, $ONNO_2$, of this molecule [13]. The two ^{14}N signals observed are assigned as indicated in Table 6.4 on the basis of comparison with reported resonance positions in other nitroso- and nitro-species. The ^{14}N spectrum of the pure liquid at 223K was obtained from a sample under excess ($ca.$ 1700 kNm^{-2}, i.e. 17 atmospheres) pressure of NO generated by a 1 : 1.2 molecular mixture of NO_2 and NO. With equimolar proportions of NO_2 and NO at 223K the

TABLE 6.4. ^{14}N Chemical Shifts of some Nitrogen–Oxygen Species

Molecule or ion	Solvent (TK)	^{14}N shift (p.p.m.)[a]	Half-height width (Hz)	References
NNO	neat (193)	+138		43
NNO	neat (193)	+222		43
ONNO$_2$	neat (223)	−67 ± 10[b]	300	13, 53
ONNO$_2$	neat (223)	−302 ± 10[b]	745	13,14
N$_2$O$_4$	neat (263)	+11 ± 10		13, 53
	CCl$_4$ (257 to 241)	+15 ± 3		44
	CFCl$_3$ (213)	+13 ± 10	315	53
	CFCl$_3$ (203)	+18 ± 10	360	53
	CHCl$_3$ (243 to 215)	+15.5 to +17.2		44
	C$_6$H$_6$ (253 to 231)	+17.0 to +18.0		44
	CH$_3$NO$_2$ (254 to 217)	+18.7 to +20.3		44
	(CH$_3$)$_2$SO (254)	+18.2		44
	CH$_3$COOH (246 to 231)	+13.9 to +17.6		44
	HSO$_3$F (298)	+89 ± 5		44
N$_2$O$_5$	CCl$_4$	+60 ± 1[c]	80	33
	CCl$_4$	+62 ± 1	40	5
	CH$_3$NO$_2$	+58 ± 1	18	54
	CHCl$_3$ (248)	+48 ± 2		44
	HNO$_3$[e]	+47.5 ± 0.5	30	5
	(CH$_3$CO)$_2$O (CH$_3$COONO$_2$)	+64 ± 0.5	31	54
HNO$_3$	97%	+47.5 ± 0.5[c]	17	33
	100%	+43 ± 0.5	10	5
	H$_2$O	+43 to 0	10 to 8	5, 113
N$_2$O$_2^{2\ominus}$	1.0MNaOH	−83 ± 10[b]	3000	9
H$_2$N$_2$O$_2$	1.0MHClO$_4$	−86 ± 10[b]	3750	9
N$_2$O$_3^{2\ominus}$	0.1MNaOH	−15 ± 10[b]	3540	10
	0.5MNaOH	+43	1100	55
NO$_2^{\ominus}$	H$_2$O	−247 ± 20		7
	H$_2$O	−237 ± 4[d]	520	33
	H$_2$O (298)	−250 ± 10		44
NO$_3^{\ominus}$	H$_2$O	0	12	33
NO$^{\oplus}$	H$_2$SO$_4$	0 ± 10	very broad	14
NO$_2^{\oplus}$	H$_2$SO$_4$[f]	+126 ± 1	29	5
	H$_2$SO$_4$	+125 ± 10		46
	HSO$_3$F (298)	+129 ± 5	very broad	44

[a] Referred to external NO$_3^{\ominus}$ unless otherwise indicated.
[b] Referred to external CH$_3$NO$_2$.
[c] Referred to internal CH$_3$NO$_2$.
[d] Referred to internal NO$_3^{\ominus}$.
[e] 3 : 1 mole ratio HNO$_3$: N$_2$O$_5$
[f] 80% H$_2$SO$_4$, 20% HNO$_3$

single N$_2$O$_4$ signal, consistent with its symmetrical, planar O$_2$NNO$_2$ structure, is also present at +11 p.p.m. On further increasing the NO$_2$: NO ratio to 1.1 : 1 and then 3 : 1, the intensity of the N$_2$O$_4$ signal increases at the expense of the N$_2$O$_3$ signals in the expected proportions. In a given sample, no change in the ratio of tetroxide to trioxide intensities has been observed within experimental error, over the temperature range 243–193°K, and the pure tetroxide yields a single resonance with about the same chemical shift as in the mixture with N_2O_3. The spectra of N_2O_3 and N_2O_4 give no indication of the involvement of significant amounts of the ONONO, $ONONO_2$ or ONNO species in the various equilibria in these liquids at the temperatures studied [13]. ^{14}N resonances have also been observed [44] (Table 6.4) for mixtures of N_2O_4 with various solvents at several temperatures and concentrations in an attempt to study solute–solvent interactions [45]. Unfortunately the changes in chemical shift observed with variation of temperature, concentration and the nature of the solvent (i.e. 'onium' or π-donor [45]) are within the limits of error, and ^{14}N NMR spectra are evidently rather insensitive to such interactions.

An exchange equilibrium, presumably between N_2O_4 and $NO_2^{\oplus}$, is postulated to account for the high chemical shift observed for N_2O_4 in HSO_3F solution. The absence of a ^{14}N signal from the $NO^{\oplus}$ cation, which could also be a likely product of tetroxide solvolysis in this medium, is attributed to broadening of the signal beyond detectability limits by quadrupole effects. The fact that $NO^{\oplus}$ in H_2SO_4 gives a very broad resonance (Table 6.4) supports this view.

^{14}N NMR spectra were first reported for the system N_2O_5–$NO_2^{\oplus}$–$NO_3^{\ominus}$–HNO_3–H_2O by Ogg and Ray [46] who recognized that the single resonance displayed by solutions of N_2O_5 in HNO_3 is attributable to rapid chemical exchange between $NO_2^{\oplus}$, HNO_3 and $NO_3^{\ominus}$. They also recorded the spectrum of N_2O_5 in CCl_4. Unfortunately, Ogg and Ray did not tabulate their chemical shifts, which are difficult to ascertain with any degree of precision from the illustrations in their paper. The later chemical shift values quoted in Table 6.4 are, however, in broad agreement with the results of Ogg and Ray.

The single resonance at ca. 60 p.p.m. observed for solutions of N_2O_5 in CCl_4 is consistent with the presence of the symmetrical, molecular form O_2NONO_2, of the compound indicated by other physical measurements [47]. This form is also present in $CHCl_3$ [47], in the light of which the significant difference in chemical shifts (Table 6.4 [44]) compared to that in CCl_4 is somewhat surprising, assuming that ingress of moisture into the N_2O_5–$CHCl_3$ system did not occur. The position of the single resonance line found for solutions of dinitrogen pentoxide in 100% nitric acid is a function of the pentoxide concentration, shifting upfield from the value of +43 p.p.m. for the pure acid. An early Raman study [48] of a 6% solution of N_2O_5 in HNO_3 indicated the complete dissociation of N_2O_5 into $NO_2^{\oplus}$ and $NO_3^{\ominus}$ ions. However, a proportion of undissociated N_2O_5 may be present at higher concentrations [49]. The chemical shift for the dinitrogen pentoxide molecule is seen to occur (Table 6.4), within experimental error, almost exactly midway between the values for the nitronium and nitrate ions assuming negligible solvent shifts for these three species. No doubt $NO_2^{\oplus}$ and $NO_3^{\ominus}$ undergo rapid exchange in solution leading to an average chemical shift value close to that of unionized N_2O_5. Ionization of the pentoxide in solution is not therefore detectable by ^{14}N NMR data. An added complication arises from the self-dissociation [50] of pure nitric acid itself:

$$2HNO_3 \rightleftharpoons H_2O + NO_2^{\oplus} + NO_3^{\ominus}.$$

Solutions of dinitrogen pentoxide in this acid will thus contain the species HNO_3, $NO_2^{\oplus}$ and $NO_3^{\ominus}$ and possibly molecular N_2O_5. Rapid exchange between all of these can be envisaged, the precise position of the single resonance line in the range +43 to +62 p.p.m. then being determined by their relative concentrations. In solvents of relatively high dielectric constant, e.g. nitromethane, $\epsilon = 37$, partial ionization of the pentoxide is likely to occur, but rapid exchange phenomena involving the solvent are not anticipated. The ^{14}N resonance of N_2O_5 in CH_3NO_2 (Table 6.4) is, however, practically coincident with that observed in CCl_4, in which N_2O_5 is known to be entirely molecular [47].

The shifts of nitric acid solutions in water are concentration dependent and appear in the range $0 \rightarrow 43$ p.p.m. from aqueous nitrate ion. The position of the sharp resonance line reflects the extent of ionization of the acid:

$$HNO_3 + H_2O \rightleftharpoons NO_3^{\ominus} + H_3O^{\oplus}$$

(see also Section 6.5). The discrepancy in the shifts quoted for the 100% and 97% acid solutions (Table 6.4) is believed to arise from the use of CH_3NO_2 as internal standard in the latter case. Work in the author's laboratory has shown that nitromethane displays a small shift to low field of the nitrate ion, ca. −5 p.p.m. when in pure nitric acid, due presumably to some protonation of the nitromethane. The self-dissociation of analytically 100% nitric acid means that its ^{14}N chemical shift is an average value resulting from the rapid exchange of the nitrogen-containing species in the equilibrium. The true resonance position of the HNO_3 molecule is therefore unknown, but is expected to be only slightly affected, probably < 0.5 p.p.m., by the low concentrations, *ca.* 1–2% by weight, of $NO_2^{\oplus}$ and $NO_3^{\ominus}$.

The effectively identical chemical shifts of hyponitrous acid and its dianion agree with the evidence from vibrational spectra that the ONNO structure is unchanged on pro- tonation [51]. The $^{\ominus}O_2NNO^{\ominus}$ structure of the oxyhyponitrite (nitrohydroxamate) ion, has now been confirmed by asym- metric isotopic labelling and decomposition [52]. However, in spite of the presence of two environmentally non-equivalent nitrogen atoms, only one exceedingly broad ^{14}N resonance signal is observed.

6.2.1.5   *Nitrogen Fluorides, Oxo-fluorides and other Oxo-halides.*

The nitrogen fluorides and oxo-fluorides present an oppor- tunity for the direct detection of fine structure in their ^{14}N signals due to $^{14}N-^{19}F$ spin–spin interaction. Fine structure in ^{14}N NMR spectra has been resolved for three N—F molecules; F_3NO and NF_3 (C_{3v} symmetry) show $1 : 3 : 3 : 1$ quartets and FNO_2 (C_{2v} symmetry) shows a $1 : 1$ doublet [44]. The coupling constants determined in this ^{14}N study [44] are listed in Table 6.5, together with complementary values obtained from ^{19}F NMR studies on the above com- pounds and other N—F species [57–65]. Spin–spin coupling effects involving nuclei other than protons are not well understood, but it is possible to comment in broad terms on the implications of the data in Table 6.5 so far as electronic and molecular structure are concerned.

Evidence for the formation of the fluorodiazonium cation, $FNN^{\oplus}$, by reaction of *cis*-difluorodiazine, FNNF, with the powerful Lewis acid AsF_5 was obtained from ^{19}F NMR spectroscopy [57]. FNN is isoelectronic with CO_2, $NO_2^{\oplus}$ and N_2O and is thus expected to be linear. This geometry implies that of all the species listed in Table 6.5, $FNN^{\oplus}$ should possess the highest degree of *s*-character in the nitrogen hybrid orbital, *sp*, involved in the N—F bond. Therefore, if the Fermi contact interaction is still of major importance in N—F systems, this should in turn result in the one-bond coupling constant, $J(N-F)$, for $FNN^{\oplus}$ being the

TABLE 6.5.   Some ^{14}N–^{19}F Coupling Constants (Hz)

Species	^{14}N NMR data	^{19}F NMR data		References
	$^1J_{NF}$	$^1J(N-F)$	$^2J(N-F)$	
FNN$^\oplus$		328		57
F$_2$NO$^\oplus$		250 ± 3		58
NF$_4^\oplus$		234		59
		231		60
NF$_3$	158 ± 5			44
		160		61
		155		62
FNNF (cis)		±145	±37	62
F$_3$NO	134 ± 2			44
		136		63
FNNF (trans)		±136	±73	62
FNO$_2$	109 ± 5			44
		113		64
FCN			33	65

largest of those listed in Table 6.5, as actually observed. However, other data in Table 6.5 is not so readily rationalized. For example, the large differencies in coupling constant values between $F_2NO^\oplus$ and FNO_2, both trigonal planar sp^2 and isoelectronic with $NO_3^\ominus$, on the one hand and between $NF_4^\oplus$ and F_3NO, both tetrahedral sp^3 and isoelectronic, on the other, are not easily interpreted in detail. However, it is seen that replacement of F by $O^\ominus$ brings about an approximate halving of the coupling constant in each case.

A ^{19}F NMR study on F_3NO has yielded $^1J(N-F) = 136$ Hz. A 1 : 1 : 1 triplet was observed over the temperature range 163–208K, each component having a linewidth of less than 10 Hz. This is taken to indicate a high degree of symmetry around the nitrogen in F_3NO in contrast to NF_3, F_2NNF_2 and FNO whose ^{19}F signals are quadrupole broadened by the ^{14}N nuclei [63].

Double irradiation studies [62] have shown that the two-bond coupling constants, $^2J(N-F)$, for cis and trans FNNF are of opposite sign to the one-bond coupling constants, $^1J(N-F)$. The cis-isomer was originally thought to be 1,1-difluorodiazine, $F_2\overset{\oplus}{N}=\overset{\ominus}{N}$, but the double irradiation behaviour of the molecule does not support this structure since only a single ^{14}N resonance frequency is observed [62]. Furthermore, from a comparison of the two-bond coupling constants for difluorodiazine (Table 6.5) it may be inferred that the F–N–N bond angles are probably quite different in the *cis-* and *trans-*isomers.

^{19}F NMR work has played a major role in the identification of the fluoro-cations FNN$^{\oplus}$, NF$_4^{\oplus}$ and F$_2$NO$^{\oplus}$ in compatible solvents, e.g. HF. However, no ^{14}N resonance due to the F$_2$NO$^{\oplus}$ cation has been observed for solutions of F$_2$NO$^{\oplus}$AsF$_6^{\ominus}$ in HSO$_3$F or HF [44] nor do ^{14}N resonances for FNN$^{\oplus}$, NF$_4^{\oplus}$ and FCN appear to have been recorded as yet. The ^{14}N shifts which have been reported for O—N—F and N—F species are collected in Table 6.6, together with values for other nitrogen oxo-halides. These observed shifts relative to aqueous NO$_3^{\ominus}$ cover a fairly wide range for compounds possessing a lone pair of electrons on nitrogen, from −352 p.p.m. for BrNO to +35 p.p.m. for N$_2$F$_4$. Current interpretations of ^{14}N chemical shifts stress the dominance of the *paramagnetic* contribution, $\sigma_N{}^P$ to the chemical shift which depends upon the average excitation energy, $\Delta E_{av.}$ as discussed in Chapter 1.

The observed order of chemical shifts BrNO < ClNO < FNO < *trans*-FNNF < *cis*-FNNF < F$_2$NNF$_2$ corresponds to an

TABLES 6.6. ^{14}N Chemical Shifts of some Nitrogen Fluorides, Oxo-fluorides and other Oxo-halides

Molecule or ion	Solvent ($T°$K)	^{14}N shift (p.p.m.)[a]	Half-height width (Hz)	References
FNO	neat (195)	−116 ± 5		44
	neat (205)	−101 ± 10[b]	270	14
F$_3$NO	neat (163)	+131 ± 5		44
		+134		66
ClNO	neat (253)	−224 ± 5		44
BrNO	neat (253)	−352 ± 5		44
FNO$_2$	neat (163)	+82 ± 5		44
	neat (208)	+69 ± 10[b]	155	53
ClNO$_2$	neat (253)	+68 ± 5		44
	neat (273)	+63 ± 10[b]	95	53
ClONO$_2$	neat (195)	+43.6		44
FNNF (*cis*)		−5.3		62
FNNF (*trans*)		−67.2		62
F$_2$NNF$_2$	neat (163)	+35		44
		+47.8		62
F$_2$NNF$^{\oplus}$	anhydrous HF (298)	−22		44
NF$_3$	neat (121)	+8		44
		+6		66

[a] Referred to external NO$_3^{\ominus}$ unless otherwise indicated.
[b] From external nitromethane.

increase in ΔE_{av}. resulting from an increasing stabilization of the nitrogen lone pair with increasing electronegativity of the substituents around nitrogen. When the lone pair is substituted by an oxygen atom the ^{14}N chemical shift is moved upfield and the range for XNO_2 compounds is narrower than for XNO systems (X = F, Cl). The general trend whereby the resonance moves upfield with increasing electronegativity of X, observed previously for alkyl and aryl derivatives, is apparently found for inorganic derivatives also. Witanowski [33] has given an empirical rationalization of this observation. Two further noteworthy shifts are included in Table 6.6. Firstly, that of $ClONO_2$ not unexpectedly falls in the region observed for covalent nitrates, as shown by the shifts for N_2O_5 and HNO_3 in Table 6.4 and by those for some anhydrous covalent metal nitrates and nitrato-complexes in Table 6.14. Secondly, the trifluorodiazenium cation $F_2N^{\oplus}=NF$, obtained by reaction of F_2NNF_2 with AsF_5 [67, 68], shows a comparable shift to that of the isoelectronic and presumably isostructural oxy-hyponitrite anion, $^{\ominus}O_2N=NO^{\ominus}$, (Table 6.4) which likewise displays only a single ^{14}N resonance in spite of the presence of non-equivalent nitrogen atoms [44]. No ^{14}N–^{19}F coupling constants are resolvable from the ^{19}F NMR study of $F_2NNF^{\oplus}$ [67].

6.2.1.6   Nitrogen–Sulphur Compounds

Cyclic inorganic systems containing nitrogen atoms appear to be potentially ideal candidates for study by nitrogen nuclear magnetic resonance methods. The first reported investigation of this type on a sulphur–nitrogen ring [69] involved a solution of 97.2% ^{15}N enriched $S_4N_3^{\oplus}Cl^{\ominus}$ in concentrated (70%) nitric acid. In this somewhat unusual solvent it is possible to obtain stable 3–4M solutions which display two sharp and well resolved ^{15}N resonances, a high-field triplet ($\delta_{NO^{\ominus}}$ = +12 p.p.m.) and a lower field doublet ($\delta_{NO_3^{\ominus}}$ = 0 p.p.m.) in the intensity ratio 1 : 2, consistent with the presence of the $S_4N_3^{\oplus}$ cation of structure (I) in 70% HNO_3. The observed ^{15}N–^{15}N coupling

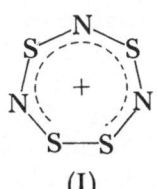

(I)

constant is 7 Hz. This work provided the first example of the use of ^{15}N NMR spectroscopy as a structural tool for the detection of non-equivalent nitrogen atoms in an inorganic molecule. However, solutions of $S_4{}^{14}N_3^{\oplus}Cl^{\ominus}$ in HNO_3 (3M) and H_2SO_4 (1M) give no observable ^{14}N resonances, presumably due to a combination of quadrupole broadening effects, too great a dilution in H_2SO_4 and obscuring of the $S_4N_3^{\oplus}$ signal(s) by the solvent resonance in the case of HNO_3. Nevertheless, ^{14}N NMR has made a valuable contribution towards the establishment of structure (II) for tetrasulphur dinitride [70]. A 3M solution of S_4N_2 in CS_2 gives a

(II)

single resonance of half-width 15 p.p.m. (109 Hz) at +105 ± 1 p.p.m. from aqueous nitrate ion as external reference. For purposes of comparison, CS_2 solutions of the two 'saturated' sulphur–nitrogen ring compounds N-methylheptasulphurimide, (III) and N,N'-dimethyl-1,4-hexasulphurdi-imide (IV) have also

(III)   (IV)

been examined. Both give single resonances at $\delta_{NO_3^{\ominus}}$ = +355 ± 5 p.p.m. The band width of the S_4N_2 signal is not unusually broad for sp^2-hybridized nitrogen in an unsymmetrical electronic environment and does not suggest two unresolved signals. The width is, in fact, smaller than for S_4N_4 [71] or S_7NMe where only one signal is possible. The ^{14}N NMR spectrum of S_4N_2 is therefore indicative of nitrogen atoms in similar environments. The chemical shift and linewidth are consistent with nitrogen multiply bonded either to nitrogen or to sulphur. The singly-bonded nitrogen in the two N-methyl sulphur imides examined gives resonances in the same region as ammonia, amines and the ammonium ion. The ^{14}N signal for S_4N_2 [70] is far downfield from this region and the chemical shift is much nearer to $\delta_{NO_3^\ominus}$ values for two compounds known to contain N=S bonds, viz., bis(phenylimino)sulphur, +69 p.p.m. [71] and isothiazole, +89 p.p.m. [25]. The shift to low-field accompanying double bonding to nitrogen is caused by deshielding[25] which results from a magnetic excitation of the low-energy $n \to \pi^*$ transition, leading to a circular motion of charge. S_4N_4 is somewhat anomalous, in that it contains S—N multiple bonds but gives a ^{14}N NMR signal at $\delta_{NO_3^\ominus} = +244$ p.p.m. This, however, is an untypical case in which the $n \to \pi^*$ transition happens to produce no charge rotation owing to the spherical symmetry of the π-electron distribution [71].

6.2.1.7 Nitrogen–Phosphorus and Nitrogen–Boron Compounds

The ^{14}N NMR technique has not yet found wide application to studies of nitrogen–phosphorus compounds, e.g. cyclophosphazenes, and nitrogen–boron compounds have only recently commanded attention in this respect. This neglect of the ^{14}N nucleus in these promising areas may be attributable to the presence of rather more easily studied nuclei, viz. ^{31}P and ^{11}B in these compounds. Some ^{14}N results on a number of pyridine–borane adducts are available [72]. The ^{14}N shifts in these adducts are observed to lower field than those of the corresponding pyridinium hydrochlorides which may signify that the *paramagnetic* contribution to the nitrogen screening tensor, σ_N^p is greater in the former than in the latter. Some preliminary ^{14}N measurements have been made on adducts of triethylamine with BH_3, BF_3, BCl_3, BBr_3, $BHBr_2$ and BH_2Br, but no obvious trends were observed. The ^{14}N resonances of these compounds, relative to $NO_3^\ominus$ mostly measured on CH_2Cl_2 solutions fall within the range +315 to +333 ± 3 p.p.m. (half-height linewidths *ca.* 100 Hz). The resonance of liquid Et_3N appears at +327 ± 3 p.p.m. (linewidth *ca.* 250 Hz) [73]. In the most recent study of nitrogen–boron compounds [123] ^{14}N chemical shifts and linewidths were reported for a considerable number of amino-boranes of the types $B(NR_2)_3$, $B(NHR)_3$, $R'B(NR_2)_2$, $R'B(NHR)_2$, $R_2'BNHR$, R_2BNH_2 and $(R_2'B)_2NR$, together with some hydrazino-boranes (for each of which two ^{14}N resonances were observed) and dialkylaminohalogeno-boranes. The results have been discussed in connection with the [11]B chemical shift data of these compounds. A decrease in shielding of the ^{14}N nucleus is observed with increasing B—N bond order. Neighbouring group anisotropy effects, especially of the heavier halogens, amplify this trend.

6.2.2 Complexes of Metals with Nitrogen-Containing Ligands

The effect on the ^{14}N chemical shift of bonding to a diamagnetic metal ion has been investigated for the following ligands: ammonia [6, 12, 25, 29] (Table 6.7), cyanide [12, 25, 75] (Table 6.8), isocyanate [29] (Table 6.9), fulminate [56] (Table 6.10), thiocyanate [76] (Table 6.11), isoselenocyanate [124] (Table 6.11), azide [39] (Table 6.12), nitric oxide [12, 56, 77] (Table 6.13), nitrite [12, 77] (Table 6.13), nitrate [5, 7, 54, 77] (Table 6.14), ethylenediamine, 2,2'-dipyridyl and some amino acids [78] (Table 6.15), isohydrocyanic acid, and a substantial number of organic nitriles [79, 125] and isonitriles [79]. The chemical shifts in complexes containing these ligands span a range in excess of 700 p.p.m. The chemical shift differences between free and complexed ligand have been interpreted in terms of changes in σ_N^p on complex formation [12]. The high-field shifts observed for ammonia and nitrite ligands (Tables 6.7 and 6.13) are found to be essentially independent of the lowest energy electronic absorption maxima. This has been attributed to the constraint placed upon the lone pair electrons of the ligand when a metal-

TABLE 6.7.   [14]N Chemical Shifts of some Ammino-complexes

Compound	Solvent	[14]N shift (p.p.m.)[a]	References
NH_3	liquid	$+376 \pm 1$	74
$[Co(NH_3)_6]Cl_3$		$+418 \pm 6^b$	12
$[Co(NH_3)_6]^{3\oplus}$	$H_2O(Cl^\ominus)$	$+425 \pm 10$	25
$[Co(NH_3)_5Cl]^{2\oplus}$	$H_2O(Cl^\ominus)$	$+420 \pm 20$	25
$[Co(NH_3)_5(H_2O)]^{3\oplus}$	$H_2O(Cl^\ominus)$	$+414 \pm 20$	25
$[Ru(NH_3)_6]Cl_2$	H_2O	$+448 \pm 4^b$	12
$[Rh(NH_3)_6]Cl_3$	H_2O	$+427 \pm 4^b$	12
$[Pt(NH_3)_4]^{2\oplus}$	$H_2O(Cl^\ominus)$	$+408 \pm 10$	25
$[Ag(NH_3)_2]^{\oplus}$		$+384 \pm 6$	6
$[Ag(NH_3)_2]^{\oplus}$	H_2O	$+382$	29

a Referred to external $NO_3^\ominus$ unless otherwise indicated.
b Referred to external CH_3CN; recalculated to external $NO_3^\ominus$.

nitrogen bond is formed. These electrons are then less susceptible to the magnetic field (lowered ground state energy) and contribute less to σ_N^p which is correspondingly decreased, resulting in a high-field shift. In the cyano-complexes (Table 6.8) the ligand nitrogen atom is not directly bound to the metal atom and the ^{14}N chemical shifts of these complexes are reported [12] to be to low field of that of the free ligand by amounts which vary roughly with the wavelength of the lower-lying excited electronic states of the complexes, comparing complexes of the same symmetry. This has been accounted for in terms of metal–carbon σ-bonding accompanied by d-electron delocalization into the π-molecular orbital set of the ligand. However, the chemical shift value for KCN quoted by Bramley *et al.* [12] as -5 p.p.m. relative to CH_3CN, appears to be in error. Recalculation of the chemical shift values for the complexes to external nitrate (Table 6.8) and comparison with the Herbison-Evans and Richards value for KCN [25] shows that the chemical shifts for cyano-complexes actually differ rather little from that of the cyanide ion itself. This may still be accounted for in terms of compensation of σ-electron loss from ligated cyanide by π-electron gain from the metal,

TABLE 6.8. ^{14}N Chemical Shifts of some Cyano-complexes

Compound	Solvent	^{14}N shift (p.p.m.)[a]	References
KCN	H_2O	$+96 \pm 2$	25
$K_2[Cr(CN)_5NO]$	H_2O	$+102 \pm 5$	25
$K_4[Fe(CN)_6]$	H_2O	$+111 \pm 15$	25
		$+83$[b,c]	75
$Na_4[Os(CN)_6]$	H_2O	$+101 \pm 13$[b]	12
$Na_3[Co(CN)_6]$		$+98 \pm 8$[b]	12
$K_3[Co(CN)_6]$	H_2O	$+94 \pm 12$	25
		$+76$[b,c]	75
$K_2[Ni(CN)_4]$		$+86 \pm 5$[b]	12
		$+111$[b,c]	75
$K_2[Pd(CN)_4]$		$+99 \pm 8$[b]	12
$K_2[Pt(CN)_4]$		$+112 \pm 6$[b]	12
$K_2[Zn(CN)_4]$		$+97 \pm 4$[b]	12
$K_2[Cd(CN)_4]$		$+87 \pm 5$[b]	12
$K[Cu(CN)_2]$		$+111$[b,c]	75
$K[Ag(CN)_2]$		$+111$[b,c]	75
$Na[Au(CN)_2]$		$+116 \pm 2$[b]	12

a Referred to external $NO_3^{\ominus}$ unless otherwise indicated.
b Referred to external CH_3CN; recalculated to external $NO_3^{\ominus}$.
c Approximate values only.

leading to very similar resultant nitrogen shielding in the cyano-complexes studied (Table 6.8) compared to that in the free cyanide ion.

High-field shifts relative to the free ion are again found in isocyanato-complexes (Table 6.9) and serve as a diagnostic feature of N-bonding, although no definitively established cyanato- (O-bonded) complex has yet been reported [29]. The positions of the ^{14}N resonances of the metal isocyanato-complexes depend on the degree of polarization of the nitrogen lone-pair electrons, but partial dissociation in solution leading to exchange between isocyanate ligands and uncoordinated ions may also contribute to the rather small positive shifts observed for certain of these species. There are

TABLE 6.9. ^{14}N Chemical Shifts of some Isocyanato-complexes [29]

Compound	Solvent	^{14}N shift (p.p.m.)a
NCO$^\ominus$	H$_2$O	+300 ± 1
Et$_4$N$^\oplus$Ag(NCO)$_2^\ominus$	Me$_2$CO	+344 ± 2
Me$_4$N$^\oplus$Ag(NCO)$_2^\ominus$	Me$_2$CO	+342 ± 2
Ph$_4$As$^\oplus$Ag(NCO)$_2^\ominus$	Me$_2$CO	+342 ± 2
(Et$_2$NH$_2^\oplus$)$_2$Zn(NCO)$_4^{2\ominus}$	Me$_2$CO	+325 ± 2
(Et$_3$NH$^\oplus$)$_2$Zn(NCO)$_4^{2\ominus}$	Me$_2$CO	+322 ± 4
(Me$_4$N$^\oplus$)$_2$Hg(NCO)$_4^{2\ominus}$	Me$_2$CO	+317 ± 2
(Et$_4$N$^\oplus$)$_2$Hg(NCO)$_4^{2\ominus}$	MeNO$_2$	+312 ± 2
(Et$_4$N$^\oplus$)$_2$Sn(NCO)$_6^{2\ominus}$	Me$_2$CO	+308 ± 1
EtOCN		+222 ± 1

a Referred to external NO$_3^\ominus$.

significant shift differences between complexes of M(NCO)$_2^\ominus$, M(NCO)$_4^{2\ominus}$ and M(NCO)$_6^{2\ominus}$ stoichiometry and this phenomenon should permit ready distinction to be made between such species in solution. The back-donation of electron density from filled metal d-orbitals into empty antibonding (π^*) orbitals of the NCO ligands, although expected to occur to only a limited extent, may nevertheless exert some influence on the ^{14}N chemical shifts of such isocyanato-metallates. Certain metal complexes containing coordinated fulminate, i.e. the isomeric —CNO system, have also been studied by ^{14}N NMR spectroscopy [56]. These complexes exhibit resonances 10–20 p.p.m. to low field of the free ligand (Table 6.10), such downfield shifts being commonly observed for linear triatomic groups in which a nitrogen atom is not directly bonded to an organic group or a metal as in

ROCN [29, 56],   RSCN [25, 76],   MSCN [76],   RNNN [39] and MNNN [39] systems. The presence of a metal–carbon bond in the fulminate complexes is also supported by the ^{14}N shift of +165 p.p.m., referred to external $NO_3^{\ominus}$, observed [56] for the compound 2,4,6-trimethylbenzo-nitrileoxide (V).

$$\underset{\text{(V)}}{\underset{\text{Me}}{\overset{\text{CNO}}{\text{Me}\diagdown\diagup\text{Me}}}}$$

On inspection of Tables 6.1, 6.9 and 6.10, it may be seen that the ^{14}N shifts of –CNO, –OCN and –NCO compounds are respectively ca. +170 p.p.m., ca. +210 p.p.m. and ca. +350

TABLE 6.10. ^{14}N Chemical Shifts of some Fulminato-complexes [56]

Compound	Solvent	^{14}N shift (p.p.m.)[a]
$CNO^{\ominus}$	H_2O	+176
$Na_2[Ni(CNO)_4]$	H_2O	+163
$(Pr_4{}^nN)_2[Ni(CNO)_4]$	CH_2Cl_2	+162
$[(Ph_3P)_2Pd(CNO)_2]$	$CHCl_3$	+165
$PhHgCNO$	Dioxan	+155
$(Et_4N)[Ag(CNO)_2]$	CH_2Cl_2	+164

[a] Referred to external $NO_3^{\ominus}$.

p.p.m. relative to $NO_3^{\ominus}$. This allows a ready distinction to be made between these isomeric systems and is of particular interest in relation to the isomerizations fulminate → cyanate or isocyanate [80] and nitrileoxide → isocyanate [81]. Also, following the trend found by Kent and Wagner [28, 43], for some linear triatomic molecules and ions, increasing π-electron density on the nitrogen atom in the order –CNO < –OCN < –NCO is suggested by the ^{14}N chemical shifts.

The use of ^{14}N NMR spectroscopy as a means of distinguishing alternative bonding modes of a nitrogen-containing ambidentate ligand was first demonstrated by a study on metal thiocyanate complexes [76]. It was observed (Table 6.11) that there is no appreciable nitrogen chemical shift relative to thiocyanate ion if sulphur-bonding occurs,

TABLE 6.11.　　^{14}N Chemical Shifts of some Thiocyanato-,Isothiocyanato-
and Isoselenocyanato-complexes

Compound	Solvent	^{14}N shift (p.p.m.)[a]	Reference
SCN$^\ominus$	H$_2$O	+166	76
	H$_2$O	+170	124
SeCN$^\ominus$	H$_2$O	+136	124
Na$_2$[Pd(SCN)$_4$]	H$_2$O	+148	76
K$_2$[Pt(SCN)$_4$]	H$_2$O	+166	76
K$_3$[RH(SCN)$_6$]	H$_2$O	+158	76
Na$_3$[Ir(SCN)$_6$]	H$_2$O	+163	76
(NH$_4$)$_2$[Hg(SCN)$_4$]	H$_2$O	+146	124
Na$_2$[Hg(SCN)$_4$]	H$_2$O	+157	76
Sr(SCN)$_2$	H$_2$O	+173	124
EtSCN[a]	CHCl$_3$	+98.5	76
K[Ph$_3$Sn(NCSe)$_2$]	MeOH	+215	124
cis-[Pt(NCS)$_2$(Bu$_3$P)$_2$]	CHCl$_3$	+302	76
cis-[Pt(NCS)$_2$(Bu$_2$PhP)$_2$]	CHCl$_3$	+249	76
trans-[PtH(NCS)(Et$_3$P)$_2$]	CHCl$_3$	+239	76
cis-[Pt(NCS)$_2$(Bu$_3$As)$_2$]	CHCl$_3$	+303	76
trans-[Ni(NCS)$_2$(Et$_3$P)$_2$]	CHCl$_3$	+293	76
trans-[Ni(NCS)$_2$(Bu$_2$PhP)$_2$]	CHCl$_3$	+291	76
Na$_2$[Cd(NCS)$_4$]	H$_2$O	+178	76
Na$_2$[Cd(NCS)$_4$]	MeOH	+220	76
K$_4$[Cd(NCS)$_6$]	H$_2$O	+183	124
Na$_2$[Zn(NCS)$_4$]	EtOH	+255.5	76
K$_2$[Zn(NCS)$_4$]	H$_2$O	+238	124
K$_2$[Zn(NCS)$_4$]·2Me$_2$CO	H$_2$O	+220	124
K$_2$[Ru(NCS)$_5$(NO)]	H$_2$O	+245	76
Ph$_3$Sn(NCS)	MeOH	+253	124
BH[Ph$_3$Sn(NCS)$_2$][c]	MeOH	+209	124
EtNCS[b]	neat	+268	76
PhNCS	neat	+267	124

　[a] Referred to external NO$_3^\ominus$ (ref. 76).
　　Referred to internal NO$_3^\ominus$ (ref. 124).
　[b] See also ref. 25.
　[c] B = benzylamine.

whereas bonding *via* nitrogen produces a significant high-field
shift. The authors have related these effects to the $n \to \pi^*$
transition of the lone-pair electrons of the nitrogen
atom [76]. The thiocyanate ligands in K$_2$[Ru(NCS)$_5$NO]
have been shown to be N-bonded, the effect of the NO$^\oplus$
ligand on the ruthenium atom being analogous to that of the
phosphine and arsine ligands on palladium and platinum. The
very large solvent effect on the ^{14}N shift observed for
Na$_2$[Cd(NCS)$_4$] is attributed to the existence in solution of

both N- and S-bonded species in kinetic equilibrium. Studies of the effect of added HCl on aqueous KSCN and of $HClO_4$ on aqueous ethanolic KSCN have led to the conclusion that the main tautomer present in such solutions is HNCS, albeit appreciably dissociated, and that the ^{14}N shift of the pure acid must be in the region of $\delta_{SCN}\ominus = +100$ p.p.m. [76]. This was confirmed recently for solutions of HNCS in diethylether [5] and benzene [5, 30] (Table 6.2). A further notable feature of the results summarized in Table 6.11 is the observation of ^{195}Pt—^{14}N coupling for the complex cis-[Pt(NCS)$_2$(Bu$_3$P)$_2$], with a coupling constant of 430 Hz, the only value known for these two atoms. This may be compared with values of 2–6 kHz for Pt—P coupling constants [82].

A further ^{14}N NMR study [124] of diamagnetic thiocyanate and isothiocyanate species reported recently, includes chemical shifts for the selenocyanate compounds KSeCN and K[Ph$_3$Sn(NCSe)$_2$] (Table 6.11). The low-field shift observed for K$_4$[Cd(NCS)$_6$] is again attributed to an equilibrium between N- and S-bonded species, whereas the low value for the pentacoordinate anion, [Ph$_3$Sn(NCS)$_2$]$^\ominus$, is considered to arise from some dissociation of the isothiocyanate ligands in methanol. This is thought to occur also with [Ph$_3$Sn(NCSe)$_2$]$^\ominus$. Dependences of the linewidths for aqueous KSCN solutions upon different concentrations and temperatures, have also been studied.

The study of covalent azides [39] discussed in Section 6.2.1.3 and Table 6.3 has included measurements on a number of metal azido-complexes (Table 6.12), each of which display the three ^{14}N resonances expected for coordinated azide ligands, and the observed chemical shifts indicate that the bonding is very similar to that in organic azides (cf. Table 6.3). In common with the nitrogen-bonded ligands already discussed, azide shows a chemical shift for the bonded atom, N_a, to high field of that observed for the free ligand. This may be referred to as the 'coordination shift'. Only one N_a signal is observed for the complexes Pd$_2$(N$_3$)$_6^{2\ominus}$ and (Ph$_3$P)$_2$Pd$_2$(N$_3$)$_4$ which contain both terminal and bridging azide ligands. Also, the N_a coordination shift found for (Ph$_3$P)$_4$Pd(N$_3$)$_2^{2\ominus}$ which possesses only azide bridges is close to that of the terminal azide groups in (Ph$_3$P)$_2$Pd(N$_3$)$_2$. In fact, a recent X-ray study [83] of the (N$_3$)$_2$Pd(N$_3$)$_2$Pd(N$_3$)$_2^{2\ominus}$ anion has brought to light a remark-

TABLE 6.12. ^{14}N Chemical Shifts of some Azido-complexes [39]

$$M-\overset{\ominus}{N}_a-\overset{\oplus}{N}_b\equiv N_c$$

Compound	Solvent	^{14}N shifts (p.p.m.)[a]		
		N_a	N_b	N_c
$N_3^\ominus$	H_2O	+277 (55)[b]	+128 (22)	+277 (55)[b]
$(Ph_4As)_2Sn(N_3)_6$	CH_2Cl_2	+293 (800)	+143 (90)	+225 (90)
$(Ph_4As)_2Pb(N_3)_6$	CH_2Cl_2	+280 (328)	+143 (185)	+200 (185)
$(Ph_4As)Au(N_3)_4$	CH_2Cl_2	+310 (410)	+140 (55)	+183 (127)
$(Ph_4As)Au(N_3)_2$	CH_2Cl_2	+336 (180)	+140 (75)	+248 (75)
$(Ph_4As)_2Pd(N_3)_4$	CH_2Cl_2	+334 (245)	+143 (110)	+225 (125)
$(Ph_4As)_2Pt(N_3)_4$	CH_2Cl_2	+345 (180)	+130 (110)	+237 (250)
$(Ph_4As)_2Pd_2(N_3)_6$	CH_2Cl_2	+355 (218)	+135 (145)	+230 (324)
Ph_3PAuN_3	CH_2Cl_2	+315 (330)	+132 (110)	+229 (290)
cis-$(Bu_3P)_2Pt(N_3)_2$	CH_2Cl_2	+326 (220)	+135 (145)	+226 (110)
cis-$(PhBu_2P)_2Pt(N_3)_2$	CH_2Cl_2	+344 (180)	+130 (290)	+230 (360)
cis-$(Ph_2BuP)_2Pt(N_3)_2$	CH_2Cl_2	+365 (73)	+135 (270)	+230 (270)
cis-$(Ph_3P)_2Pt(N_3)_2$	CH_2Cl_2	+351 (110)	+131 (186)	+225 (250)
$(Ph_3P)_2Pd(N_3)_2$	CH_2Cl_2	+363 (145)	+140 (110)	+230 (75)
$(Ph_3P)_2Pd_2(N_3)_4$	CH_2Cl_2	+374 (218)	+134 (145)	+255 (320)
$(Ph_3P)_4Pd_2(N_3)_2(BF_4)_2$	CH_2Cl_2	+365 (110)	+129 (40)	+192 (92)

[a] Referred to external $NO_3^\ominus$; linewidths (Hz) at half peak height in parentheses.
[b] Resonance of intensity 2 relative to the N_b signal.

able similarity of the N–N distances and Pd–N_3 angles in both terminal and bridging N_3 ligands. The azide bridges in these complexes appear to be of the N-diazonium type (VI) [83, 84] and evidently attachment of a second metal atom to the lone pair of the N_a atom does not significantly change the net shielding of the N_a nucleus.

$$\begin{matrix} M \\ {}\diagdown \\ {}\diagup \\ M \end{matrix} N-N\equiv N$$

(VI)

The results presented in Table 6.13 show that the chemical shifts of nitrosyl complexes fall into two distinct groups. For those complexes incorporating the $RuNO^{3\oplus}$ unit, moderate high-field shifts relative to $NO^\oplus$ are seen whilst for the low oxidation-state iron and cobalt compounds, in which the nitrosyl group competes for metal electrons with other π-acceptor ligands, similarly modest shifts to low field are displayed. The authors of a paper [56] dealing with the latter compounds have quoted chemical shifts for $NO^{\oplus}$ in its $HSO_4^\ominus$ and $PF_6^\ominus$ salts, in H_2SO_4 and nitromethane respectively, as −694 and −693 p.p.m. relative to $NO_3^\ominus$. Subsequently [79]

TABLE 6.13. ^{14}N Chemical Shifts of some Nitrosyl and Nitro-complexes

Compound	Solvent	^{14}N shift (p.p.m.)[a]	References
NO$^{\oplus}$	H_2SO_4	0 ± 10	14
$[RuNOCl_3(H_2O)_2] \cdot 3H_2O$	H_2O	$+36$[b]	77
$[RuNO(NO_3)_3(H_2O)_2] \cdot xH_2O$	T.B.P.	$+43$[b]	77
$Na_2[RuNO(NO_2)_4OH]$	H_2O	$+10$[b]	77
$H_2[RuCl_5NO]$	1MHCl	$+25 \pm 2$[c]	12
$[Fe(CO)_2(NO)_2]$		-35 ± 1[c]	12
$[Fe(CO)_2(NO)_2]$	C_6H_6	-30	56
$[Co(CO)_3NO]$		-17 ± 4[c]	12
$[Co(CO)_3NO]$	C_6H_6	-15	56
$Na[Fe(CO)_3NO]$	Et_2O	-22	56
$[Co(NO)_2Cl]_2$	EtOH	-44	56
$[Co(NO)_2Br]_2$	EtOH	-44	56
$[Co(NO)_2(PPh_3)Cl]$	Me_2CO	-66	56
$[Co(NO)diars_2Br]Br$	MeOH	-36	56
NO$_2^{\ominus}$	H_2O	-237 ± 4	33
$Na_2[RuNO(NO_2)_4OH]$	H_2O	$+10$[b]	77
$Na_3[Co(NO_2)_6]$	H_2O	-3 ± 2[c]	12
$Na_2[Pd(NO_2)_4]$	H_2O	-2 ± 4[c]	12
$Na_2[Pt(NO_2)_4]$	H_2O	-4 ± 4[c]	12

[a] Referred to external NO$_3^{\ominus}$ unless otherwise indicated.
[b] Referred to external NO$_2^{\ominus}$; recalculated to external NO$_3^{\ominus}$.
[c] Referred to external CH_3CN; recalculated to external NO$_3^{\ominus}$.

they have recognized the possibility that these inordinately low values may have resulted from the presence of paramagnetic nitric oxide arising as a decomposition product in their solutions. The ^{14}N chemical shift value [14] for NO$^{\oplus}$ included in Tables 6.4 and 6.13 is much more acceptable, although rather inaccurate, since the resonance is very broad. In nitrosyl complexes, the nitrogen atom of the ligand is adjacent to the metal and it is well known that coordinated NO$^{\oplus}$ behaves as both a σ-donor and a strong acceptor of metal d_π-electrons. These effects are expected to be largely compensatory as far as nitrogen shielding is concerned [12] and to give a range of ^{14}N resonances in nitrosyl complexes spanning the free ligand resonance as observed for the ^{13}C resonances of some carbonyl complexes [85]. Table 6.13 shows that these expectations are fulfilled and it seems reasonable to suppose that the gradation from positive to negative shifts is a reflection of decreasing importance of σ-bonding and increasing transfer of metal d-electrons into π^* orbitals of NO.

A qualitative interpretation [12] of the high field shifts of nitro-complexes relative to the nitrite ion (Table 6.13) was mentioned briefly earlier in this section. The similarity in the chemical shifts of metal nitro-complexes, nitroalkanes, such as nitromethane, and the nitrate ion is striking.

In the early work on the ruthenium nitrosyl complexes [77] the proximity of the ^{14}N NMR signals for coordinated nitrosyl, nitro- and nitrato-groups was noted and a ^{14}N shift of +4 p.p.m. relative to $NO_3^{\ominus}$ was reported for the nitrate ligands in the complex $[RuNO(NO_3)_3-(H_2O)_2] \cdot xH_2O$ in tributylphosphate solution. Schmidt, Brown and Williams [7] measured the ^{14}N NMR spectra of a series of diamagnetic metal nitrates in aqueous solution and not surprisingly reported shifts and linewidths identical to those in saturated NH_4NO_3. Also predictably, large shifts from $NO_3^{\ominus}$ were found for aqueous and ethanolic solutions of paramagnetic nitrates, e.g. $Cu(NO_3)_2 \cdot 3H_2O$ [7]. More recently, a chemical shift ($\delta_{NO_3^{\ominus}}$) value of +4 p.p.m. for a solution of uranyl nitrate in methanol has been reported [25]. Further to this work, an extensive series of ^{14}N NMR measurements on diamagnetic, anhydrous covalent metal nitrates and nitrato-complexes in non-aqueous solvents has been carried out recently in the author's laboratory [5, 54] and the results are summarized in Table 6.14. The existence of metal–nitrate covalent bonding in a given solid compound may be readily established by vibrational spectroscopy which, in favourable circumstances, will also allow the mode of nitrate attachment, i.e. *via* unidentate, $MONO_2$ or bidentate, MO_2NO bonding of oxygen atoms, to be determined [86]. In addition, an appreciable number of X-ray crystal structure determinations on metal nitrates and nitrato-complexes have been published [86], as a result of which some understanding of the structural principles operative in such compounds is being gained. Vibrational spectroscopic data indicate the presence of covalent nitrate bonding in all the compounds listed in Table 6.14 and the designations ONO_2 or O_2NO are used in the formulae of compounds for which an X-ray structure of the solid is additionally available; ref. 86 gives further details. The solution behaviour of such covalent nitrates is a feature of major interest in their chemistry and the nature of solution species in non-interacting, weakly interacting and strongly interacting, or reacting, solvents has been studied by ^{14}N NMR spectroscopy [5, 54]. For this purpose, ^{14}N NMR is superior to both vibrational spectroscopy, which is prone to interference by solvent absorption and electronic spectroscopy, which for nitrates of non-transition metals, gives broad, featureless absorption bands which are difficult to interpret.

The results in Table 6.14 clearly demonstrate that metal–nitrate covalent bonding in diamagnetic compounds is readily detectable by ^{14}N NMR, resulting in an upfield shift from $NO_3^{\ominus}$ in all cases. It is also seen that the highest shifts yet observed for covalent metal nitrates fall 10 p.p.m. below those of organic nitrates such as $MeONO_2$ and $EtONO_2$ and are appreciably lower than the shifts for 100% HNO_3 and N_2O_5 (Table 6.4). For those compounds which could be

TABLE 6.14. ^{14}N Chemical Shifts of some Anhydrous Covalent Metal Nitrates and Nitrato-complexes [5, 54]

Compound	Solvent	Line-width (Hz)	^{14}N shift (p.p.m.)[a]
$MeONO_2$	neat	10	$+35 \pm 0.5$
$MeONO_2$	CCl_4	10	$+36 \pm 0.5$
$EtONO_2$[b]	neat	15	$+36 \pm 0.5$
$EtONO_2$[b]	CCl_4	15	$+37 \pm 0.5$
$Ti(O_2NO)_4$	CCl_4	28	$+25$
$Ti(O_2NO)_4$	$MeNO_2$	58	$+25$
$NO^{\oplus}[Au(ONO_2)_4]^{\ominus}$	$MeNO_2$	51	$+25 \pm 2$
$Sn(O_2NO)_4$	CCl_4	35	$+24$
$Sn(O_2NO)_4$	$MeNO_2$	72	$+24$
$CrO_2(NO_3)_2$	neat	32	$+23$
$CrO_2(NO_3)_2$	CCl_4	18	$+23$
$NO_2^{\oplus}[Ga(NO_3)_4]^{\ominus}$	$MeNO_2$	38	$+22$ (NO_3)
$NO_2^{\oplus}[Ga(NO_3)_4]^{\ominus}$	$MeNO_2$	72	$+125$ ($NO_2^{\oplus}$)
$VO(NO_3)_3$	neat	80	$+21 \pm 1$
$VO(NO_3)_3$	CCl_4	18	$+21 \pm 1$
$VO(NO_3)_3$	MeCN	110	$+21 \pm 1$
$Co(O_2NO)_3$	CCl_4	28	$+11 \pm 0.5$
$Al(NO_3)_3 2MeCN$	EtOAc	100	$+22 \pm 1$
$MeHgNO_3$	C_6H_6	30	$+14 \pm 2$
$(Ph_4As)_2[Zn(O_2NO)_4]$	MeCN	22	$+9 \pm 1$
$(NH_4)_2[Ce(O_2NO)_6]$	H_2O	86	$+8 \pm 1$
$UO_2(NO_3)_2$[c]	MeOH		$+4 \pm 1$
$[RuNO(NO_3)_3(H_2O)_2] \cdot xH_2O$[d]	TBP		$+4$
$Cd(NO_3)_2 \cdot MeCN$	MeCN	100	$+2 \pm 1$

[a] Referred to external $NO_3^{\ominus}$.
[b] See also ref. 33.
[c] Ref. 25.
[d] Ref. 77.

measured as pure liquids, or as solutions in CCl_4, presumed non-interacting, or CH_3NO_2, presumed weakly interacting, the shifts occur in the very narrow range +21 to +25 p.p.m., with the exception of that for $Co(NO_3)_3$. Figure 6.4 illustrates the ^{14}N NMR spectrum of a solution of nitronium tetranitratogallate(III), $NO_2^{\oplus}[Ga(NO_3)_4]^{\ominus}$, in nitromethane as solvent and internal standard. The two resonances in 1 : 4 intensity ratio, in addition to that of the solvent, confirm the presence of $NO_2^{\oplus}$ and $Ga(NO_3)_4^{\ominus}$ ions in nitromethane

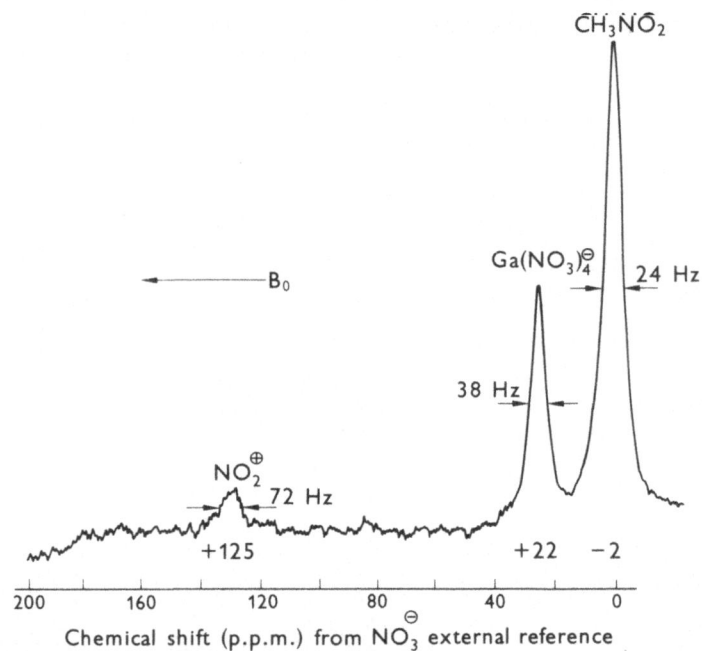

Fig. 6.4. ^{14}N NMR spectrum at 7.226 MHz of nitronium tetranitratogallate(III) in nitromethane.

solution (see also Tables 6.4 and 6.14). Comparison of the shifts for $Ti(O_2NO)_4$, $NO^{\oplus}[Au(ONO_2)_4]^{\ominus}$ and $Sn(O_2NO)_4$ (Table 6.14) shows that either bidentate or unidentate covalent interaction can lead to strikingly similar nitrogen shielding effects, so that unfortunately ^{14}N NMR does not allow a distinction between bonding modes to be made.

    The extent of 'perturbation' of the nitrate group by the metal depends on the oxidation state of the latter as well as on the bonding mode. Partial dissociation of nitrate ligands may occur in donor and/or high dielectric solvents leading to exchange between 'free' and coordinated nitrate. Such factors may account for some of the lower shifts recorded in Table 6.14. The effect of changing solvent and thus of changing dielectric constant, solvating power and donor power on the ^{14}N chemical shift has been studied [5] in the case of $[Ce(O_2NO)_6]^{2\ominus}$. The species present in solutions of metal nitrates in 100% HNO_3 have also been investigated by ^{14}N NMR [5, 54]. Furthermore, the great value of the ^{14}N NMR technique in following reactions of nitrates in solution has now become evident [54]. A brief summary of current investigations is given in Section 6.5.

As found for ammonia, (Table 6.7), the ^{14}N shifts of glycine, alanine, ethylenediamine and 2,2′-dipyridyl co-ordinated in Co(III) complexes are upfield of those of the free ligands [78] (Table 6.15). The most significant feature of these results is the high-field shifts of the trans-glycinato- and -alaninato-complexes compared to their cis-isomers. This observation has been interpreted in terms of the splitting of molecular orbital energy levels on reduction of symmetry from C_3 to C_i in going from the cis- to the trans-complex [78].

^{14}N chemical shifts relative to $NO_3^{\ominus}$ have been reported for a large number of isonitrile complexes [79] of Mn, Cr, Fe, Co, Ni, Cu, Mo, W and Au. ^{14}N 'coordination shifts' to low field of the free isonitriles are found for all those complexes in which the nitrogen atom of the ligand is not directly bound to the metal atom. Coordination shifts to increasingly

TABLE 6.15. ^{14}N Chemical Shifts of some Co(III) Complexes and Bidentate Ligands [78]

Compound	Solvent	^{14}N shift (p.p.m.)[a]
$H_3N^{\oplus}CH_2COOH$	20% HCl	$+331 \pm 5$
$H_3N^{\oplus}CH_2COO^{\ominus}$		$+335 \pm 3$
cis-Co$(H_2NCH_2COO)_3$	20% HCl	$+340 \pm 5$
trans-Co$(H_2NCH_2COO)_3$		$+399 \pm 7$
$H_3N^{\oplus}CH_2CH_2COOH$	20% HCl	$+301 \pm 6$
$H_3N^{\oplus}CH_2CH_2COO^{\ominus}$		$+310 \pm 5$
cis-Co$(H_2NCH_2CH_2COO)_3$		$+331 \pm 5$
trans-Co$(H_2NCH_2CH_2COO)_3$		$+371 \pm 7$
$H_2NCH_2CH_2NH_2$		$+358 \pm 2$
Co$(H_2NCH_2CH_2NH_2)_3$		$+373 \pm 5$
dipy(2,2′-dipyridyl)		$+48 \pm 3$
Co(dipy)$_3$		$+165 \pm 8$

[a] Referred to external $NO_2^{\ominus}$; recalculated to external $NO_3^{\ominus}$.

low field are correlated with increasing metal-to-ligand π-bonding, as indicated by infra-red investigations. Thus, the ^{14}N measurements are taken to indicate an increase in the importance of π-bonding from cationic isonitrile complexes, e.g. $[Fe(CH_3NC)_6]Br_2 \cdot 2H_2O$, in which it is believed to be insignificant, through isonitrile carbonyl complexes, e.g. $[Fe(CO)_3(CNCH_3)_2]$, with an intermediate degree of π-back bonding to compounds in which isonitriles are the only ligands bound to zerovalent metals, e.g. $[Cr(CNPh)_6]$. In the latter, π-bonding is of major significance. The ^{14}N resonances for certain isohydrocyanic acid complexes [79], e.g. $[Cr(CO)_5CNH]$ are found to lie in the same chemical shift region as those of the isonitrile complexes. From infra-red studies it is known that isohydrocyanic acid behaves only as a σ-donor; however, the measurement of a coordination shift relative to the free CNH ligand has not been possible in this case. As with all complexes involving direct metal–nitrogen bonding, the ^{14}N signals of compounds containing nitrile ligands, e.g. $[Cr(CO)_5NCCH_3]$, are to high field of the free ligand resonances [79, 125].

6.3   Studies on Paramagnetic Metal Complexes

Large chemical shifts and broad absorption bands are expected from ^{14}N nuclei in paramagnetic environments [3, 87]. Either high- or low-field isotropic shifts may occur and these may result either from the presence of electron spin density at the nucleus in question (contact shifts) or from direct dipole–dipole interactions between the magnetic dipoles of the unpaired electrons and those of the nuclear spins, involving no delocalization of the electron spin (pseudo-contact shifts) [87]. Paramagnetic line broadening is a manifestation of short nuclear relaxation times brought about by the rapidly fluctuating local magnetic fields produced by the unpaired electrons. In general, for paramagnetic species containing resonating nuclei, these line shifting and broadening effects can provide much information concerning electronic and geometric structures and modes of re-action [87]. However, since many ^{14}N resonance signals are also quadrupolar broadened it is not to be expected that ^{14}N nuclei will prove to be as useful in studying paramagnetic species as are nuclei with $I = \frac{1}{2}$. Nevertheless, ^{14}N NMR studies of paramagnetic inorganic species, in particular transition metal complexes, can in favourable circumstances provide several kinds of physicochemical information [3]. This will now be illustrated by reference to the four phenomena which have been most fruitfully investigated by ^{14}N NMR work on paramagnetic transition metal complexes, viz. ion association, kinetics of electron transfer, ligand exchange processes, and spin density distribution.

6.3.1 Ion Association

Ion association between the diamagnetic tetra-n-butyl-ammonium and phosphonium cations and the paramagnetic anions $[Ph_3PCoI_3]^\ominus$, $[Ph_3PNiI_3]^\ominus$ and $[Fe(CN)_6]^{3\ominus}$ has been studied by ^1H, ^{14}N and ^{31}P NMR spectroscopy [88–93]. The ^1H spectra of the tetra-n-butylammonium protons have been interpreted in terms of a pseudo-contact interaction with the unpaired electrons on the metal [90]. However, low-field shifts have been reported for the ^{14}N and ^{31}P signals and it has been argued that this observation supports the postulate of direct transfer of positive spin density from the metal to the nitrogen and phosphorus atoms of the cation, rather than a pseudo-contact interaction [91, 92]. The direct spin transfer mechanism predicts that the resonance of the cation protons should undergo a high-field shift irrespective of the particular paramagnetic anion present in solution. However, proton shifts in opposite senses are observed for the tetra-n-butylammonium cation in the presence of $[Ph_3PCoI_3]^\ominus$ or $[Ph_3PNiI_3]^\ominus$. Recent ^{14}N NMR data has been interpreted in favour of the pseudo-contact model with an additional small contact interaction attributed to weak covalent interaction between the anion and cation in the ion pair [93].

6.3.2 Kinetics of Electron Transfer

^{14}N and ^{13}C resonances have been studied in a number of paramagnetic and diamagnetic cyanometallate complexes [75]. ^{14}N signals of comparable width can be observed from diamagnetic hexacyanoferrate(II), $[Fe(CN)_6]^{4\ominus}$, and from paramagnetic hexacyanoferrate(III), $[Fe(CN)_6]^{3\ominus}$, but the resonance of the latter is shifted 724 p.p.m. to low field of the resonance of the former, this shift being attributed to a contact interaction. Aqueous solutions containing both

$K_3[Fe(CN)_6]$ and $K_4[Fe(CN)_6]$ display a single peak, intermediate in position between the resonances of each constituent, indicating the presence of a fast exchange process [94]. Measurement of ^{14}N NMR linewidths has enabled the kinetics of the electron-transfer reaction between the two species to be studied. The reaction has been found to obey the rate equation:

$$\text{rate} = k[Fe(CN)_6^{3\ominus}][Fe(CN)_6^{4\ominus}]$$

with $k = 9.2 \pm 1.3 \times 10^4$ sec^{-1} mol^{-1} at 305°K and an Arrhenius activation energy of 17.6 kJ mol^{-1}. The effect of replacing the potassium ion by other alkali and alkaline earth cations has also been studied. The cations are found to have a marked catalytic effect on the reaction rate, which increases from $H^{\oplus}$ to $Cs^{\oplus}$ and $Mg^{2\oplus}$ to $Sr^{2\oplus}$. Possible mechanisms of electron transfer, in relation to the catalytic effect of cations, have been considered [94]. The authors have concluded that the most likely mechanism is one in which the cation transfers the electron by serving as a bridge between the two anions. The Arrhenius plot for the reaction rate appears to be non-linear and is tentatively interpreted as being due to a change in the reaction mechanism with the temperature. In a later note [126], remeasurement of rate constants as a function of temperature with higher precision and over a much larger range of concentrations has been reported. The new Arrhenius plot shows that the dependence of log K on $1/T$ is linear in the temperature range 298–338K and gives $E = 17.6 \pm 2.5$ kJ mol^{-1}, a pre-exponential factor of 1.2×10^8 sec^{-1} and activation parameters $\Delta H\ddagger = 15.0$ kJ mol^{-1} and $\Delta S\ddagger = -99.1$ J K^{-1} mol^{-1} at 312K. The significance of E is discussed and an attempt has been made to measure the effect upon this parameter of substituting $K^{\oplus}$ by $H^{\oplus}$ or $Cs^{\oplus}$. The results have led to the tentative conclusion that the catalytic effect of the cations on the electron transfer reaction between $[Fe(CN)_6]^{3\ominus}$ and $[Fe(CN)_6]^{4\ominus}$ is not expressed in their E values, as might be expected if the capability of ion-pair formation is the rate-determining factor in the reaction. Ion-pair formation is confirmed by cation resonance measurements [127].

6.3.3   Ligand Exchange Processes

^{14}N NMR has been quite extensively used in kinetic studies of exchange processes involving paramagnetic M(II) ions of the

$3d$ transition series and nitrogen-containing ligands, for example, ammonia exchange between nickel ammine complexes and ammonia in both aqueous and anhydrous ammonia solutions [95]. Since proton NMR studies [96] of this system in aqueous ammonia signify similar exchange rates, it is concluded that ammonia molecules are undergoing exchange with the $[Ni(NH_3)_6]^{2\oplus}$ ion and that individual protons are exchanging at a much slower rate. Similar conclusions are reached from ^1H NMR data on liquid ammonia solutions of $[Ni(NH_3)_6]^{2\oplus}$ [97-99]. Ammonia exchange has also been examined by ^{14}N NMR for Co(II) in aqueous ammonia [128] and for Co(II) [129] and Mn(II) [130] in anhydrous ammonia. Other nitrogen ligands and metal cations for which exchange processes have been studied by ^{14}N NMR include ethylenediamine—Cu(II) [100]; thiocyanate—Ni(II) and Co(II) [101, 102, 131]; acetonitrile—Mn(II) and Fe(II) [103, 132]; and pyridine—Co(II) [104].

The results for Cu(II) in ethylenediamine solutions, which also include proton NMR measurements, are again consistent with the view that protons and ^{14}N nuclei both experience the Cu(II) environment through exchange of the ethylenediamine molecule as a whole between the bulk solvent and the Cu(II) first coordination sphere [100].

In the case of aqueous Co(II)–thiocyanate solutions [102], use of both ^{17}O and ^{14}N chemical shift data permit a unique assignment of coordination numbers of $NCS^\ominus$ and H_2O in all the species involved in the stepwise equilibria. Complex species found to be present in significant amounts over the concentration range 0–7.98M are $[Co(H_2O)_6]^{2\oplus}$, $[Co(NCS)(H_2O)_5]^\oplus$, $[Co(NCS)_2(H_2O)_2]$, $[Co(NCS)_3(H_2O)]^\ominus$ and $[Co(NCS)_4]^{2\ominus}$.

The ^{14}N nuclear relaxation times of acetonitrile solutions of $Mn(ClO_4)_2$ have been studied as a function of temperature [103]. The kinetic parameters for the exchange of acetonitrile from the first coordination sphere are derived from the equations of Swift and Connick [87, 105] describing the broadening of the NMR lines of nuclei in dilute solutions of paramagnetic ions, where chemical exchange is a factor.

A similar ^{14}N NMR study of solvent exchange in the system $[Fe(CH_3CN)_6]^{2\oplus}/CH_3CN$ has been made [132] and a comparison of the exchange parameters given for $Mn^{2\oplus}$, $Fe^{2\oplus}$, $Co^{2\oplus}$ and $Ni^{2\oplus}$ in the solvents CH_3CN, NH_3, H_2O and CH_3OH. These results are discussed in terms of crystal field considerations. From the ^{14}N NMR study [104] of the

$CoCl_2$-pyridine system, the equilibrium constant $[Co(py)_2 Cl_2] [py]^2 / [Co(py)_4 Cl_2]$ has been determined, together with ΔH and ΔS for this equilibrium. The mean lifetime of pyridine on $[Co(py)_4 Cl_2]$ has also been calculated, together with $\Delta H^{\ddagger}$ and $\Delta S^{\ddagger}$ values for the exchange process.

6.3.4　Spin Density Distribution

Complexes of bis(acetylacetonato)nickel(II), $[Ni(acac)_2]$, with organic ligands have been widely used to study the delocalization of the unpaired electron spin density by proton magnetic resonance [133, 134]. Comparable ^{14}N NMR studies have now been carried out for a number of organic nitrogen donor molecules [135] in the presence of $[Ni(acac)_2]$ in order to obtain information on the delocalization of the unpaired electrons at the site of the coordinating ligand atoms. The nitrogen ligands L used are: aniline, diethylamine, dipropylamine, tert-butylamine, piperidine and pyridine and the composition of the resulting complexes is expected to be $[Ni(acac)_2 L_2]$ with *trans* L ligands. Large downfield ^{14}N shifts, relative to the free ligand are observed, e.g. aniline, -10150 ± 500; pyridine, -15400 ± 300 p.p.m. The rather large, positive spin densities observed are interpreted in terms of occupancy by one of the unpaired electrons of an antibonding orbital consisting of the d_{z^2} orbital of the nickel atom and hybrid orbitals of the nitrogen atom. The results are taken to indicate that piperidine, the most basic, gives the greatest degree of covalency to the Ni—N bond, while aniline, the least basic, gives the least covalency. However, a comparison of spin densities and basicity constants for the other compounds investigated shows no clear connection between these parameters. More recently, measurement of contact shifts for ^1H, ^{13}C and ^{14}N nuclei have been used to give a 'map' of the distribution of spin density in the complexes $[Ni(en)_2]^{2\oplus}$ (en = ethylenediamine) and $[Ni(pn)_2]^{2\oplus}$ (pn = 1,3-diaminopropane) [136]. Large downfield ^{14}N shifts relative to the free ligand are again observed, i.e. $[Ni(en)_2]^{2\oplus}$, $-13,050 \pm 400$, $[Ni(pn)_2]^{2\oplus}$, $-15,400 \pm 400$ p.p.m. and attributed to extensive delocalization of the unpaired $d_{x^2-y^2}$ electron of Ni(II) on to the nitrogen atoms of the ligands. The role of spin polarization, hyperconjugation and the influence of the geometry of the ring on the spin density distribution are discussed. LCAO—MO calculations of the distribution of spin density on the ligands of the complex $[Ni(en)_2]^{2\oplus}$ have been performed and considered in relation to the experimentally determined spin populations.

[14]N chemical shifts have also been measured for a series of octahedral chromium(III) complexes and d^5 pentacyano strong-field complexes [137]. With the exception of $[Cr(CN)_6]^{3\ominus}$, the Cr(III), t_{2g}^3, complexes are shown to have the same chemical shifts within experimental error as their diamagnetic Co(III), t_{2g}^6, counterparts. In earlier work [25] no isotropic hyperfine shift was observed for the cyano resonance in $[Cr(CN)_5NO]^{2\ominus}$ (Table 6.8). This effect is attributed to a negligible pseudo-contact term, since the $^4A_2$ ground state of Cr(III) is expected to have little magnetic anisotropy. Also, the results show the contact term to be of small magnitude, which is not surprising since there can be no nitrogen s-character associated with the t_{2g} orbitals. It is not clear why only hexacyanochromate(III) shows an appreciable contact interaction of cyanide ligand with the electron spins on chromium. Sizeable paramagnetic shifts are observed for the d^5 complexes. The magnitude of the pseudo-contact shift has been estimated for $Na_2[Fe(CN)_5NH_3]$ and $Na_3[Fe(CN)_5NO_2]$ and proved to have a sign consistent with a $^2B_2$ ground state, but too large a magnitude for explanation by traditional theoretical treatments.

6.4 Studies on Solids

A nucleus possessing an electric quadrupole moment will experience an interaction of the nuclear quadrupole and an electric field gradient at the nucleus. The NMR spectrum of the nucleus can show splitting due to this interaction, if the nucleus is located in a crystalline solid. Examination of the NMR spectrum of a single crystal containing such a nucleus allows information to be obtained concerning the electric field gradient in the vicinity of the nucleus, and its orientation with respect to the crystallographic axes. In favourable cases, inferences can be drawn concerning crystal structure and chemical bonding. However, even with polycrystalline powders, some chemical bonding information is still available, although structural information is lost. [14]N NMR is a potentially more powerful method than nuclear quadrupole resonance in studying phase transitions, crystal structure, symmetry and other solid state phenomena which reflect

changes in nuclear quadrupole interactions. The only requirement is that sufficiently large and well-defined single crystals
must be available [3].

The first determination of ^{14}N quadrupole coupling constants in solids by the use of ^{14}N NMR was reported by
Forman [106] for potassium azide. By the use of a single
crystal, the relative orientation of molecular and crystallographic axes can be found and it was shown that the azide
groups lie along [110] axes and are not bent. More recently, a
^{14}N NMR investigation of some solid metal nitrates [107] has
further highlighted the potentialities of ^{14}N studies on solids.
Single crystal spectra of the compounds $UO_2(NO_3)_2 \cdot 6H_2O$
and $RbUO_2(NO_3)_3$ have again allowed determination of
quadrupole coupling constants and asymmetry parameters of
the nitrogen nuclei, as well as the orientation of the nitrate
groups with respect to the crystallographic axes. In the case
of uranyl nitrate hexahydrate the results are used to decide
between conflicting crystal structures and the existence of two
different crystal structures for rubidium uranyl nitrate is
confirmed. Powdered samples of anhydrous zinc, cadmium
and mercury(II) nitrates have also been investigated by ^{14}N
NMR. While the results of these studies are not of sufficiently
high accuracy to permit the extraction of reliable values for
the quadrupole coupling constants and asymmetry parameters,
it is possible to plot the relationship between them for each
nitrate. This plot clearly shows that the nitrate environment
in $Cd(NO_3)_2$ is markedly different from, and more symmetrical than, that in $Zn(NO_3)_2$ and $Hg(NO_3)_2$. It was
concluded from earlier infra-red studies that the nitrate groups
in $Zn(NO_3)_2$ and $Hg(NO_3)_2$ are coordinated to the metal
cation [108], and in $Zn(NO_3)_2$ they are probably bidentate
or bridging [109], whereas ionic nitrate groups were indicated
in $Cd(NO_3)_2$ [108]. These conclusions are strongly supported
by the nitrogen magnetic resonance results.

^{14}N data for several solid mononitrides [110–112] possessing NaCl-type lattice structures have been interpreted in terms
of Knight shifts and line widths. These parameters are
independent of temperature and applied magnetic field in the
case of the diamagnetic nitrides, but depend on both of these
experimental variables in the case of the paramagnetic
compound UN [112]. ^{15}N NMR measurements have also
been made on enriched $U^{15}N$ [138]. The Knight shift, K, and
the linewidth have been measured under similar conditions to

those in the previous NMR measurement [112] on $U^{14}N$, except that in $U^{15}N$ the observed dispersion lines are distorted and broadened experimentally. The K values and their dependence upon molar susceptibility are very similar for both $U^{14}N$ and $U^{15}N$.

6.5   Miscellaneous Applications

6.5.1   The Dissociation of Nitric Acid

An early study utilized ^{14}N NMR to determine the degree of dissociation of aqueous nitric acid [113] over a range of concentrations up to *ca.* 15M. The values obtained by ^{14}N NMR are rather higher than those obtained from Raman intensities [114] and the ^{14}N NMR measurements could no doubt be profitably repeated using modern equipment. In a recent investigation, ^{14}N NMR spectroscopy has been used to study the equilibrium between HNO_3 and $NO_2^{\oplus}$ in nitric acid–sulphuric acid mixtures [139].

6.5.2   The Knight Shift of the Sodium–Ammonia System

The temperature dependence $(240\text{–}295°K)$ of the Knight shift of the sodium–ammonia system has been studied [115] by means of ^{14}N and ^{23}Na NMR in the concentration range corresponding to mole ratios (R) of $5.7\text{–}700$ (NH_3/Na), and kinetic data has been obtained. At high sodium concentrations $(R = 5.7)$, the electron density at the sodium atom is 9% of the free atom value but falls on dilution $(R \geqslant 300)$, to 0.13%. This is accompanied by a parallel increase in the electron density at the ^{14}N nucleus, clearly indicating that with increasing dilution, the electrons increasingly occupy expanded orbitals in the dielectric medium. A theoretical model involving the $Na \cdot 6NH_3$ species yields electron densities which are an order of magnitude too high. Other models produce more plausible values, but it is felt that more refined wavefunctions are required to determine the extension of the expanded orbitals in the dielectric medium.

6.5.3   Studies of Donor–Acceptor Interaction

The donor properties of nitrobenzene and pentafluoronitrobenzene towards the boron halides BF_3, BCl_3, BBr_3, $PhBCl_2$ and Bu^nBCl_2 in CH_2Cl_2 have been investigated by ^{11}B, ^{14}N and ^{19}F NMR [116]. The ^{14}N chemical shifts on complex formation are small and do not bear any simple relationship to the donor or acceptor properties of the system. However, the ^{14}N linewidths are related to the ^{11}B chemical shifts between coordinated and uncoordinated boron halide and increase with increasing acceptor properties of the latter.

6.5.4   Identification of Reaction Products

Recent work in the author's laboratory [54] has shown ^{14}N NMR to be an excellent technique for the identification of reaction products, e.g. of anhydrous covalent nitrates or nitrato-complexes and reactive organic solvents. The reaction between dinitrogen pentoxide and acetic anhydride may be cited by way of illustration. It has been known for many years that this reaction yields the somewhat unstable (and explosive!) compound acetyl nitrate, $AcONO_2$ [117].

It has been shown that the production of tetranitromethane, $C(NO_2)_4$, in the further slow reaction of acetyl nitrate with excess acetic anhydride can be monitored by ^{14}N NMR spectroscopy [54]. Spectra obtained 15 min, 17 hr and 30 days after mixing N_2O_5 and excess Ac_2O are illustrated in Fig. 6.5. The signal at +64 p.p.m. is the only one observed immediately after preparation of a fresh solution of N_2O_5 in excess Ac_2O. The position of this resonance is practically identical to that of molecular N_2O_5 in CCl_4 (Table 6.4). It is attributed to acetyl nitrate but would also be consistent with fast exchange between the nitrogen-containing species in the following equilibria

$$2AcONO_2 \rightleftharpoons Ac_2O + N_2O_5 \rightleftharpoons NO_2^{\oplus} + NO_3^{\ominus}$$

Whatever the precise origin of the peak at +64 p.p.m. may be, the diminution in its intensity with time is accompanied by a corresponding increase in intensity of the characteristically sharp resonance of $C(NO_2)_4$ at +43 p.p.m., linewidth 7 Hz. The potentialities of ^{14}N NMR as a method of following the course, and kinetics, of such reactions are evident and similar studies on a range of metal nitrates and organic reactants are currently in progress [54].

This brief survey has indicated many areas and aspects of Inorganic Chemistry to which ^{14}N NMR studies have already been advantageously applied. The range of chemical information available is encouraging and a continuing and increasing employment of the technique by Inorganic Chemists can be

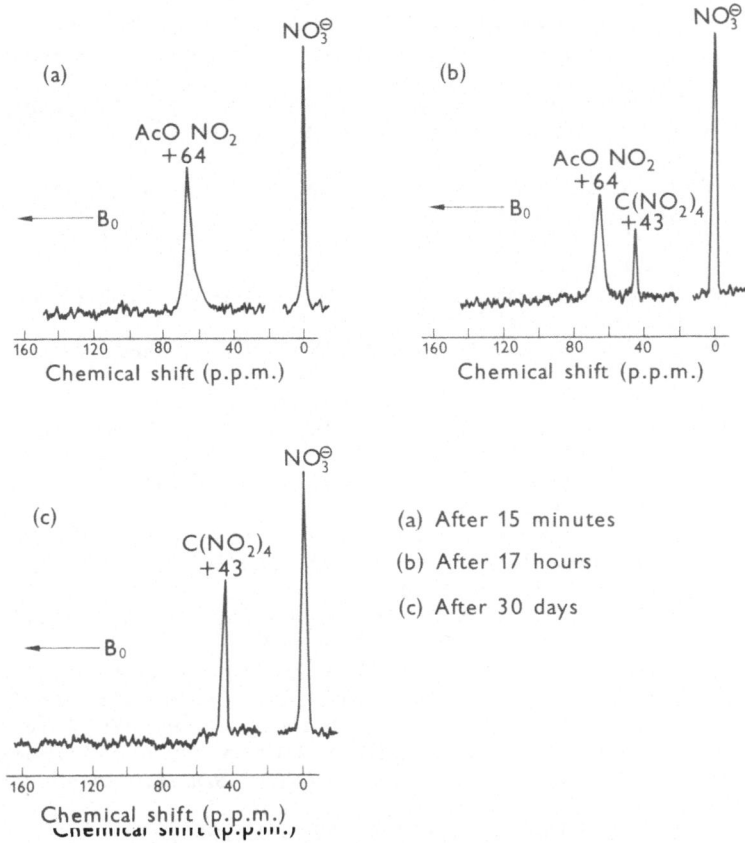

Fig. 6.5. ^{14}N NMR spectra at 7.226 MHz of dinitrogen pentoxide in excess acetic anhydride at *ca.* 300 K; NO$_3^{\ominus}$ as external reference.

confidently predicted. In particular, ^{14}N NMR studies on solids, preferably single crystals, are worthy of wider application. The advent of Fourier pulsed ^{14}N studies (Chapter 2) holds promise from the experimental viewpoint, especially for sparingly soluble compounds.

References

1. E. F. MOONEY and P. H. WINSON, in *Annual Reports on N.M.R. Spectroscopy*, Vol. 2 (E. F. MOONEY, Ed.), Academic Press, London and New York, 1969, p. 125.
2. E. W. RANDALL and D. G. GILLIES, in *Progress in N.M.R. Spectroscopy*, Vol. 6 (J. W. EMSLEY, J. FEENEY, and L. H. SUTCLIFFE, Eds.), Pergamon Press, Oxford, 1971, p. 119.

3. M. WITANOWSKI and G. A. WEBB, in *Annual Reports on N.M.R. Spectroscopy*, Vol. 5 (E. F. MOONEY, Ed.), Academic Press, London and New York, 1972, 395.
4. W. G. PROCTOR and F. C. YU, *Phys. Rev.*, 77, 717 (1950).
5. K. F. CHEW, W. DERBYSHIRE, and N. LOGAN, unpublished results.
6. B. M. SCHMIDT, L. C. BROWN, and D. H. WILLIAMS, *J. Mol. Spectrosc.*, 2, 539 (1958).
7. B. M. SCHMIDT, L. C. BROWN, and D. H. WILLIAMS, *J. Mol. Spectrosc.*, 2, 551 (1958).
8. B. M. SCHMIDT, L. C. BROWN, and D. H. WILLIAMS, *J. Mol. Spectrosc.*, 3, 30 (1959).
9. J. MASON and W. van BRONSWIJK, *J. Chem. Soc. A*, 791 (1971).
10. J. MASON, Personal Communication.
11. M. WITANOWSKI and H. JANUSZEWSKI, *J. Chem. Soc. B*, 1062 (1967).
12. R. BRAMLEY, B. N. FIGGIS, and R. S. NYHOLM, *J. Chem. Soc. A*, 861 (1967).
13. L.-O. ANDERSSON and J. MASON, *Chem. Comm.*, 99 (1968).
14. L.-O. ANDERSSON, J. MASON, and W. van BRONSWIJK, *J. Chem. Soc. A*, 296 (1970).
15. P. HAMPSON and A. MATHIAS, *Mol. Phys.*, 11, 541 (1966).
16. A. MATHIAS, *Mol. Phys.*, 12, 381 (1967).
17. P. HAMPSON and A. MATHIAS, *Mol. Phys.*, 13, 361 (1967).
18. P. HAMPSON and A. MATHIAS, *Chem. Comm.*, 371 (1967).
19. P. HAMPSON and A. MATHIAS, *Chem. Comm.*, 825 (1968).
20. P. HAMPSON and A. MATHIAS, *J. Chem. Soc. B*, 673 (1968).
21. E. B. BAKER and W. L. BURD, *Rev. Sci. Instruments*, 28, 313 (1957).
22. E. B. BAKER, *J. Chem. Phys.*, 37, 911 (1962).
23. W. McFARLANE and R. R. DEAN, *J. Chem. Soc. A*, 1535 (1968).
24. G. A. OLAH and T. E. KIOVSKY, *J. Amer. Chem. Soc.*, 90, 4666 (1968).
25. D. HERBISON-EVANS and R. E. RICHARDS, *Mol. Phys.*, 8, 19 (1964).
26. A. H. NORBURY and A. I. P. SINHA, *Quart. Rev.*, 24, 69 (1970).
27. D. MARTIN, *Angew. Chem., Eng. Ed.*, 3, 311 (1964).
28. E. L. WAGNER, *J. Chem. Phys.*, 43, 2728 (1965).
29. K. F. CHEW, W. DERBYSHIRE, N. LOGAN, A. H. NORBURY, and A. I. P. SINHA, *Chem. Comm.*, 1708 (1970).
30. K. M. MACKAY and S. R. STOBART, *Spectrochim. Acta*, 27A, 923 (1971).
31. J. NELSON, R. SPRATT, and S. M. NELSON, *J. Chem. Soc. A*, 583 (1970).
32. N. GROVING and A. HOLM, *Acta. Chem. Scand.*, 19, 1768 (1965).
33. M. WITANOWSKI, *J. Amer. Chem. Soc.*, 90, 5683 (1968).
34. J. N. SHOOLERY, *J. Chem. Phys.*, 31, 1427 (1959).
35. *Tables of Interatomic Distances and Configurations in Molecules and Ions*, The Chemical Society, London, 1958.

36. J. E. GRIFFITHS, *J. Chem. Phys.*, **48**, 278 (1968).
37. K. M. MACKAY and S. R. STOBART, *Spectrochim. Acta*, **26A**, 373 (1970).
38. J. N. SHOOLERY, R. G. SHULMAN, and D. M. YOST, *J. Chem. Phys.*, **19**, 250 (1951).
39. W. BECK, W. BECKER, K. F. CHEW, W. DERBYSHIRE, N. LOGAN, D. M. REVITT, and D. B. SOWERBY, *J. Chem. Soc. Dalton*, 245 (1972).
40. T. KANDA, Y. SAITO, and K. KAWAMURA, *Bull. Chem. Soc. Japan*, **35**, 172 (1962).
41. R. A. FORMAN, *J. Chem. Phys.*, **39**, 2393 (1963).
42. D. HERBISON-EVANS and R. E. RICHARDS, *Mol. Phys.*, **7**, 515 (1964).
43. J. E. KENT and E. L. WAGNER, *J. Chem. Phys.*, **44**, 3530 (1966).
44. A. M. QURESHI, J. A. RIPMEESTER, and F. AUBKE, *Canad. J. Chem.*, **47**, 4247 (1969).
45. C. C. ADDISON and J. C. SHELDON, *J. Chem. Soc.*, 3142 (1958).
46. R. A. OGG and J. D. RAY, *J. Chem. Phys.*, **25**, 1285 (1956).
47. C. C. ADDISON and N. LOGAN, in *Developments in Inorganic Nitrogen Chemistry*, Vol. 2 (C. B. COLBURN, Ed.), Elsevier, Amsterdam, in the press.
48. C. K. INGOLD and D. J. MILLEN, *J. Chem. Soc.*, 2612 (1950).
49. E. BERL and H. H. SAENGER, *Monatsh. Chem.*, **53**, 1036 (1929).
50. W. H. LEE, in *The Chemistry of Non-Aqueous Solvents*, Vol. 2 (J. J. LAGOWSKI, Ed.), Academic Press, London and New York, 1967, p. 151.
51. G. E. McGRAW, D. L. BERNITT, and I. C. HISATSUNE, *Spectrochim. Acta*, **23A**, 25 (1967).
52. D. N. HENDRICKSON and W. L. JOLLY, *Inorg. Chem.*, **8**, 693 (1969).
53. J. MASON and W. van BRONSWIJK, *J. Chem. Soc. A*, 1763 (1970).
54. M. I. KHALIL and N. LOGAN, unpublished results.
55. W. L. JOLLY and N. LOGAN, unpublished results.
56. W. BECKER and W. BECK, *Z. Naturforsch.*, **25b**, 101 (1970).
57. D. MOY and A. R. YOUNG, *J. Amer. Chem. Soc.*, **87**, 1889 (1965).
58. C. A. WAMSER, W. B. FOX, B. SUKORNICK, J. R. HOLMES, B. B. STEWART, R. JUURIK, N. VANDERKOOI, and D. GOULD, *Inorg. Chem.*, **8**, 1249 (1969).
59. K. O. CHRISTE, J. P. GUERTIN, A. E. PAVLATH, and W. SAWODNY, *Inorg. Chem.*, **6**, 533 (1967).
60. W. E. TOLBERG, R. T. REWICK, R. S. STRINGHAM, and M. E. HILL, *Inorg. Chem.*, **6**, 1156 (1967).
61. E. L. MUETTERTIES and W. D. PHILLIPS, *J. Amer. Chem. Soc.*, **81**, 1084 (1959).
62. J. H. NOGGLE, J. D. BALDESCHWEILER, and C. B. COLBURN, *J. Chem. Phys.*, **37**, 182 (1962).
63. N. BARTLETT, J. PASSMORE, and E. J. WELLS, *Chem. Comm.*, 213 (1966); W. B. FOX, J. S. MACKENZIE, E. R. McCARTHY, J. R. HOLMES, R. F. STAHL, and R. JUURIK, *Inorg. Chem.*, **7**, 2064 (1968).

64. R. A. OGG and J. D. RAY, *J. Chem. Phys.*, **25**, 797 (1956).
65. F. S. FAWCETT and R. D. LIPSCOMB, *J. Amer. Chem. Soc.*, **82**, 1509 (1960).
66. J. MASON and W. van BRONSWIJK, *Chem. Comm.*, 357 (1969).
67. A. R. YOUNG and D. MOY, *Inorg. Chem.*, **6**, 178 (1967).
68. J. K. RUFF, *J. Amer. Chem. Soc.*, **87**, 1140 (1965).
69. N. LOGAN and W. L. JOLLY, *Inorg. Chem.*, **4**, 1508 (1965).
70. J. NELSON and H. G. HEAL, *J. Chem. Soc. A*, 136 (1971).
71. J. MASON, *J. Chem. Soc. A*, 1567 (1969).
72. E. F. MOONEY and M. A. QASEEM, *J. Inorg. Nuc. Chem.*, **30**, 1439 (1968).
73. K. F. CHEW, W. DERBYSHIRE, M. F. LAPPERT and N. LOGAN, unpublished results.
74. R. A. OGG and J. D. RAY, *J. Chem. Phys.*, **26**, 1339 (1957).
75. M. SHPORER, G. RON, A. LOEWENSTEIN, and G. NAVON, *Inorg. Chem.*, **4**, 358 (1965).
76. O. W. HOWARTH, R. E. RICHARDS, and L. M. VENANZI, *J. Chem. Soc.*, 3335 (1964).
77. B. B. MURRAY, U.S.A.E.C. Report DD-391 (1959).
78. B. M. FUNG and S. C. WEI, *J. Mag. Res*, **3**, 1 (1970).
79. W. BECKER, W. BECK, and R. RIECK, *Z. Naturforsch.*, **25b**, 1332 (1970).
80. W. BECK and E. SCHUIERER, *J. Organomet. Chem.*, **3**, 55 (1965).
81. C. GRUNDMANN and J. M. DEAN, *J. Org. Chem.*, **30**, 2809 (1965).
82. A. PIDCOCK, R. E. RICHARDS, and L. M. VENANZI, *Proc. Chem. Soc.*, 184 (1962).
83. W. P. FEHLHAMMER and L. F. DAHL, *J. Amer. Chem. Soc.*, **94**, 3377 (1972).
84. R. MASON, G. A. RUSHOLME, W. BECK, H. ENGELMANN, K. JOOS, B. LINDENBERG, and H. S. SMEDAL, *Chem. Comm.* 496 (1971).
85. R. ETTINGER, P. BLUME, P. C. LAUTERBUR, and A. PATTERSON, *J. Chem. Phys.*, **33**, 1597 (1960).
86. C. C. ADDISON, C. D. GARNER, N. LOGAN, and S. C. WALLWORK, *Quart. Rev.*, **25**, 289 (1971).
87. G. A. WEBB, in *Annual Reports of N.M.R. Spectroscopy*, Vol. 3 (E. F. MOONEY, Ed.), Academic Press, London and New York, 1970, p. 211.
88. G. N. LaMAR, *J. Chem. Phys.*, **41**, 2992 (1964).
89. G. N. LaMAR, *J. Chem. Phys.*, **43**, 235 (1965).
90. R. H. FISCHER and W. D. HORROCKS, *Inorg. Chem.*, **7**, 2659 (1968).
91. P. K. BURKERT, H. P. FRITZ, W. GRETNER, H. J. KELLER, and K. E. SCHWARZHANS, *Inorg. Nucl. Chem. Letts.*, **4**, 237 (1968).
92. H. P. FRITZ, W. GRETNER, H. J. KELLER, and K. E. SCHWARZHANS, *Z. Naturforsch.*, **25b**, 174 (1970).
93. D. G. BROWN and R. S. DRAGO, *J. Amer. Chem. Soc.*, **92**, 1871 (1970).
94. M. SHPORER, G. RON, A. LOEWENSTEIN, and G. NAVON, *Inorg. Chem.*, **4**, 361 (1965).

95. H. H. GLAESER, G. A. LO, H. W. DODGEN, and J. P. HUNT, *Inorg. Chem.*, **4**, 206 (1965).
96. A. L. van GEET, *Inorg Chem.*, **7**, 2026 (1968).
97. T. J. SWIFT and H. H. LO, *J. Amer. Chem. Soc.*, **88**, 2994 (1966).
98. B. B. WAYLAND and W. L. RICE, *Inorg. Chem.*, **6**, 2270 (1967).
99. W. L. RICE and B. B. WAYLAND, *Inorg. Chem.*, **7**, 1040 (1968).
100. M. ALEI, W. B. LEWIS, A. B. DENISON, and L. O. MORGAN, *J. Chem. Phys.*, **47**, 1062 (1967).
101. R. B. JORDAN, H. W. DODGEN, and J. P. HUNT, *Inorg. Chem.*, **5**, 1906 (1966).
102. A. H. ZELTMANN and L. O. MORGAN, *Inorg. Chem.*, **9**, 2522 (1970).
103. W. L. PURCELL and R. S. MARIANELLI, *Inorg. Chem.*, **9**, 1724 (1970).
104. G. D. HOWARD and R. S. MARIANELLI, *Inorg. Chem.*, **9**, 1738 (1970).
105. T. J. SWIFT and R. E. CONNICK, *J. Chem. Phys.*, **37**, 307 (1962).
106. R. A. FORMAN, *J. Chem. Phys.*, **45**, 1118 (1966).
107. B. A. WHITEHOUSE, J. D. RAY, and D. J. ROYER, *J. Mag. Res.*, **1**, 311 (1969).
108. C. C. ADDISON and B. M. GATEHOUSE, *J. Chem. Soc.*, 613 (1960).
109. B. O. FIELD and C. J. HARDY, *J. Chem. Soc.*, 4428 (1964).
110. R. G. SHULMAN and B. J. WYLUDA, *J. Phys. Chem. Solids*, **23**, 166 (1962).
111. M. KUZNIETZ, *J. Chem. Phys.*, **49**, 3731 (1968).
112. M. KUZNIETZ, *Phys. Rev.*, **180**, 476 (1969).
113. Y. MASUDA and T. KANDA, *J. Phys. Soc. Japan*, **9**, 82 (1954).
114. O. REDLICH and J. BIGELEISEN, *J. Amer. Chem. Soc.*, **65**, 1883 (1943).
115. J. V. ACRIVOS and K. S. PITZER, *J. Phys. Chem.*, **66**, 1693 (1962).
116. E. F. MOONEY, M. A. QASEEM, and P. H. WINSON, *J. Chem. Soc. B*, 224 (1968).
117. A. PICTET and E. KHOTINSKY, *Ber. Dtsch. Chem. Ges.*, **40**, 1163 (1907).
118. J. W. LEHMAN and B. M. FUNG, *Inorg. Chem.*, **11**, 214 (1972).
119. P. CLIFTON and L. PRATT, *Proc. Chem. Soc.*, 339 (1963).
120. W. L. JOLLY, A. D. HARRIS, and T. S. BRIGGS, *Inorg. Chem.*, **4**, 1064 (1965).
121. J. S. GRIFFITH and L. E. ORGEL, *Trans. Faraday Soc.*, **53**, 601 (1957).
122. R. FREEMAN, G. R. MURRAY, and R. E. RICHARDS, *Proc. Roy. Soc. Ser. A*, **242**, 455 (1957).
123. W. BECK, W. BECKER, H. NÖTH, and B. WRACKMEYER, *Chem. Ber.*, **105**, 2883 (1972).
124. H. BÖHLAND and E. MÜHLE, *Z. Anorg. Allg. Chemie*, **379**, 273 (1970).
125. J. SADLEJ and Z. KEÇKI, *Roczniki Chem.*, **45**, 445 (1971).
126. A. LOEWENSTEIN and G. RON, *Inorg. Chem.*, **6**, 1604 (1967).

127. A. LOEWENSTEIN and G. RON, *Proceedings of the XIVth Colloque Ampere*, Ljubljana, Yugoslavia, Sept. 1966.
128. R. MURRAY, S. F. LINCOLN, H. H. GLAESER, H. W. DODGEN, and J. P. HUNT, *Inorg. Chem.*, **8**, 554 (1969).
129. H. H. GLAESER, H. W. DODGEN, and J. P. HUNT, *Inorg. Chem.*, **4**, 1061 (1965).
130. M. GRANT, H. W. DODGEN, and J. P. HUNT, *J. Amer. Chem. Soc.*, **91**, 6318 (1969).
131. R. MURRAY, H. W. DODGEN, and J. P. HUNT, *Inorg. Chem.*, **3**, 1576 (1964).
132. R. J. WEST and S. F. LINCOLN, *Austral. J. Chem.*, **24**, 1169 (1971).
133. J. A. HAPPE and R. L. WARD, *J. Chem. Phys.*, **39**, 1211 (1963).
134. E. E. ZAEV and Yu. N. MOLIN, *Zh. Strukt. Khim.*, **7**, 680 (1966).
135. Yu. N. MOLIN, *Zh. Strukt. Khim.*, **10**, 932 (1969).
136. Yu. N. MOLIN, P. V. SCHASTNEV, and N. D. CHUVYLKIN, *Zh. Strukt. Khim.*, **12**, 403 (1971).
137. B. R. McGARVEY and J. PEARLMAN, J. Mag. Res., **1**, 178 (1969).
138. M. KUZNIETZ and D. O. van OSTENBURG, Phys. Rev (B), **2**, 3453 (1970).
139. F. SEEL, V. HARTMANN, and W. GOMBLER, *Z. Naturforsch.*, **27b**, 325 (1972).

Author Index

Subject Index